Molecular Biology, Biochemistry and Biophysics
Molekularbiologie, Biochemie und Biophysik

8

Molecular Biology
Biochemistry and Biophysics

Molekularbiologie
Biochemie und Biophysik

8

Chapman & Hall Limited · London
Springer-Verlag Berlin · Heidelberg · New York

1970

Protein Sequence Determination

A Sourcebook of Methods and Techniques

Edited by Saul B. Needleman

With 77 Figures

Chapman & Hall Limited · London
Springer-Verlag Berlin · Heidelberg · New York
1970

Saul B. Needleman, Ph. D.

Assistant Professor of Biochemistry and Neurology

Department of Biochemistry, Northwestern University Medical School
and Radioisotope Service V. A. Research Hospital
Chicago, Illinois, USA

Chapman & Hall Limited ISBN 0412-09990-X

 Library of Congress Catalog Card Number 79-117944. Printed in Germany. Typesetting and printing: Carl Ritter & Co., Wiesbaden. Bookbinding: Großbuchbinderei Karl Hanke, Düsseldorf

Title No. 3808

For Sondra

and

Marty, Arthur, Beth, Heidi

Preface

"All the King's horses and all the King's men
couldn't put Humpty Dumpty together again."

It is entirely possible that the difficulty facing "all the King's men" was principally the lack of a sufficient guide to the techniques of reassembling from a series of small components, the original entity. It is the sincere hope of the editor of the present work and of each of the contributing authors that the modern researcher will not face a similar predicament in his endeavours to reconstruct the complete primary sequence of a protein from the array of component amino acids. Rather, it is the intent that, with this volume, he may proceed untimorously if not with outright confidence toward achieving his purpose.

To the newcomer in protein sequencing, compelled by necessity, or fascination, to investigate the exact order of amino acids in proteins, the question of "where to begin" — or "how to do it" is urgent. To those more skilled, a ready source of additional techniques should nevertheless be of value. This volume attempts to present in a single source a discussion of the methods and techniques useful to the determination of the primary structures of proteins and peptides. Hopefully, this book will tell the reader "how to do it".

Theoretical considerations, the derivation of mathematical concepts, discussions of active sites and of mechanisms of action are held to a minimum. While such topics are of considerable interest to many, they would burden the newcomer with details beyond his immediate needs. Ample references to the literature have been included to provide an extensive bibliography for the advanced worker and to satisfy the curiosity of the graduate student. Although not intended primarily for this purpose, the diversity of topics included make this volume a conceivable choice as a text for a senior or graduate course in biochemical techniques and methodology.

The contributors to this volume have been carefully selected to insure a balance of ideas from established leaders in the field and from "second-generation" workers who are even now in the process of making their primary scientific contributions. The international origin of the authors reflects the widespread interest in primary structure and conformational studies of proteins. The selection of authors and of subject matter was made solely by the editor and any reproach resulting from these selections must be directed towards him.

As is usual in such cases, credits must be given to many persons who directly or indirectly have contributed to the creation of this work. Perhaps the greatest vote of thanks must be given to every worker in the field who has come before, for it is they who made possible, who provided the means for that which appears in this volume, to aid those who will follow. In a more immediate sense, however, the editor must specifically mention with thanks and gratitude Dr. ALFRED GOLDSTONE, who was more a friend and advisor than co-author on this undertaking, and Mr. THEODOR K. MAYER who shared a load not really his. Finally, but not in the least, how does one thank SONDRA, my wife, who spent the many hours listening patiently to words beyond her ken but who gave them meaning.

SAUL B. NEEDLEMAN

Highland Park, Illinois
July, 1970

Contents

Contributors

Pehr Edman	St. Vincent's School of Medical Research, Victory Parade, Fitzroy, N. 6, Victoria, Australia
John W. Eveleigh	Instrument and Controls Division, Oak Ridge National Laboratory, Oak Ridge, Tennessee, USA
Angelo Fontana	Istituto di Chimica Organica, Universita di Padova, Padova, Italy
Duane Gish	Upjohn Company, Kalamazoo, Michigan, USA
Alfred D. Goldstone	Neurology Department, Northwestern University Medical School and Veterans Research Hospital, Chicago, Illinois, USA
Charles B. Kasper	McArdle Laboratories, University of Wisconsin, Madison, Wisconsin, USA
Amos Linenberg	Hydronautics Inc., Rehovat, Israel
Kozo Narita	Institute for Protein Research, University of Osaka, Osaka, Japan
Saul B. Needleman	Department of Biochemistry, Northwestern University Medical School and Veterans Research Hospital, Chicago, Illinois, USA
Ernesto Scoffone	Istituto di Chimica Organica, Universita di Padova, Padova, Italy
Ken E. Van Holde	Department of Biochemistry and Biophysics, Oregon State University, Corvallis, Oregon, USA
George D. Winter	Technicon Corporation, Ardsley, New York

Chapter 1

General Considerations

Saul B. Needleman

Protein sequence determinations have reached new levels of interest in recent years such that these studies no longer are confined to a few major protein laboratories. Now, researchers in widely diverse areas of biochemistry and the medical sciences face the requirements for knowing the primary sequence of a protein. Indeed, the time is approaching, perhaps, when the true essence of many research programs will necessitate the complete characterization of the amino acid sequence of the protein of central interest. The reasons for wanting to know the sequence of a protein probably are as great and as complex as the number of research problems dealing with proteins.

Unfortunately, wanting to know the sequence — or even requiring that information — does not accomplish it. Perhaps even more unfortunate is the prevailing idea that sequencing is difficult to do. It must be said that this is only partially the case. From the fundamental techniques of the early workers have evolved a diversity of approaches to sequence studies which have become standard procedure in the protein laboratory. In the 25 years since Sanger's revolutionary work on the structure of insulin, it is safe to say that the development of new techniques or modifications of the old ones have increased in direct proportion to the number of workers in the field. While this might appear to add to the difficulties or complexity attending a detailed examination of the amino acid sequence of a protein, this diversity in available techniques have, instead, made the basic resolution of the problem attainable in the moderately equipped laboratory — when provided with sufficient literature sources to guide the way.

What does this book intend to accomplish? It is the intention of the contents of this volume to provide a quick but detailed discussion of the techniques involved in sequencing a protein — to provide a handy and fairly complete reference to the literature on this subject — to tell you "how to do it".

The difficulties involved in the final steps of primary sequence determination dictate the absolute requirement for having a single homogeneous substance upon which to perform the various experimental procedures. The reconstruction into a meaningful array — the complete protein sequence — from the collection of bits and pieces of short peptides is described in Chapter 9. One might imagine that the presence of even small amounts of contaminating materials would create considerable confusion in putting together the final jigsaw puzzle, even with the use of computer programming. Nothing can be more disconcerting than to have some small peptide unit remain which can not be conveniently fit into the final structure. Towards this end, in Chapter 2, is presented a discussion of analytical procedures used in establishing the purity and homogeneity of a protein. The physical characterization of the protein molecule is examined according to a series of convenient parameters — the mass, charge and molecular size of the molecule. Thus the application of sedimentation methods in the analytical ultracentrifuge provides one means for establishing the homogeneity of a protein. Electrophoresis differentiates members of a series according to charge and gel filtration is discussed as a technique for separation according to molecular size. The problems of purification are approached in terms of the criteria to be applied to the selection of the most reasonable procedures for the protein of interest and for establishing that purification has proceeded to the point of readiness for further analysis. Included in this chapter is a discussion of the special problems introduced by the presence of subunits in the protein.

Once the protein has been obtained in a satisfactory state of purity, it is of interest to determine which amino acid residues occupy the amino-terminal and carboxyl-terminal positions. The techniques used in obtaining this information are described in Chapter 3. The interpretation of terminal group analysis is presented in relation to the presence of free and masked terminal sites in the protein. Thus it is possible to establish the presence of multi-chain systems, of unusual configurations in the protein or, of blocked terminal residues. This is useful infomation in selecting the analytical procedures to be followed later.

In the two chapters which follow are discussed details for amino acid composition analysis by the most commonly used procedures currently available, automated ion exchange column chromatography in the automatic amino acid analyzer and by the rapidly developing technique of gas chromatography of volatile amino acid derivatives. It is from the amino acid composition analysis that one establishes a material balance during sequencing. Obviously, the sum of the amino acids present in the individual small peptides cannot exceed the total composition of the original protein — or be less than that amount. Knowledge of the presence of specific amino acids or even of the number of certain residues in the protein often suggests the application of special simplifying procedures in establishing the sequence of the individual small peptides.

With few exceptions, the protein as isolated from its native source is too large, too complex, for direct application of sequencing techniques. While it might be possible to establish the order of the first half dozen or so amino acids in the protein with little difficulty, further progress would be impeded by the increased resistance with each succeeding step to undergo reaction. Thus, at each step, incomplete release of amino acids would lead to the accumulation of proteins in varying stages of stepwise degradation. Interpretation at each additional step would become increasingly difficult and meaningless. A much simpler approach to sequence determination involves cleavage of the original protein in a known manner into a group of short peptides. These can be fractionated into the component peptides each of which in turn can be subjected to detailed sequence analysis with considerably less difficulty than could the original intact protein. In Chapter 6 are outlined a number of non-enzymic and enzymic methods in common use for converting the protein into a reproducible group of short peptides of manageable size. Since overlapping peptides (see Chapter 9) are needed in order to unequivocably establish the sequence of a protein, it is generally necessary to provide for analysis two sets of short peptides obtained by different procedures. The advantages and disadvantages for the several methods are presented. Details are described for the fractionation of the peptide mixture and for obtaining pure single components for sequence study.

The analysis of each small peptide is aided by knowing whether specific amino acids are present in the peptide. Special color reactions are known for several of the amino acids and the necessary reagents and details for performing the color analysis are given in Chapter 7. Similar color reactions may be used on the intact protein to guide in the selection of proper non-enzymic procedures for the chemical cleavage of the protein to smaller units prior to enzymic hydrolysis to the peptide level.

With the total amino acid composition in hand, together with the composition of key short peptides, one is ready to begin elucidation of the amino acid sequence of each of the necessary peptides. The method used almost exclusively for sequence analysis is that of Edman using phenylisothiocyanate for the derivatization of the N-terminal amino acid residue. The procedure for conducting the Edman reaction is discussed in great detail in Chapter 8.

As the sequence is completed for each member of at least two separate sets of peptides, it is possible to begin to reconstruct the total and exact primary structure of the original protein. While a few computer programs have been designed for collating the data, manual procedures involving overlapping peptides are more generally used. The computer programs usually need know only details of the composition of a set of randomly derived peptides from a protein as well as the N-terminal amino acid of each peptide. Some of the problems of computer programming are discussed in Chapter 9. In addition, the principles of overlap analysis and protein reconstruction by the manual operation are given in this chapter.

Primary sequence analysis is essentially a degradative process. The large protein is reduced to a set of smaller peptides each of which is then analyzed at the amino acid level in a stepwise manner from which is reconstructed information about the complete amino acid arrangement within the original protein. Yet, methods for the chemical syntheses of peptides or proteins are well within the range of intent of this book. Not infrequently, it is only by comparison against authentic synthetic models that fine structure analysis can be achieved in difficult or controversial cases. Towards this end, many of the methods used in preparing peptides of known structure are delineated in Chapter 10. The problems of synthesis are as varied as those of sequence analysis. Thus, the choice of functional group blocking agents is discussed with reference to the nature of the final structure of the desired peptide and to the methods available to release the free peptide from its masked intermediate form.

It is evident then, that in this volume, in the short span of a few 100 pages, one can find the answer to almost every problem that can arise during sequence analysis in the hands of the novice. The "how to do it" is here but, where details prove to be insufficient in the text, it is possible to turn to the almost 1200 references to the original literature. Each chapter stands alone as a primary reference to one aspect of the overall problem, that of sequencing a protein. Yet, the whole can only be the sum and continuity of its parts and represents in fact, an integrated presentation. This book serves as a road map through a challenging and thought-provoking adventure.

Humpty Dumpty sits on the wall. The egg is not yet broken. And so, at the end, it is time to begin.

Chapter 2

Physical Characterization of the Protein Molecule

Ken E. van Holde

Expending research effort on impure materials is a gross waste of time and talent. Yet there always exists the danger that a protein chemist, working with materials that are intrinsically hard to purify, may carry out detailed chemical studies with contaminated samples. It is the purpose of this chapter to outline some techniques by which at least the physical homogeneity of a protein may be judged. Such studies are a necessary prelude to detailed chemical investigations of proteins.

It must be understood at the outset that physical homogeneity is not synonomous with chemical homogeneity, nor *vice versa*. Polypeptide chains may differ in chemical composition in ways which cannot be detected by physical techniques now at our disposal; for example, two chains which differ only in a valine substitution for a leucine are likely to be indistinguishable by any physical method. Conversely, differences in state of aggregation which are easily resolved by the ultracentrifuge may result if a given polypeptide chain is placed in different solvent environments.

In this chapter, there is not space to describe the theoretical bases for the various techniques. To fully understand any of these methods, the reader must turn to the cited references. We shall be most concerned here with the applicability and reliability of various techniques and in particular to the kinds of heterogeneity which they detect.

I. The Accessible Physical Parameters

At the present time, there are a limited number of physical parameters which may be determined for macromolecules in solution, or which may be used for separation. These include molecular weight, density, charge, and in a somewhat less precise way, molecular dimensions. A macromolecular preparation can be heterogeneous in any one, or all, of these parameters. As an example, let us consider molecular weight. Three different situations may be distinguished:

(a) Homogeneity: All of the macromolecules in the sample have exactly the same molecular weight.

(b) Paucidispersity: There exists a small number of well defined molecular species. Impure protein preparations, or self-associating substances are frequently of this type.

(c) Polydispersity: The sample contains a wide range of molecular species. This kind of system is frequently found with synthetic (and some natural) polymers. Species may differ from one another only by one monomer molecular weight.

The way in which a heterogeneous system is best described will depend upon whether it is paucidisperse or polydisperse. In the former case, a statement of the weight or number fraction of the various species is often feasible. For a truly polydisperse substance, the *molecular weight distribution* provides the only complete description; one defines $W(M)dM$ as the weight fraction of material between M and $M+dM$. Alternately, one may recognize that most polydisperse systems are mixtures of polymers of a given monomer and give W_i the weight fraction of i-mer, as a function of i.

More frequently, a characterization of polydispersity (or paucidispersity) is given by the statement of several average molecular weights. Averages may be defined with respect to the weight fraction (W_i) of various species, as shown in Table 1.

Table 1. *Average molecular weights*

Designation	Defining equation	Measured by
Number average	$M_n = \Sigma W_i / \Sigma (W_i/M_i)$	Osmometry, vapor pressure, sedimentation equilibrium
Weight average	$M_w = \Sigma W_i M_i / \Sigma W_i$	Sedimentation equilibrium, light scattering
Z-Average	$M_z = \Sigma W_i M_i^2 / \Sigma W_i M_i$	Sedimentation equilibrium
$Z+1$ Average	$M_{z+1} = \Sigma M_i M_i^3 / \Sigma W_i M_i^2$	Sedimentation equilibrium
etc.		

It can be easily shown that for a molecularly homogeneous substance, $M_n = M_w = M_z = M_{z+1} = \ldots$, whereas for a heterogeneous material, $M_n < M_w < M_z < M_{z+1} < \ldots$. The identity of several properly measured molecular weights is a critical test for homogeneity.

Heterogeneity in molecular weight is by no means the only kind. In every measurable physical parameter, a sample may or may not be homogeneous. For example, preparations may be obtained which are roughly homogeneous in molecular weight, but heterogeneous in density. Two polypeptide chains may have almost exactly the same weight, but differ in charge, because of amino acid substitution. Finally, it is possible, at least in principle, that a molecularly homogeneous protein might exhibit conformational heterogeneity. It is because of these possibilities that more than one physical technique is usually needed to establish the purity of a protein preparation.

We shall not be much concerned in the chapter with grossly heterogeneous substances, since such are not likely to be subjected to the kind of analysis described in later chapters. Rather, we shall emphasize the study of paucidisperse materials.

II. Analysis of Mass: Sedimentation Methods

A. Sedimentation Velocity

Nowadays, the first test to which a "purified" protein preparation is usually put is a sedimentation velocity experiment. The analysis is all too often incomplete; in some cases only a single experiment at a single rotor speed and concentration is reported. A typical (if hypothetical) result is shown in Fig. 2-1. In this example, two boundaries are observed. The major component, we shall say, sediments with a sedimentation coefficient (s) of 5.4×10^{-13} sec. The sedimentation coefficient has

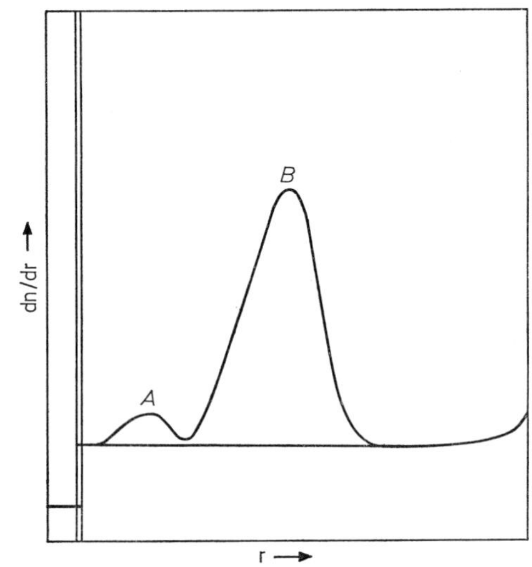

Fig. 2-1. A schlieren pattern from sedimentation of a hypothetical protein. The main component (B) is contaminated with a small amount (ca 10%) of a more slowly sedimenting impurity. The main component may or may not be homogeneous; the slight skewing of the boundary could result from concentration dependence of s

been calculated [874] by measuring the distance from the center of rotation of the maximum in the concentration gradient curve at various times:

$$s = \frac{dr_b/dt}{\omega^2 r} = \frac{\text{velocity of boundary motion}}{\text{centrifugal field strength}} \tag{1A}$$

or

$$s = \frac{\ln r_b/r_o}{\omega^2 t} \quad \text{the integrated form of (1A)}\,. \tag{1B}$$

Here r is the distance from the center of rotation, t is time, and ω is the angular velocity of the rotor. Following such an analysis, the statement will frequently be made that "with the exception of a small amount of slowly sedimenting material, the protein appeared to be homogeneous in the ultracentrifuge."

A statement of this kind is nearly devoid of meaning. In the first place, what is meant by "a small amount" of slowly sedimenting material? To give a more precise statement, one might determine the fraction of small component by measurement of the areas of the two peaks. This will give (approximately, see below) the *weight* fractions of the component. (More exactly, it will give the refractometric fraction, probably about the same thing.) How is this to be determined? It must be remembered [814] that the area under each peak will diminish with time because of the radial dilution effect. Precisely, the apparent concentration (c_p) will vary as

$$c_p = c_0 \exp(-2s\omega^2 t) \tag{2}$$

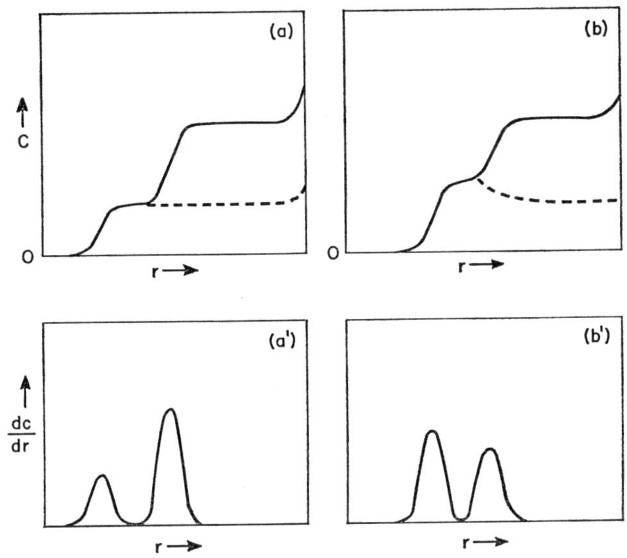

Fig. 2-2. A schematic representation of sedimentation of two-component mixtures without (a, a′) and with (b, b′) the Johnson-Ogston effect. In each case the upper figure represents concentration as a function of distance, the lower the concentration gradient

where the s value is for the boundary in question, and c_0 the initial concentration. Boundaries with different velocity will vary differently in apparent area with time; the area of each should be extrapolated back to zero time. However, this may still not yield an accurate representation of the relative amounts of the two components. If the sedimentation coefficients of these materials vary with protein concentration, as is usually the case, an irksome complication called the "Johnson-Ogston" effect will be involved. This is illustrated in Fig. 2-2. The slow component will sediment more rapidly in the region behind the fast boundary, where it is not interacting with the fast component. It will therefore "pile up" behind the fast boundary, leading to an overestimation of the amount of slow component and an underestimation of the amount of fast component. This effect can be corrected for [see 874] but the correction requires data not usually available at this level of analysis.

In many cases (especially if the boundaries differ greatly in s) the Johnson-Ogston effect is not too serious. In this event, correction for, or extrapolation for the effect of radial dilution leads to a reasonable analysis. Suppose the weight fraction of slowly sedimenting material is small (say 10%). Is this significant? The answer depends upon what kind of studies are to be made with the mixture. Suppose that end groups are to be determined, and that the molecular weight of the slow component is only $^1/_4$ that of the fast. Then about 31% of the *molecules* (number fraction) will be of the smaller species. Such contamination might be disastrous for studies of this kind. It would be essential to subject the sample to some kind of fractionation procedure (gel permeation chromatography, for example — see section IV) before continuing study. A "minor" contaminant may not be inconsequential!

Let us analyze the statement further. It is implied that the major component is homogeneous, simply because a single boundary is observed. *There are many ways in which a heterogeneous substance can yield a single, fairly sharp boundary.* To list a few:

1. The substance may be pauci- or polydisperse, with a number of components of very similar sedimentation coefficient, not resolved because of diffusion broadening of the boundary. This danger is particularly acute with low molecular weight ($M < 20{,}000$) substances, for diffusion will be rapid and sedimentation slow. Such mixtures will not give boundaries of the same shape as would be expected for a single component. However, there is no easy test to apply, for the shape of the boundary for even a homogeneous material is not strictly a Gaussian diffusion boundary [335]. Furthermore, even the sum of two Gaussian curves *looks* Gaussian if the peaks are not resolved. Problems of boundary resolution are considered in some detail in Svedberg and Pedersen [1023].

2. Because of concentration-dependent sedimentation, the boundary may be sharpened. A word of explanation will suffice here: In general, s will decrease with increasing solute concentration. This means that molecules on the low concentration edge of a boundary will tend to catch up with molecules on the high concentration side. In extreme cases, this can cause a very heterogeneous substance to appear homogeneous if the concentration used to load the cell is too high.

3. Substances which are in rapidly reversible equilibrium may not be resolved in sedimentation. For example, a substance in rapid monomer-dimer equilibrium will yield a single boundary with s varying between that of monomer and dimer as the concentration is increased [345].

The pitfalls described above could be largely avoided by more experiments. To critically test the homogeneity of a low molecular weight substance, the highest rotor speeds and lowest practical concentrations should be employed. Variation of the concentration over wide limits will usually detect concentration-dependent boundary sharpening and rapid equilibria. For these reasons, the frequent admonitions that sedimentation velocity data be extrapolated to zero concentration are not nit-picking.

Even after the best has been done, the sedimentation velocity method does not easily give *assurance* of homogeneity. The very best would be this: to use a known diffusion coefficient *and* concentration dependence of s to predict the boundary shape at different times [335], and compare with observation. This is almost never done. However, intelligent use of the sedimentation velocity method, utilizing variation of both rotor speed and concentration, can at least provide evidence against gross paucidispersity.

B. Sedimentation Equilibrium

For critical tests of homogeneity in mass, the sedimentation equilibrium method is powerful. This is because it allows (with proper care) the unambiguous determination of several molecular weight averages.

Let us consider what is observed in such an experiment. We shall not be concerned here with either the theory of sedimentation equilibrium or details of experimental procedure. For these, the reader is referred to SCHACHMAN [874], FUJITA [335], or to a number of review articles. (WILLIAMS *et al.* [1138], VAN HOLDE [1076], TRAUTMAN [1066].) Instead, we shall emphasize the problems and potentialities in the analysis of the sedimentation equilibrium results.

Suppose that the ultracentrifuge has been operated, at a relatively slow speed, for a time sufficient for equilibrium to be attained. Then for a homogeneous solute, or for each component in a mixture, the concentration will be distributed exponentially in the cell. At a point r cm from the center of rotation the concentration will be:

$$c(r) = c(a) \exp\{M(1 - \bar{v}\varrho)\, \omega^2(r^2 - a^2)/2\,RT\}. \tag{3}$$

Here a is the distance of the meniscus from the center of rotation, $\bar{v}$ is the partial specific volume of the solute, ϱ the solution density, R the gas constant, and T the absolute temperature. This equation assumes that the solution behaves ideally. In practice ideal behavior can only be approached, (although often very closely) in experiments at finite concentration. The use of this equation requires that the concentration, or some quantity proportional to the concentration, be measured at different points in the ultracentrifuge cell. This will be easy to do if a scanning absorption optical system is employed; it may even be possible with such an optical system to measure independently the concentrations of different components. More commonly, however, one uses the schlieren or Rayleigh optical systems. The former yields the refractive index gradient, the latter the refractive index difference between any two points. Obviously, the schlieren patterns can be integrated to yield concentration differences between points, as well. Note, however, that neither yields the *absolute* concentration, needed for Eq. (3). To cope with this problem, two different lines of approach have been used :

1. In the "high speed" method (also called "Yphantis method" or "meniscus depletion method" [1167] the rotor velocity is chosen sufficiently high that the concentration near the meniscus becomes negligeably small (see Fig. 2-3A). Then concentration differences between point a and any other point (r) yield the total concentration at r.

2. In the "low speed" method [1077], (Fig. 2-3B) the concentration is not zero at the meniscus; ideally the concentration ratio between the ends of the cell is about 3:1. To obtain the meniscus concentration (and thus the concentration at other points) one makes use of the conservation of mass, which yields

$$c(a) = c_0 - \frac{1}{b^2 - a^2} \int_a^b (b^2 - r^2) \frac{dc}{dr}\, dr \tag{4}$$

where b is the bottom of the solution column, and c_0 is the initial concentration. The latter is frequently determined in terms of area on a schlieren diagram, or number of interference fringes, in a "synthetic boundary" experiment in which solvent is layered over solution.

With $c(r)$ known as a function of r, a substance may be tested for homogeneity by use of a form of Eq. (3). Taking natural logarithms, Eq. (3) converts to the form

$$\ln c(r) = \text{constant} + M(1 - \bar{v}\varrho)\,\omega^2 r^2/2\,RT\,. \tag{5}$$

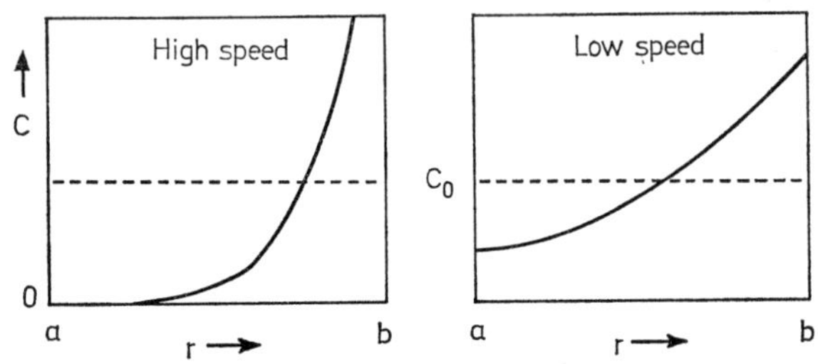

Fig. 2-3. The difference between the "low speed" and "high speed" sedimentation equilibrium methods. See text

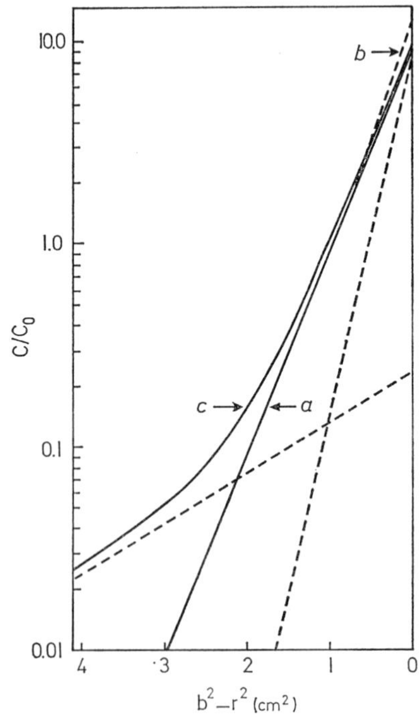

Fig. 2-4. Approximate graphs of ln c vs r^2 for hypothetical mixtures. In each case the high speed method has been used. Curve (a) represents a homogeneous protein. In curve (b), 10% of this material has irreversibly dimerized; in curve (c) 10% has irreversibly dissociated to 1/4 molecules. The dotted lines show expected slopes for dimers and 1/4 molecules. For (b), $M_z/M_w = 1.18$; for (c), $M_z/M_w = 1.06$

Thus, for a homogeneous substance, a graph of ln c vs r^2 should be a straight line. If the substance is not homogeneous in molecular weight, the total concentration (which is what will be determined by most optical methods) will be a sum of exponential functions, and ln $c(r)$ vs r^2 will show upward curvature. The observation of such curvature is direct evidence for heterogeneity but appreciably heterogeneous substances may give, under some circumstances, nearly straight lines. The low speed and high speed methods differ somewhat in sensitivity to heterogeneity.

Methods for treating the data so as to most easily detect heterogeneity are discussed by FUJITA and WILLIAMS [336]. In general, the high speed technique, which produces greater resolution of components in the cell, will be more sensitive. In Fig. 2-4 are shown the kinds of ln c vs r^2 curves to be expected for several mixtures. Note that the high speed technique is capable of revealing a fairly small amount of low-molecular weight contaminant. With high molecular weight impurities it does not fare quite so well, since it may not be possible to resolve data all of the way to the bottom of the cell. This can be remedied by carrying out experiments at very low initial concentrations, but under these conditions the ultracentrifuge must be operating very smoothly to avoid convection [1167].

A rather different way of detecting heterogeneity involves the computation of different average molecular weights. In this calculation the low-speed method has considerable advantages, for since the entire solution column can be examined, average weights for the entire sample can be determined. For example, the weight average can be easily measured from the total concentration difference across the cell

$$M_w = \frac{2\,RT}{\omega^2(1-\bar{v}\varrho)\,(b^2-a^2)}\,\frac{c(b)-c(a)}{c_0} \tag{6}$$

and the Z-average from

$$M_z = \frac{2\,RT}{\omega^2(1-\bar{v}\varrho)}\,\frac{\frac{1}{b}\left(\frac{dc}{dr}\right)_b - \frac{1}{a}\left(\frac{dc}{dr}\right)_a}{c(b)-c(a)}\,. \tag{7}$$

If this calculation yields $M_z > M_w$, the sample must be heterogeneous. Values of M_z/M_w for the mixtures shown in Fig. 2-4 are given in the legend. Since, with careful work, M_w can be determined to better than 1%, and M_z to a few percent, relatively small heterogeneity can be detected. Under some circumstances, other averages may be obtained, to confirm the analysis.

In a single experiment, neither identity of two averages, nor apparent lack of curvature in ln c vs r^2 graphs *prove* homogeneity. This is because macromolecular solutions may be non-ideal. In general, non-ideality will tend to give apparent molecular weight values which decrease with increasing concentration. This will tend to produce downward curvature in the ln c vs r^2 graph; such downward curvature may nearly compensate for the upward curvature due to heterogeneity (see Fig. 2-5). Similarly, non-ideality will usually depress the apparent Z-average molecular weight more than the weight average. It is possible, then, to have $M_z^{\mathrm{app}} = M_w^{\mathrm{app}}$ for a non-ideal, heterogeneous substance at some finite concentration. Such pitfalls can usually be avoided by carrying out experiments at several concentrations, and extrapolating apparent molecular weights to zero concentration.

Usually, aqueous solutions of globular proteins are not too non-ideal, so the above dangers are not too severe. Exceptions are:

1. Highly asymmetric proteins.
2. Proteins dissolved in denaturing solvents, such as guanidine hydrochloride.

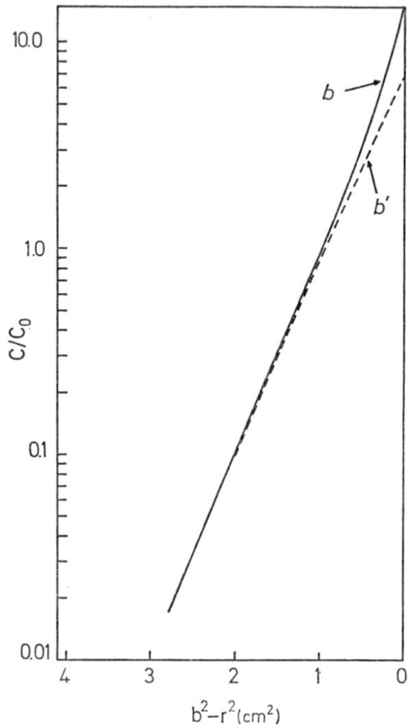

Fig. 2-5. A schematic representation of the effects of non-ideality. Curve (b) represents the same mixture as curve (b) in Fig. 4; (b′) shows the effect of appreciable non-ideality on this graph

3. Proteins carrying a high net charge, especially when dissolved in buffer of low ionic strength. It is a good general rule to have the ionic strength at least 0.1, unless the protein is isoelectric.

In such cases, investigation at more than one concentration is *imperative*.

C. Density Gradient Techniques

The development of density gradient methods has been of enormous utility to the investigation of crude and partially purified protein preparations. They have been less used, however, for the kind of problem we consider here — the proof of homogeneity in materials to be subjected to detailed chemical analysis. We shall consider the reasons for this, and some ways in which gradient methods can be useful.

Of the two kinds of density gradient methods — sedimentation equilibrium in a density gradient [682] and density gradient sedimentation velocity [157], the former has been little employed with proteins. This is probably because the breadth of the band which forms at the isodensity point is inversely proportional to $\sqrt{M}$; most proteins, being fairly low molecular weight macromolecules, will form very broad bands, making resolution difficult. Furthermore, many proteins are insoluble in the concentrated salt solutions needed for such studies.

It is quite possible, however, that this technique will become of greater importance in the study of certain kinds of conjugated proteins, such as lipoproteins and glycoproteins. Heterogeneity in such substances may reside largely in the amount of bound lipid or carbohydrate, with an accompanying heterogeneity in specific volume. Already, this technique has been used with good success in investigations of lipoprotein preparations (see, for example [1136]).

Sedimentation velocity studies in density gradients (often sucrose gradients) are customarily used for rather different purposes. By the use of isotopic labeling, or by enzymatic assay, it is frequently possible to study the sedimentation of proteins in very dilute solution or in grossly contaminated mixtures. For the kind of material which is considered in this volume — available in moderate quantities and at least fair purity, the density gradient technique has little to recommend it over the more conventional sedimentation velocity measurement. In general, s values cannot be obtained as precisely by this method, nor is the resolution markedly superior. However, one combination of techniques would seem to provide excellent evidence for homogeneity. If both enzymatic activity and total protein concentration can be determined, the demonstration of uniform *specific* activity across a sedimenting band should be a quite convincing demonstration of purity. The criterion seems to have been little used to date.

III. Analysis by Charge: Electrophoresis

All proteins are polyampholytes, and carry an electric charge, determined by the amino acid composition, pH, ionic strength, and the nature of the buffer ions. Since a homogeneous protein should have, under given conditions, a definite value for this charge, methods which can effect separation on the basis of charge are powerful tools in characterization. The simplest way in which such separation can be effected is on the basis of the migration in an electric field: *electrophoresis.*

A rigorous analysis of the process of electrophoresis is beset with difficulties [1036].

If we consider the simplest possible situation, an isolated charged particle in a non-conducting medium, the velocity of migration can be calculated by equating the Coulombic force on the particle to the force of viscous resistance to motion through the solution.

$$fv = ZeE. \tag{8}$$

Here v is the velocity, f the frictional coefficient, Z the number of electron charges (of magnitude e each) and E the electric field. The *electrophoretic mobility* is defined as the ratio of velocity to field:

$$U = \frac{v}{E} = \frac{Ze}{f}. \tag{9}$$

For spherical particles of radius a, $f = 6\,\pi\eta a$ (Stokes' law), so

$$U = \frac{Ze}{6\,\pi\eta a} \tag{10}$$

where a is the molecular radius and η the solvent viscosity. These equations embody the most important statement about electrophoresis: To a first approximation, the mobility is proportional to the charge.

Unfortunately, such a treatment misses many important complications involved in the study of electrophoresis of proteins in aqueous solutions. A protein is not an isolated particle. It is surrounded by an environment of other small ions; in general these must form an ion atmosphere, predominantly of opposite charge, which will interact in a number of ways with the motion of the particle. For these reasons, Eqs. (8 to 10) are rough approximations, and it is not a straightforward calculation to obtain the net charge Z from mobility data.

However, for purposes of separation or the determination of homogeneity, such complications are not of serious consequence. The fact remains that two protein molecules, identical except for a difference in a single charged group, may be separated and distinguished by electrophoresis. Thus, the method is remarkably sensitive to the parameter for which it is designed.

A wide variety of techniques have been used for electrophoresis. These differ greatly in their sensitivity, the amount of protein required, and the conditions under which each may be employed. A few of the more widely used methods are described below:

A. Moving Boundary Electrophoresis

This is the classical method developed by TISELIUS [1059]. While little used today, it is still probably the most accurate method for the determination of mobilities and quantitative analysis of mixtures. Since the method is described in detail in the older literature, (see, for example, the review by ALBERTY [18] a brief description here will suffice. The apparatus used, as shown in Fig. 2-6, is essentially a U-tube, fitted with sliding partitions to facilitate the formation of boundaries and side tubes to hold the electrodes. In operation, the apparatus is filled as shown, the boundaries are formed, displaced into the observation channels, and a constant current is applied. The positions of the boundaries, and the refractive index changes across them are determined by observation through a schlieren or interferometric optical system. The entire assembly is immersed in a constant temperature bath, usually kept at 4 °C to minimize the possibility of convection.

In such an apparatus, the rate of motion of the boundaries can be measured with precision. It will be found, however, that the ascending and descending boundaries (Fig. 2-7) do not move at the same rate, nor will the areas or shapes of corresponding boundaries in the ascending and descending limbs be the same. These complications occur because the electrical fields in the boundary regions will differ; this is because the current must be the same throughout the cell, whereas the conductivity differs in the various regions. Furthermore, the original protein solution, even though dialysed against the buffer, will not contain exactly the same concentration of small ions as the buffer, because of the Donnan Effect. This will give rise to the small stationary boundaries at the original boundary positions. As a result of these complications, the quantitative analysis of materials by moving boundary electrophoresis is by no means simple. The reader is referred to the review by ALBERTY [18] for more details.

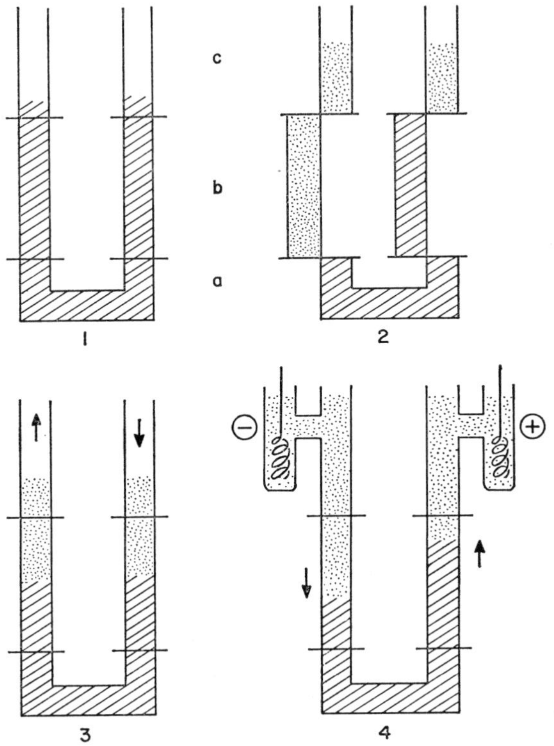

Fig. 2-6. Formation of boundaries in a Tiselius electrophoresis apparatus. In step one, both limbs of the cell are filled with protein solution (shaded). Next (step 2) protein is removed from the top of the right limb and the top and center of the left, and replaced with buffer, (dotted). Note that the partitions have been slid to allow this. In step 3 the partitions have been closed, and the boundaries formed. They have then been displaced to the center of the middle limbs for observation. Finally (step 4) electrophoresis commences. The electrodes are shown in this case

Because the method is somewhat awkward, because large quantities of protein are required, and because the separation of the materials is far from complete, the Tiselius method is no longer widely used. Rather, in recent years attention has been centered on a group of techniques in which supporting materials are utilized to stabilize the migration of bands or zones of protein components. These may be broadly described by the term *zonal electrophoresis*.

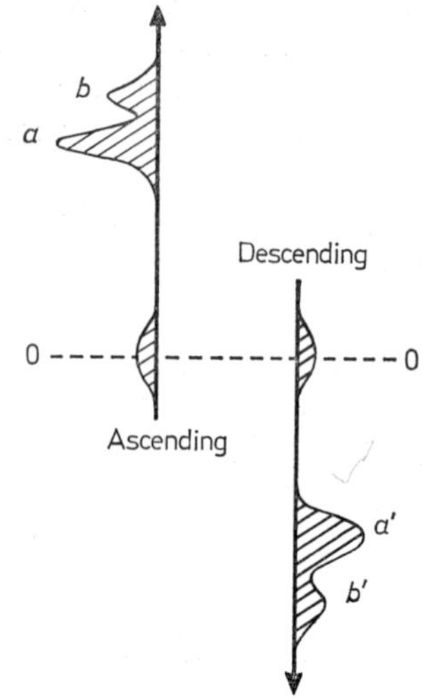

Fig. 2-7. Ascending and descending patterns for the electrophoresis of a hypothetical two-component mixture. Note that the forms of boundaries (*a*) and (*a'*, *b*) and (*b'*) are not identical, and that small stationary boundaries are left at the origin (O). See text

B. Zonal Electrophoresis on Supporting Media

The development of suitable solid, permeable supporting media has allowed the exploitation of the possibilities of zonal techniques. The general difference between these techniques and moving boundary electrophoresis is illustrated by comparing Figs. 2-7 and 2-8. Whereas in the Tiselius technique only the fastest and slowest components can be isolated, the electrophoresis of a zone through a solid medium which prevents convection can lead to a "spectral" analysis of the material into

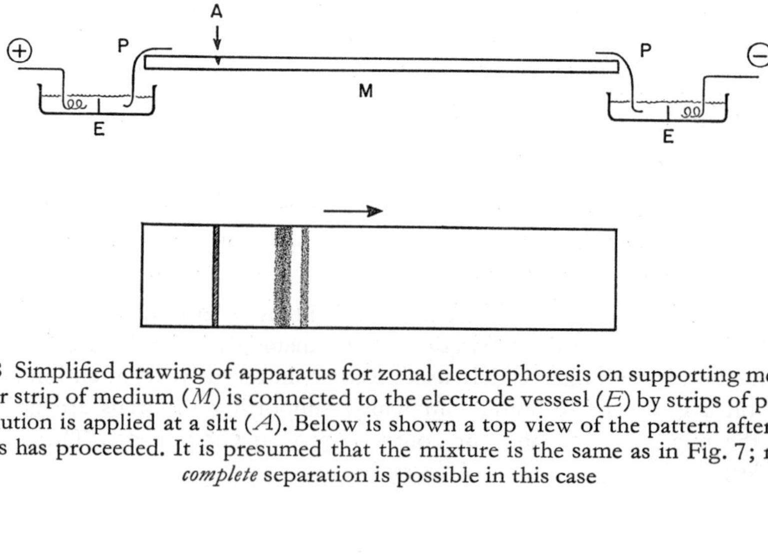

Fig. 2-8 Simplified drawing of apparatus for zonal electrophoresis on supporting media. The block or strip of medium (*M*) is connected to the electrode vessesl (*E*) by strips of paper (*P*). The solution is applied at a slit (*A*). Below is shown a top view of the pattern after electrophoresis has proceeded. It is presumed that the mixture is the same as in Fig. 7; note that *complete* separation is possible in this case

individual components. These bands may be revealed by staining, by autoradiography, by fluorescence, or by elution of strips of the medium followed by assay. The latter technique allows at least partial recovery of the individual components. Furthermore, the methods are often rapid, and may require only minute quantities of protein.

With all of these advantages, it is small wonder that zonal techniques have largely superseded the old moving boundary method. However, some price is paid in the use of solid media. In the first place, the rate of migration through such a substance as starch, polyacrylamide, or paper is even harder to account for theoretically than the electrophoretic mobility in aqueous solution. The viscosity appropriate to Eq. (10) is difficult to define, the effective cross-section of the cell is ambiguous, and interaction between the solute and the supporting media (absorption, gel filtration effects, etc.) further complicate the picture. For these reasons, analysis of zonal electrophoresis on solid media is almost always carried out empirically, by comparison of the distance of migration of the substance being investigated with that of standard substances.

Table 2. *Some solid media for zonal electrophoresis*

Medium	Description	Advantages	Disadvantages	Reference
Paper	Filter paper, moistened with buffer, held between electrode chambers.	Fast, simple, requires only small amounts of protein.	Endosmotic effects quite severe, evaporation may be a problem.	[47]
Cellulose acetate strip	Strips of thin cellulose acetate sheet, used much like paper.	Can use exceedingly small quantities of protein.	Somewhat more difficult to use than paper.	[559]
Starch block	A slurry of starch placed in a trough connected to electrodes.	Simple to prepare, can be used for fairly large scale separations. Elution of products fairly easy.	Endosmosis, impurities in starch, gel filtration effects.	[586]
Starch gel	A gel is prepared from partially hydrolysed starch, in trough or between vertical plates.	High resolving power, great versatility.	Some impediment of motion of large proteins through the gel, removal of samples more difficult.	[971]
Agar gel	Like starch gel, but more dilute gel can be employed.	Open gel structure gives conditions approaching free electrophoresis. Little absorption.	Endosmosis can be severe in alkaline solutions.	[361]
Polyacrylamide gel	Acrylamide is polymerized in small tubes to form gel. These are placed between electrode chambers.	Easy to work with, clear, controllable in porosity. No contaminants, endosmosis negligible.	Generally restricted to small scale since the gel is of low electrical resistance.	[815] [979]

A further complication, peculiar to solid media, may or may not be of importance, depending upon the chemical nature of the medium. This is endosmosis, the migration of solution through a charged solid under the influence of a field. These effects depend upon the nature and magnitude of the charge on the supporting medium, and the nature and ionic strength of the buffer. In unfavorable cases, endosmosis may be so great as to more than compensate for the electrophoretic mobility of the protein solute.

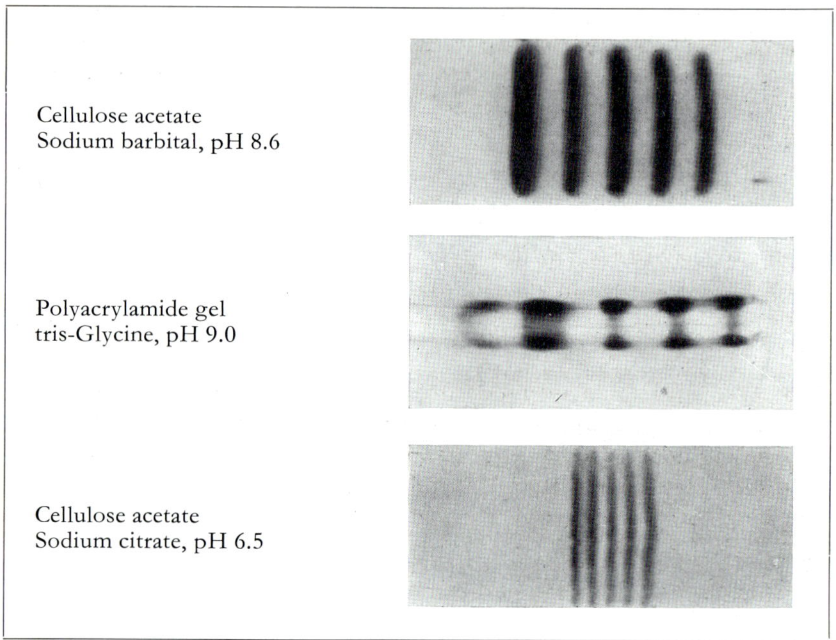

Fig. 2-9. Polyacrylamide gel electrophoresis and cellulose acetate electrophoresis pattern for the isozymes of aldolase A and C. From Penhoet *et al.* (1967)

A very wide variety of supporting media have been employed. We shall not here attempt to describe the various experimental procedures, which differ considerably in detail. Instead, general information, together with specific references, are given in Table 2. This is only a partial list; an enormous number of substances have been tried. Not listed here are a number of other media (cellulose powder, glass beads, etc.) most often used in preparative work in column apparatus. Choice of a technique for a particular problem may require considerable investigation. However, for rapid, convenient, and sensitive studies, of homogeneity, the acrylamide gel technique has a great deal to recommend it.

In Fig. 2-9 are shown patterns obtained by this method and by cellulose acetate electrophoresis with isozymes of aldolase.

IV. Analysis by Molecular Size: Gel Permeation

Recent years have seen the rapid development of a new class of methods for the determination of heterogeneity, and the rough measurement of molecular weight. These are the so-called "gel permeation" or "gel filtration" techniques. Basically, the idea is simple. Small cross-linked particles of a substance like dextran or agarose will be swollen by solvent, and permeable to small molecules. But their permeability to large molecules, such as proteins, will depend upon the "mesh" of the cross-linking. If such particles are packed into a column, we may distinguish two volumes, one accessible to all molecules (the so-called void volume, V_0, the volume between the particles) and the other accessible only to molecules able to penetrate completely into the gel particles. This latter internal volume, V_i, is determined by the dry weight of gel and its solvent uptake per gram.

Different solute molecules, which can penetrate to varying degrees into the gel particles, will thus have different volumes available to them in a column made up of such particles. Then in elution from such a column, the elution volume (V_e) will be given by

$$V_e = V_0 + K_d V_i \tag{11}$$

where K_d is a distribution coefficient ranging from zero for very large solute molecules to unity for very small ones.

There have been two general points of view as to the mechanism which accounts for the values of K_d. On the one hand, it has been assumed that equilibrium is maintained in the column, and that K_d represents that fraction of the internal volume available to a given type of molecule. The opposing view holds that the rate of diffusion of molecules into the gel particles is the controlling factor. The experiments of Ackers [5] indicate that, at least to some extent, diffusion must play a role. While equilibrium and column determination of K_d were in good agreement for tightly crosslinked gels, for the coarser gels and larger proteins considerable discrepancies were found. Furthermore, calculations based on the Stokes' radius, as given by diffusion coefficients led to good agreement with the data. The Stokes' radius (a) is an empirical measure of effective molecular size, defined by the Stokes-Einstein diffusion equation

$$D = \frac{kT}{6\pi\eta a} \tag{12}$$

where D is the diffusion coefficient, k is the Boltzman constant, T the absolute temperature, and η the solvent viscosity.

Thus, from these researches, it appears that gel filtration measures some quantity closer to a diffusion coefficient than the molecular weight itself. It should be emphasized strongly that the Stokes' radius is neither an actual molecular radius nor a direct measure of molecular weight. For example, an elongated molecule will have a Stokes' radius (as determined from D) which does not correspond to any particular dimension of the molecule itself, and which is different from the value of a for a spherical molecule of the same weight. Furthermore, it is quite possible that K_d will not measure the same value of a as does D, in the case of a very asymmetric particle. This means that the use of gel permeation as a measure of molecular weight is no more straightforward than the use of either s or D alone for this purpose. The good correlations which have been presented between K_d and M (see Fig. 2-9) derive in part from the fact that series of globular proteins, of roughly comparable hydration and asymmetry, have been used.

While these remarks are meant to caution against uncritical reliance on gel permeation as a method for accurate molecular weight determination, they are not meant to detract from the importance of the technique. As a preparative method, as a way for testing for size heterogeneity, and as a rapid means for rough M determination, the method is superb.

Two general techniques have been employed. In the column method [795], the gel beads are equilibrated with buffer and packed in a chromatographic column. A band of the protein solution is applied to the top, and eluted with the same solvent. Marker substances, such as low molecular weight compounds and large colloids (india ink or "blue dextran") are often added to define elution volumes corresponding to $K_d = 1$ and 0 respectively. In some applications, recycling is employed, to increase resolution [795]. Typical results, showing the heterogeneity of a ribonuclease sample caused by irreversible aggregation, are shown in Fig. 2-11.

In the second method, a layer of the gel is laid down (by use of a special applicator) to uniform thickness on a glass plate. The solution to be tested is placed on a spot, and elution carried out along the length of the plate. Staining then reveals the migration of the protein solutes. This method requires, of course, only minute quantities of the sample. Its resolution does not appear to be as good as that of the column methods, especially since recycling can be employed in the latter.

V. Subunits: The Special Problems of Dissociating Proteins

It has become recognized that many, if not perhaps the majority of proteins exist and function as multi-subunit structures. This has both complicated and simplified sequence determinations. The complications arise from the fact that such subunit structure must be recognized, the differences or identities between the subunits established, and methods of separation must be worked out. Simplification results from the fact that sequencing for two or more small polypeptide chains is much easier than for a single long chain. In any event, it is imperative that both the number and kinds of subunits be established before extensive chemical studies are begun. In this section, we shall be concerned with two main problems; (A) the dissociation of a protein into its constituent polypeptide chains and (B) the determination of the number of chains, and whether or not they are identical.

A. Dissociation Processes

The technique to be used for the breakdown of a protein will depend upon whether the chains are covalently or non-covalently linked. In the former case, chemical methods must be employed; in the latter, any of a number of modifications in solution environment may result in dissociation. Let us consider the latter case first.

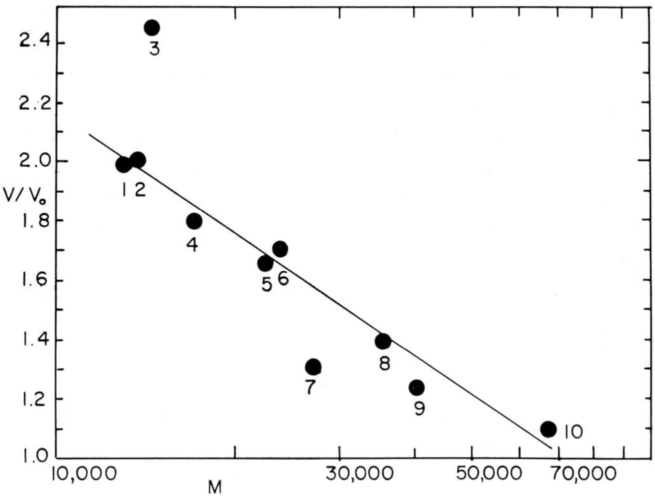

Fig. 2-10. Elution volume in gel filtration as a function of molecular weight. Conditions: 1.2 × 184 cm column of G-75 Sephadex, 25 °C to 27 °C, pH = 6, ionic strength = 0.494. Proteins are identified as follows: 1 cytochrome c, 2 ribonuclease, 3 lysozyme, 4 myoglobin, 5 α-chymotrypsin, 6 trypsin, 7 ovomucoid, 8 pepsin, 9 peroxidase, 10 serum albumin. The two which behave most anomalously are the glycoprotein, ovomucoid, and lysozyme, which has a carbohydrate substrate. The data are taken from WHITAKER (1963)

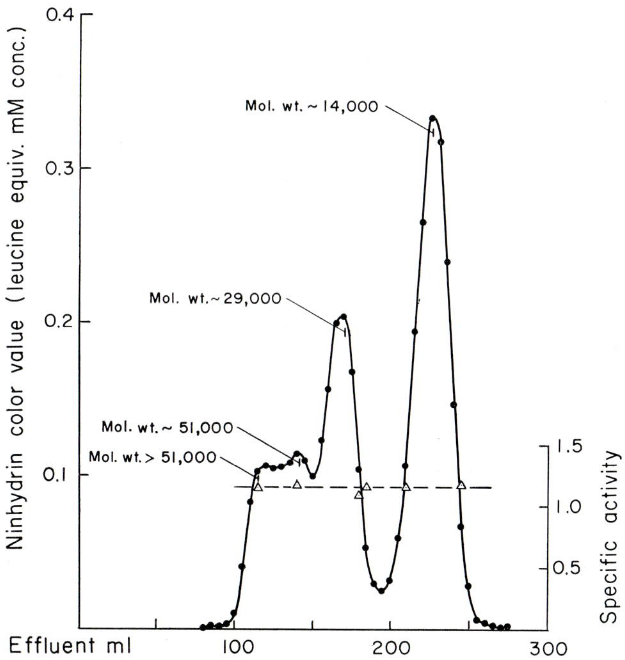

Fig. 2-11. Resolution of ribonuclease A and its aggregates on Sephadex G-75. Monomer dimer, trimer, and higher aggregates can be detected. The specific activity against cyclic CMP is also shown. Data of CRESTFIELD, STEIN, and MOORE (1962)

The non-covalently linked multi-chain structures are generally held together by relatively weak forces (hydrogen bonds, hydrophobic bonds, electrostatic interactions, etc.). For this reason even mild pertubation of the solution environment may cause dissociation. Some of the methods which have proved successful are:

1. Change of pH

This is the simplest of all procedures, and is successful in a considerable number of cases. The critical pH ranges appear to be around pH = 3 to 4 (the carboxyl titration region) and 9 to 10 (the lysine-tyrosine titration region). While simply titrating a protein to pH below 3 or above 10 to cause dissociation appears ideally simple, there are a number of complications which should be considered. In the first place, it is desirable to avoid extremes of pH, lest especially labile bonds be hydrolysed, even at moderately low temperatures. On the other hand, at moderate pH values dissociation may not be complete. This can lead to serious errors in the determination of the weights, and hence numbers of the subunits. Furthermore, physical characterization of highly charged subunits is complicated, in every case, by the existence of charge effects. These will appear, in the most part, as strong solution non-ideality. Most of these problems can be dealt with, but caution and a thorough program of experimental work are required. A case in point is the investigation of aldolase [229]. Dissociation was effected in pH 2.9 citrate buffer. The subunits appeared, from sedimentation equilibrium experiments, to be nearly homogeneous, with a molecular weight of about 50,000. Since the weight of native aldolase was then believed to be 152,000, three equal subunits were postulated. Of course, this answer is now known to be incorrect; there are clearly four units [779, 529].

Apparently, the error resulted from a combination of circumstances: (a) An incorrect weight for the native protein; it is now believed to be about 160,000. (b) Incomplete dissociation, and (c) non-ideality, which masked the heterogeneity of the "subunit" solution. More experiments should have revealed these problems.

2. Chemical Modification of the Charge

In some cases, multi-subunit proteins which do not dissociate in a practicable pH range may do so if positive or negatively charged groups are added by chemical modification. The classic example of this type is the use of succinylation to add negative charges [566].

Since such chemical modification is sometimes difficult, and complicates further chemical studies, it should perhaps only be considered as a next step if some milder methods fail. Of course, the comments made above concerning incomplete dissociation and charge effects carry equal force here.

3. Strong Denaturants (Guanidine Hydrochloride, Urea)

It has been known for a long time that concentrated solutions of materials like urea or guanidine hydrochloride denature most proteins. More recently, it has been realized that this denaturation is almost invariably accompanied by the dissociation of a protein into its constituent polypeptide chains. This is most clearly pointed out in the recent studies by TANFORD and co-workers [1037].

The principal disadvantages in employing such agents result from the high concentrations required. Solutions containing 6M guanidine hydrochloride or 8M urea are clearly unsuitable for many chemical studies, and physical investigations are seriously complicated. Removal of the denaturant often leads to re-association and/or random aggregation. Sedimentation velocity, sedimentation equilibrium, and light scattering experiments are all complicated by selective binding of either the denaturant or water by the protein. Such effects can lead to serious errors in molecular weight. To use the example of aldolase again, two competent groups have arrived at different results for the number of subunits [529, 875].

A study of these papers reveals the complexities encountered with concentrated denaturants.

4. Detergents

Substances like sodium dodecyl sulfate (SDS) act as denaturants and dissociating agents at much lower concentrations (ca 0.01 M, in many cases). However, since their action apparently involves very strong binding to the protein, with accompanying modification of the weight and net charge, they too lead to difficulties in accurate molecular weight measurements.

5. Reduction or Oxidation of Disulfide Bonds

If none of the agents listed above is successful in dissociating a protein believed to contain subunits, the possibility that these are interconnected by disulfide cross-links should be considered. In such cases reduction (with a thiol reagent such as sulfite [188] or mercaptoethanol [42], or oxidation (with performic acid, for example [413]) may be attempted. Such procedures are straightforward, but may require the accompanying use of a denaturing agent like urea or guanidine hydrochloride. In any event, the chemical study of the protein will be complicated by the fact that intrasubunit disulfides, as well as the intersubunit groups, may be destroyed in the process.

B. Determination of the Number and Kinds of Polypeptide Chains

Presuming that dissociation can be achieved, determination of the number of chains may still not be easy (as the aldolase controversy has shown). Measurement of the weight average-molecular weight alone will usually not be sufficient, for if the chains are of more than one type, an average value may be obtained which is not a simple fraction of the weight of the native protein. In fact, it is the number average molecular weight which is most useful in such circumstances. Consider the following example: a protein of weight 100,000 consists of four chains, two each of weights 10,000 and 40,000. Upon dissociation, the weight average molecular weight of the subunit mixtures would be 34,000. A premature conclusion from this data alone would be that the protein was a trimer! Of course, in such an extreme case, resolution of the two kinds of subunits by either velocity ultracentrifugation or gel filtration should be feasible. The example points up, however, the usefulness of osmometry or vapor pressure in measurements in such cases, for these techniques yield the number average molecular weight. The number average molecular weight of the products of dissociation will always give exactly the number of subunits formed [529].

In the above case, the result would be 25,000, showing that there are four subunits.

The question as to whether more than one type of subunit is involved is frequently more difficult to answer. In the example cited above the situation could be easily clarified, but more frequently the subunits are quite similar, though of more than one type. It is in this situation where electrophoretic methods become the most useful of the physical techniques, for very frequently the subunits will differ in charge.

In summary, a general protocol for the characterization of multi-subunit proteins might be described as follows:

1. Establish molecular weight and homogeneity of native protein. See Parts I to IV.
2. Attempt dissociation, using mildest methods first, but employing more than one technique. If chemical evidence shows the presence of -S-S-bonds, reductive or oxidative procedures should be employed as well, even if some dissociation is obtained by physical means.
3. Determine average molecular weight, and molecular weight homogeneity of the dissociated protein. More than one average should be obtained, and the number average is especially useful. Since dissociating conditions frequently lead to non-ideal behavior, extrapolate results to infinite dilution.
4. Use sedimentation velocity and gel filtration to look for marked paucidispersity in the subunit mixture. The latter may be a good way to effect separation for further studies. If this is possible, characterization of separated subunits is to be preferred.
5. Use electrophoretic methods to look for more subtle heterogeneity in the subunits.

Of course, the program will have to be varied for individual cases and the studies in many cases can be supplemented by hybridization experiments, etc. Nevertheless, with a thorough examination of this kind, the number and types of subunits should be determinable.

Chapter 3

End Group Determination

Kozo Narita

I. Introduction

Most polypeptides and proteins occurring in nature have been shown to consist of a single polypeptide chain, but several proteins, insulin, chymotrypsin, immunoglobulin, hemoglobin, and lactic dehydrogenase, for example, have been shown to be composed of more than a single chain. In protein molecules consisting of two or more polypeptide chains, the chains are held together either by covalent interpeptide-chain disulfide bonds of cystine residues (insulin, α-chymotrypsin, immunoglobulin etc.) or by the cumulative forces of ionic interaction, hydrogen bond and hydrophobic interaction between side chain groups of amino acid residues belonging to the neighboring different polypeptide chains (hemoglobin, lactic dehydrogenase etc.). Each polypeptide chain usually possesses an unsubstituted α-amino group at one end of the chain (amino terminal group or N-terminal group) and an unsubstituted α-carboxyl group at the other end (carboxyl terminal group or C-terminal group).

Quantitative analysis of both end groups provides valuable information concerning the purity of a protein in question, the number of polypeptide chains contained in one molecule of a protein, the molecular weight of which has previously been determined by physicochemical techniques, and branching or ramification of a polypeptide chain. Careful quantitative end group analysis on a protein whose molecular weight was not previously known allows the calculation of a minimum molecular weight.

Each analytical method for the terminal groups, to be described later in this chapter, has certain characteristic merits on one hand but also intrinsic disadvantages based upon the chemical principle involved. The use of at least two methods, which are governed entirely by different chemical principles, is recommended in order to draw a valid conclusion based upon the results of end group analyses.

When a protein is shown to contain cystine or cysteine residues by amino acid analysis, it is desirable to reduce disulfide bonds with β-mercaptoethanol in the presence of 8 to 10 M urea or 4 to 6 M guanidine hydrochloride and EDTA followed by alkylation of the resultant sulfhydryl groups prior to the end group determinations. A low concentration of the denaturants is not enough to cleave reductively all of the disulfide bonds with the thiol compound. For instance, a part of the disulfide bonds in immunoglobulin was reported to be resistant to β-mercaptoethanol in 8 M urea but was cleaved in 10 M urea or 6 M guanidine hydrochloride [166]. The chelating agent EDTA is required to minimize the unfavorable contamination by heavy metal ions which can catalyze re-oxidation of the resultant sulfhydryl groups. The sulfhydryl groups usually are alkylated with monoiodoacetic acid [218] (or its amide) or ethyleneimine [805, 806]. The former reagent converts the sulfhydryl group to an anionic carboxymethylcysteine residue while the latter converts it to a cationic β-aminoethylcysteine residue, which has a side chain length similar to that of the lysine residue and which becomes susceptible to trypsin at the carboxyl side in the peptide linkage. For this reason the aminoethylation reaction currently finds much favor. Alternatively, the performic acid oxidation of a cystine residue to two residues of cysteic acid has been used extensively since its introduction by SANGER in his study of insulin [857]. As an undesirable side reaction, the oxidation of cystine residues is accompanied by the oxidation of methionine to methionine sulfone and of tryptophan, partly to N^1-formylkynurenine and partly to unknown substances. The resulting oxidized protein sometimes is sparingly soluble in water. For a small peptide, devoid of methionine and tryptophan residues, for which the primary structure already has been elucidated the performic oxidation procedure is effectively used to determine the location of the disulfide bonds in the polypeptide chain.

Some difficulties are to be expected if reduction and subsequent alkylation of a cystine-containing protein is not performed prior to end group analyses. The N-terminal group of chymotrypsinogen, for example, is a half cystine residue which is linked to a second half cystine located in a position much further along the polypeptide chain [88]. In such a case, when the dinitrofluorobenzene method of SANGER [855] is used, monodinitrophenyl (DNP)-cystine will first be formed by acid hydrolysis but two molecules of the monosubstituted cystine can reversibly interchange half cystine residues as shown in reaction (1). Thus, half the expected amount

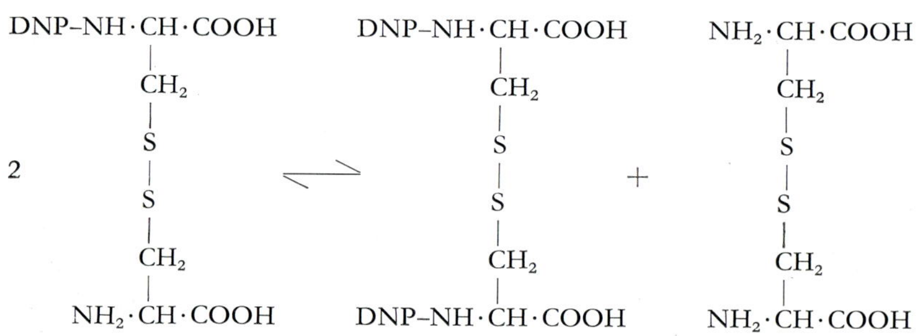

(1)

of bis-DNP-cystine will at best be determined (DNP-cystine is rather labile to acid hydrolysis) [844]. If the phenylisothiocyanate method of EDMAN [263] were used, no N-terminal phenylthiohydantoin derivative would be formed.

When a single linear polypeptide chain containing no disulfide bridges is being examined, the difficulties described above are not encountered during the determination of the primary structure of proteins. Methods that cleave disulfide bonds in proteins are described by KASPER in Chapter 6. In attempting to elucidate the structure-function relationship of a protein, however, the native protein with intact disufide bonds must of course be used.

Principles and methods of the various end group analyses have been outlined and discussed recently by BAILEY [47], and a detailed description of the techniques of some of the typical end group analyses was made by HIRS [442].

II. N-Terminal Group Determination

The general principle for amino end group determination is based upon the introduction of a marker group (colored, fluorescent, ultra-violet absorbing etc.) onto the amino function followed by the quantitative isolation and characterization of the derivatized terminal amino acid.

Although all of the terminal analysis methods described below can be used for both proteins and peptides, the cyanate method is especially suitable for proteins and large peptides while the dansyl method is preferred for oligopeptides. The phenylisothiocyanate method of Edman is described in detail by Edman in Chapter 8

A. Dinitrofluorobenzene Method of Sanger

In 1945 Sanger [855] reported that 1-fluoro-2,4-dinitrobenzene (FDNB) reacted quantitatively with amino groups of insulin under mild conditions (2) and that one mole each of dinitrophenyl (DNP)-glycine, DNP-phenylalanine and

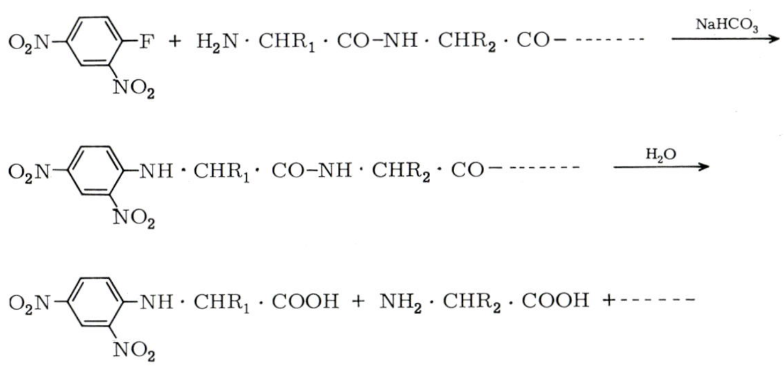

(2)

ε-DNP-lysine could be isolated from one mole of the DNP-insulin (molecular weight assumed to be 6,000) on acid hydrolysis after suitable corrections. The reagent FDNB is reactive not only with α-amino (and α-imino) and ε-amino groups but also with the sulfhydryl group of cysteine, the phenolic hydroxyl group of tyrosine and the imidazole group of histidine residues in polypeptide chains. When lysine, cysteine, tyrosine and histidine residues are located at the N-terminal positions, they are recovered as their di-DNP derivatives from DNP-proteins. One merit of the Sanger method is that α-DNP-amino acids are rather resistant to acid hydrolysis in contrast to acyl amino acids and can be recovered with yields ranging from 25 to 90% after boiling in azeotropic hydrochloric acid (5.7 N HCl) or heating with 12 N HCl at 105 °C for 12 h. Possible exceptions are DNP-glycine, DNP-cystine, DNP-proline and DNP-hydroxyproline. Stability of DNP-amino acids in hydrochloric acid was studied by Porter and Sanger [798] whose results are listed in Table 1.

Although the decomposition reaction of DNP-amino acids with acid has not been studied extensively, the breakdown products seem to be the regenerated parental amino acid, 2,4-dinitrophenol (DNP-OH), 2,4-dinitroaniline ($DNP\text{-}NH_2$)

Table 1. *Approximate breakdown of DNP-amino acids*[a]

	Hydrolysis in boiling 5.7 N HCl		Hydrolysis for 16 h in 12 N HCl at 105 °C
	Time of hydrolysis (h)	Amount unchanged (%)	Amount unchanged (%)
DNP-alanine	12	80	75
DNP-arginine	12	90	75
DNP-aspartic acid	24	60	75
Bis-DNP-cystine	12	25	0
DNP-glutamic acid	12	75	75
DNP-glycine	8	40	50
DNP-hydroxyproline	4	40	50
DNP-isoleucine	12	80	75
DNP-leucine	12	80	75
Di-DNP-lysine	8	95	75
ε-DNP-lysine	12	95	75
DNP-methionine	12	75	75
DNP-phenylalanine	12	70	50
DNP-proline	2	10	50
DNP-serine	12	90	75
DNP-threonine	24	90	75
DNP-tryptophan	12	90	0
Di-DNP-tyrosine	12	75	50
DNP-valine	12	80	75

[a] Reprinted from Ref. [798], by courtesy of the Biochemical Society.

and uncharacterized materials [817]. As the result of heating DNP-proline with hydrochloric acid, δ-chloro-α-DNP-aminovaleric acid and α-chloro-δ-DNP-aminovaleric acid are produced in addition to the regenerated parental proline [873]. The low recovery of DNP-proline is mainly attributed to this pyrolidine ring-opening reaction (3), which also may apply to the lability of DNP-hydroxyproline.

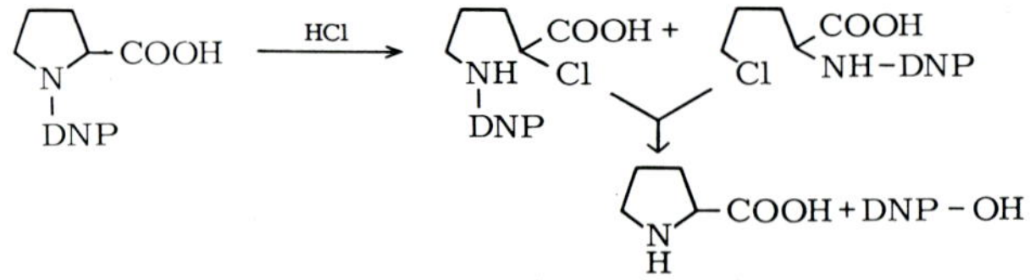

(3)

The extent of decomposition of the terminal DNP-amino acid during acid hydrolysis of a DNP-protein is not identical with that of the free DNP-amino acid alone and differs from one DNP-protein to another, depending upon the nature and amounts of constituent amino acids and upon the nature of prosthetic groups. For example, the terminal DNP-amino acid is destroyed more extensively in DNP-glycoprotein than in DNP-simple proteins. When an accurate estimate is required for a particular terminal DNP-amino acid, a recovery factor must be obtained by heating the particular DNP-amino acid with an equimolar amount of protein at a scheduled period, usually 6 to 15 h, with redistilled azeotropic hydrochloric acid in an evacuated sealed tube at 105 °C. In Fig. 3-1, the decomposition rate of DNP-alanine with 6 N HCl at 105 to 110 °C as a function of time, in the presence and absence of α-amylase from *Aspergillus oryzae* is shown. The N-terminal residue of the amylase is alanine and the corrected values of DNP-alanine released at several periods of the acid hydrolysis from DNP-amylase also are plotted. When the amounts of released DNP-amino acid at various hydrolysis times are extrapolated to zero time,

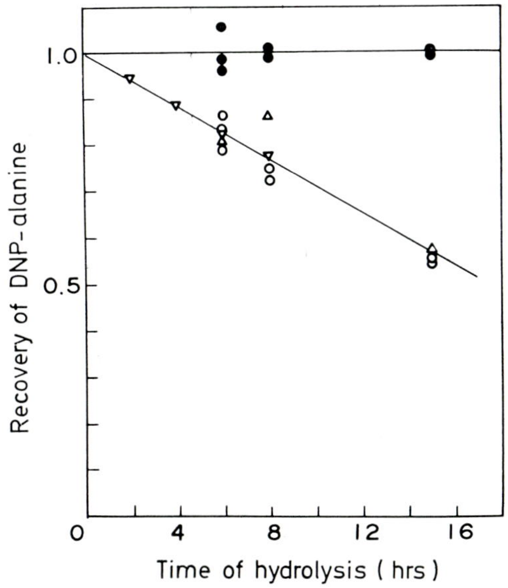

Fig. 3-1. Decomposition of DNP-alanine by heating with 6 N hydrochloric acid at 105 to 110 °C

∇, DNP-alanine alone; Δ, DNP-alanine plus equimolar amount of Taka-amylase; o, DNP-Taka-amylase, of which terminus was alanine; ●, DNP-alanine recovered from DNP-Taka-amylase after correction for decomposition during acid hydrolysis. The figure is drawn from the data in Ref. [9]

the amount of the terminal group can be estimated without obtaining a correction factor during the acid hydrolysis.

Besides the decomposition reaction of DNP-amino acids with acid, attention should be given to the photodecomposition of DNP-amino acids [10]. DNP-Amino acids are quite sensitive to light, especially to visible light. Therefore when small amount of DNP-derivatives are treated in a dilute solution, the coupling reaction, hydrolysis and subsequent chromatographic separation must be performed in the absence of light. With the exception of DNP-tryptophan, α-DNP-amino acids yield 4-nitro-2-nitrosoaniline, the corresponding aldehydes and carbon dioxide by photolysis in 1% sodium bicarbonate [841].

Generally, the α-DNP-amino acid derivatives are acids and can easily be extracted from the acid hydrolysate mixture into ether. However, DNP-cysteic acid, DNP-arginine and a part of di-DNP-histidine remain in the water phase after ether extraction together with ε-DNP-lysine, O-DNP-tyrosine and imidazole-DNP-histidine located originally in non-terminal positions in peptide chains. Generally, *imid*-DNP-histidine is very labile toward acid at high temperature and this derivative cannot be identified in the acid hydrolysate of DNP-proteins. Water-soluble DNP-amino acids are extracted with *n*- or *sec*-butanol.

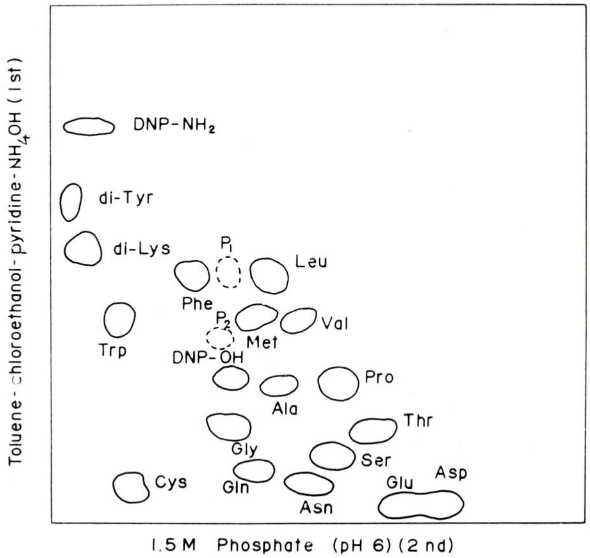

Fig. 3-2. Two-dimensional paper chromatogram of a synthetic mixture of DNP-amino acids according to the method of LEVY [609]. Development was first carried out with "toluene" of BISERTE and OSTEUX [99] and then with phosphate shown in Table 2. P_1 and P_2 respectively denote the positions of α-DNP-amino-δ-chlorovaleric acid and δ-DNP-amino-α-chlorovaleric acid

Separation and characterization of DNP-amino acids were originally carried out by chromatography on silica gel columns by SANGER [798, 855]. Modification of this procedure by using silicic acid-Celite columns was made by GREEN and KEY [370]. The use of an Amberlite CG-50 column was recommended by SEKI [938]. The mechanism of separation of DNP-amino acids with ion exchange resins seem to involve adsorption rather than an ion exchange reaction. Paper chromatography or thin layer chromatography is more convenient than column chromatography and has been widely used. Many systems for paper chromatography have been proposed, but the two-dimensional method recommended by LEVY [609] appears to be most useful. The first development is made by the toluene-chloroethanol-pyridine-aqueous ammonia solvent reported by BISERTE and OSTEUX [99] and the second by 1.5 M phosphate buffer. The locations of DNP-amino acids by this system are shown in

Fig. 3-2. The separation of DNP-aspartic acid from DNP-glutamic acid is less satisfactory and Levy recommended the use of 2.0 to 2.5 M phosphate buffer instead of 1.5 M buffer, but at this concentration the other DNP-amino acids are separated less effectively [321]. As the first solvent, *n*-butanol saturated with diluted ammonia [557] can be used in place of Biserte's "toluene" system. The pattern of distribution of the DNP-amino acids is quite similar to that obtained with the "toluene" system, but separation of di-DNP-lysine from di-DNP-tyrosine, and DNP-valine from DNP-leucine are less satisfactory. Nevertheless, the *n*-butanol-NH_4OH system is useful since it is easy to prepare and requires no equilibration of papers with the solvent in a

Table 2. *Composition of solvents for paper chromatography of DNP-amino acids*

Solvent	Composition	Remarks
Toluene-pyridine-2-chloroethanol-0.8 N NH_4OH	50 ml of toluene, 18 ml of pyridine, 30 ml of 2-chloroethanol and 30 ml of 0.8 N NH_4OH	Use the organic phase. Prepare freshly each time
n-Butanol saturated with NH_4OH	*n*-Butanol saturated with 0.1% NH_4OH. Much concentrated NH_4OH, for instance 2 N, can be used without appreciably different *Rf* values	Use the organic phase. Prepare freshly each time
tert-Amyl alcohol saturated with pH 6 phthalate	*tert*-Amyl alcohol shaken with excess pH 6 phthalate buffer, which is prepared by mixing 50 ml 0.1 M potassium hydrogen phthalate in 45.5 ml 0.1 N NaOH and made up to 100 ml	Papers are first dipped in the buffer and dried at room temperature
1.5 M Phosphate buffer	138 g (1 M) $NaH_2PO_4 \cdot H_2O$ and 71 g (0.5 M) Na_2HPO_4 are dissolved in 1 liter H_2O	Chromatogram should be developed in a warm room (about 20 °C) in winter, otherwise crystals will appear in the solvent during development
2.5 M Phosphate buffer	229 g $NaH_2PO_4 \cdot H_2O$ and 118 g Na_2HPO_4 are dissolved in 1 liter H_2O	

developing chamber, in contrast to the "toluene" system. Alternatively, *tert*-amyl alcohol-pH 6 phthalate system of Blackburn and Lowther [101] also has been used, but this system is limited either to one-dimension or to the second solvent in two-dimensional chromatography, since a paper sheet must previously be buffered with 0.1 M phthalate at pH 6. Composition of solvents and Rf values of the ether soluble DNP-amino acids are listed in Tables 2 and 3. Rf values of the acid soluble DNP-amino acids are shown in Table 4.

Two-dimensional thin layer chromatography on silica gel G plates has also been used routinely with solvent systems reported by Brenner *et al.* [154]. In Figs. 3-3 and 3-4 are shown typical two-dimensional chromatograms. Mumier and Sarrazin [709] separated all of the ether-soluble DNP-amino acids by two-dimensional thin layer chromatography on powdered cellulose without binders. The Biserte "toluene" system and saturated ammonium sulfate solution-water-sodium dodecylsulfate (100 ml:700 ml:0.576 g) are used, respectively, for the first and the second dimensions.

Table 3. *Rf values of ether extractable DNP-amino acids*[a]

	Toluene-pyridine-2-chloro-ethanol-NH_4OH[b]	*n*-Butanol saturated with 0.1% NH_4OH[c]	*tert*-Amyl alcohol saturated with pH 6 phthalate[c]	1.5 M Phosphate[c]	2.5 M Phosphate[c]	*n*-Butanol saturated with water[d]
DNP-Aspartic acid	0.08	0.08	0.03	0.62	0.44	0.12
DNP-Glutamic acid	0.09	0.10	0.06	0.56	0.37	0.14
DNP-Glycine	0.35	0.36	0.14	0.31	0.18	0.36
DNP-Serine	0.30	0.29	0.13	0.46	0.31	0.32
DNP-Threonine	0.36	0.41	0.17	0.54	0.33	0.43
DNP-Alanine	0.45	0.52	0.25	0.41	0.22	0.50
Di-DNP-Histidine	0.70	0.54	0.32	0.13	0.09	0.35
Dinitrophenol	0.43	0.62	0.45	0.31	0.20	0.56
Di-DNP-Lysine	0.72	0.65	0.49	0.11	0.06	0.72
Di-DNP-Tyrosine	0.74	0.75	0.55	0.02	0.01	0.78
DNP-Methionine	0.59	0.63	0.50	0.41	0.25	0.65
DNP-Phenylalanine	0.66	0.76	0.58	0.28	0.15	0.71
DNP-Methionine sulfoxide	0.35	—	—	—	—	—
DNP-Methionine sulfone	0.48	—	—	—	—	—
DNP-Valine	0.56	0.69	0.68	0.47	0.30	0.68
DNP-Leucine	0.64	0.78	0.70	0.40	0.30	0.74
DNP-Proline	0.43	0.53	0.50	0.52	0.41	0.48
DNP-Tryptophan	0.62	0.75	0.69	0.39	0.22	0.70
Bis-DNP-Cystine	0.32	0.24	0.28	0.16	0.10	—

[a] Dinitroaniline travels very fast in the organic solvents and very slowly in phosphate buffers. The *Rf* values of the DNP-amino acids are variable and the values listed should be considered as an indication of position.
[b] From Ref. [99].
[c] From Ref. [687].
[d] From Ref. [787].

Table 4. *Rf values of acid soluble DNP-amino acids*[a]

	Rf[b]	Color of spot	Ninhydrin reaction[c]
DNP-Arginine	0.64	Yellow	No reaction
DNP-Cysteic acid	0.25	Yellow	No reaction
ε-DNP-Lysine	0.55	Yellow	Brown-purple
O-DNP-Tyrosine	0.70	Colorless	Purple
imid-DNP-Histidine	0.33	Colorless[b] (or faint yellow[c])	Brown

[a] Solvent: *n*-Butanol-acetic acid-water (12:3:5, v/v). Prepare freshly each time.
[b] From Ref. [687], by courtesy of the Biochemical Society.
[c] From Ref. [862].

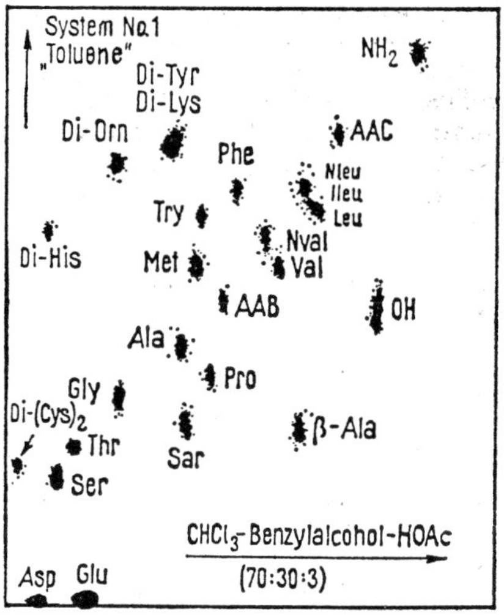

Fig. 3-3. Two-dimensional thin layer chromatogram of a synthetic mixture of DNP-amino acids (0.2 μg each). Abbreviations: OH, dinitrophenol; NH_2, dinitroaniline; AAB, α-amino-butyric acid; AAC, α-amino-caprylic acid. Reprinted from Ref. [154], by courtesy of Birkhäuser Verlag

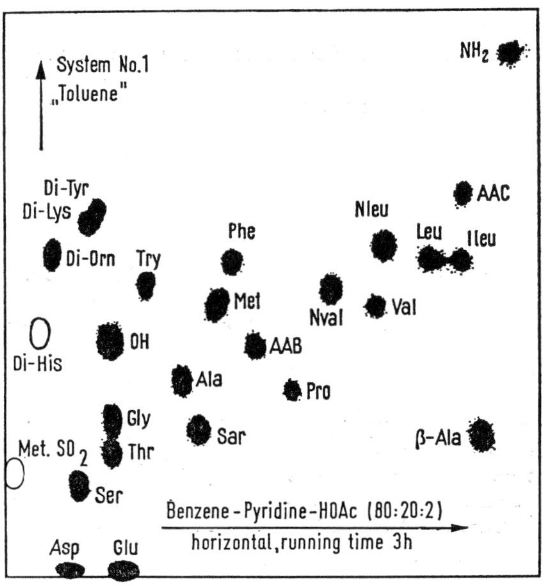

Fig. 3-4. Two-dimensional thin layer chromatogram of a synthetic mixture of DNP-amino acids (1 μg each). Abbreviations as in Fig. 3-3. Reprinted from Ref. [154], by courtesy of Birkhäuser Verlag

Most DNP-amino acids are yellow, except for O-DNP-tyrosine and *imid*-DNP-histidine which are colorless. Absorption maxima are near 360 mμ, but those of DNP-proline and -hydroxyproline are at 385 mμ in 1% $NaHCO_3$, the usual solvent for the extraction and estimation of DNP-amino acids from paper chromatograms. In 1% $NaHCO_3$ solution, the ratio of optical density of most DNP-amino acids at 390 mμ to that at 360 mμ is 0.55 to 0.65, but that of DNP-proline and DNP-hydroxyproline is 1.1 to 1.4 and that of dinitrophenol is approximately 0.8. Dinitrophenol and artifacts produced can easily be distinguished from DNP-amino acids by measuring the above ratio of optical densities at the two wave lengths. The best method is to

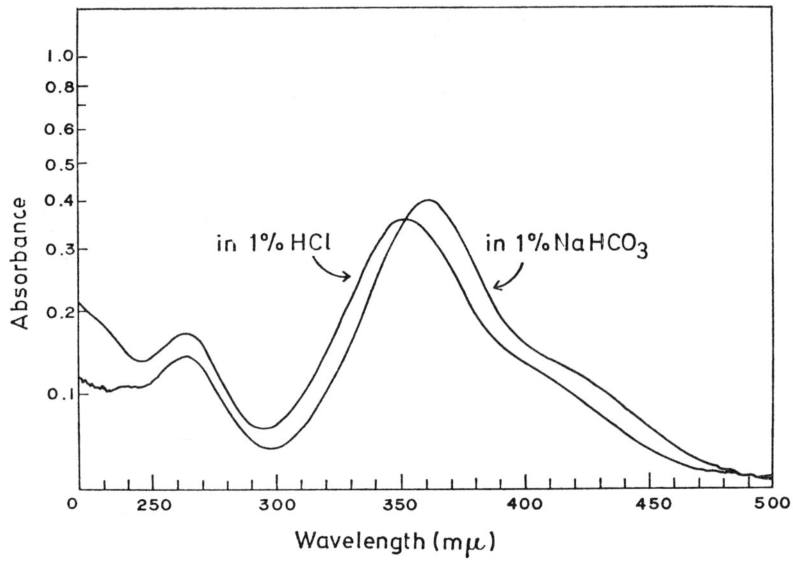

Fig. 3-5. Absorption spectrum of DNP-alanine (33.3 μM) in 1% hydrochloric acid and in 1% sodium bicarbonate solution

obtain the absorption spectra between 250 mμ and 500 mμ of the chromatographed yellow substances. Absorption of DNP-amino acids in an alkaline medium differs from that in an acidic medium as shown in Fig. 3-5.

The identified DNP-amino acids on chromatograms can be confirmed by regenerating free amino acids from the corresponding DNP-derivatives after hydrolysis with saturated barium hydroxide [688] in a sealed tube at 100 °C for 1 h or with concentrated ammonium hydroxide [629] at 100 °C for 2 h.

Methods for the preparation of the standard DNP-amino acids are not described here since these are available commercially.

1. Dinitrophenylation of Proteins and Peptides

A weighed amount of a protein (1 to 2 μmole, minimum 0.2 μmole) and an equal weight of $NaHCO_3$ are dissolved or suspended in 10 volumes of water and an additional 2 volumes of 5% ethanolic solution of 1-fluoro-2,4-dinitrobenzene (FDNB) are added. The reaction mixture is shaken for 2 h at room temperature. The reaction mixture is then acidified with a few drops of concentrated HCl. The usual DNP-proteins are insoluble in water. The precipitated DNP-protein is collected by centrifugation and washed successively with ether and ethanol to remove excess FDNB and dinitrophenol produced from the reagent, and further washed successively with water, acetone and ether. The DNP-protein thus obtained is dried in an evacuated dessicator on $CaCl_2$ or P_2O_5.

When DNP-derivatives of proteins and oligopeptides are soluble in water, the reaction mixture is diluted with two volumes of water to decrease the alcohol concentration and excess FDNB is removed several times with ether. The reaction mixture is then acidified with HCl to pH 1 and the DNP-derivatives are extracted

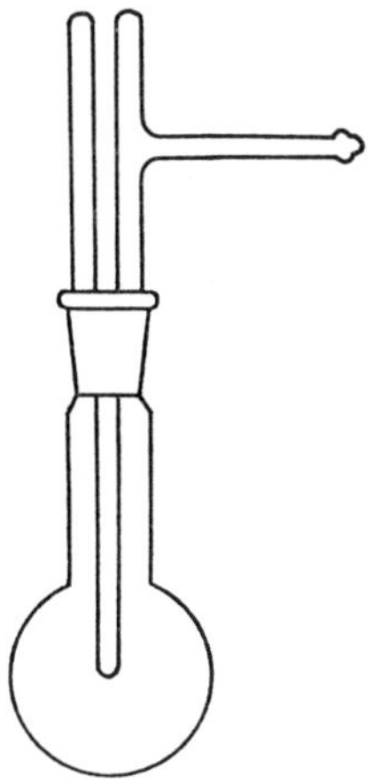

Fig. 3-6. Vacuum sublimation apparatus according to MILLS [689]

with ether, ethyl acetate, or *sec*-butanol. To prevent conversion of DNP-methionine to its sulfoxide, solvents used for the extraction of DNP-derivatives must be freed from peroxides by shaking with concentrated ferrous sulfate solution. After the organic extracts are washed with water, the DNP-derivatives are recovered on evaporation *in vacuo*. Dinitrophenol produced is removed by vacuum sublimation at around 60 °C using an apparatus suggested by MILLS [689] as shown in Fig. 3-6. Into the inner tube is placed crushed dry ice. Dinitrophenol that has sublimed on the outer surface of the inner tube after 15 min is washed away with acetone using a polyethylene flush bottle. The surface of the reaction mixture covering the bottom of the outer tube of a sublimation apparatus is renewed by dissolving the mixture in acetone and again the content is dried *in vacuo*. Sublimation is repeated three to four times until no more dinitrophenol is sublimed. The end point of sublimation can easily be recognized by observing that no yellow color appears in water, into which the acetone washings are poured.

The DNP-peptide remaining in the aqueous phase of the above extraction can be recovered by adsorbing it on the top of a talc column (1 × 5 cm), which has previously been washed with 1 N HCl. The column is washed with 1 N HCl (about 20 ml) to elute free peptides which escaped the reaction and the adsorbed DNP-peptide is eluted with 1 N HCl-ethanol mixture (1:4, v/v).

When a small amount of peptide (0.2 μmole, for instance) extracted from a paper chromatogram is treated, the peptide is dissolved in 0.1 ml of 1% triethylamine, and 0.2 ml of 5% FDNB alcohol solution is added [859]. After 2 h at room temperature, a few drops of water and triethylamine are added and the excess FDNB is extracted 3 times with ether. Salt-free DNP-peptide can be obtained on evaporation of the aqueous phase.

2. Hydrolysis of DNP-Proteins

The dried DNP-protein is weighed (minimum 0.2 μmole) into a small glass test tube and constant boiling HCl (5.7 N HCl), redistilled with a glass apparatus (to remove unfavorable metal ions) is added such that the DNP-protein is sufficiently covered with the acid. The test tube is then evacuated with a vacuum pump while cooling the contents with a dry ice bath for 5 to 10 min and sealed. In general, 80% of the weight of the DNP-protein roughly corresponds to the original protein. The molecular weight of the DNP-protein can be calculated by adding a factor, 166 X (where X is the number of DNP-groups introduced), to the molecular weight of the original protein, assuming that all of the constituent lysine, tyrosine, histidine, cysteine and the terminal amino groups have been quantitatively substituted [321]. Sanger [855] originally estimated the protein content of DNP-insulin by analyzing its amide groups since amide groups of asparagine and glutamine are known not to be substituted with nor hydrolyzed by FDNB during the reaction under the alkaline conditions used.

The optimum time of hydrolysis at 105 °C for the end group analysis differs from one DNP-protein to another and if a sufficient amount of protein is available, the best method is to follow the released amount of the terminal DNP-amino acid as a function of hydrolysis time as discussed previously. If the sample is not large enough, hydrolysis should be carried on for 6 to 8 h. This period is not enough to release DNP-amino acids quantitatively from the protein derivative but is sufficient to identify unstable DNP-amino acids during acid hydrolysis except for DNP-proline. Based on the result of a preliminary analysis, further experiments should be performed with reference to the stability of each DNP- amino acid as shown in Table 1.

3. Characterization and Estimation of the DNP-Amino Acid

The hydrolysate is diluted to about 5 volumes with water to decrease the concentration of HCl to 1 N, and the N-terminal DNP-amino acid is extracted with 5 ml of ether (should be freed from peroxide) until no more yellow color is extracted into the ether layer. This usually requires 3 to 5 extractions. The ether extracts are washed with small portions of water and the washings are returned to the aqueous phase. DNP-Arginine, DNP-cysteic acid and di-DNP-histidine cannot be completely extracted into the ether layer since these DNP-amino acids partition between the two phases. The use of *n*- or *sec*-butanol is recommended to extract them, but ε-DNP-lysine, O-DNP-tyrosine and DNP-derivatives of oligopeptides are also extractable with these solvents.

The above ether extracts are evaporated *in vacuo* and the residue is dissolved in 0.1 ml of acetone. Fifty μl of the solution is transferred quantitatively with the aid of a micropipet to a sheet of filter paper (18 × 22.5 inch, Whatman No. 1) to be used for two-dimensional chromatography. Two paper chromatograms can therefore be prepared from a single sample hydrolysate. The first development is usually carried out with Biserte's "toluene" system in an ascending manner. The paper sheet is curled into a cylinder and placed in a dish of suitable size (e.g. 7.75 inches in diameter). The paper is equilibrated for 4 h with 200 ml of 0.8 N ammonia in a small dish placed concentrically inside the larger dish in which the chromatographic solvent is added later. The entire assembly is placed in a cylindrical glass jar with a well fitting plate glass lid (e.g. 8.75 inches in diameter and 18 inches in height). An air-tight seal of the rim of the outer cylinder is maintained with silicone grease and by placing a weight on top of the lid. Biserte's "toluene" solvent is prepared according to the composition shown in Table 2. The mixture is allowed to stand for 4 h (duration of the equilibration period) and the lower aqueous layer is withdrawn. The volume of the upper layer is enough to develop two paper sheets. The organic layer is added through a hole in a glass lid to the annular space between the inner and outer dishes through a funnel fitted with a long stem. Development of DNP-amino acids is then started and continued overnight (about 15 to 16 h) in the dark; during this period the solvent rises to within about one inch of the top of the paper cylinder. The paper is then removed and dried in an air oven at 40 °C for at least 3 to 4 h; insufficient removal of the first solvent causes trailing when chromatography in the second dimension is performed.

In place of the "toluene" system, *n*-butanol saturated with 0.1% NH_4OH can be used as discussed previously.

The dried paper cylinder is then unfolded and suitable controls (5 μl of M/100 solutions of DNP-amino acids in acetone) including dinitrophenol are then spotted on either side of the unknown yellow spot. Second development is then carried out overnight in the dark with 1.5 M phosphate buffer of pH 6 listed in Table 2 either by the ascending or descending technique in a chromatographic apparatus containing a tray of water at the bottom. In the second development with phosphate buffer, no equilibration is necessary, so no air-tight chamber is required. When ascending development with phosphate is performed, care must be taken to prevent weakening of the strength of the filter paper due to absorption of the dense buffer solution.

After the paper chromatogram is air-dried, yellow spots (ultraviolet-absorbing spots under ultraviolet lamp) are marked with a pencil about 0.1 inch beyond their visible peripheries. Characterization of the yellow spots are made from the positions in relation to the dinitrophenol spot comparing with a standard map prepared from a mixture of the authentic DNP-amino acids and dinitrophenol (Fig. 3-2). The marked spots are cut out, placed in test tubes, and 4 ml of 1% $NaHCO_3$ is added. The tubes are placed in a water bath at 55 to 60 °C for 15 min to allow elution of the yellow color. A few blanks cut from each sheet are treated similarly. The extinction values of all of the solutions are read at 360 mμ and 390 mμ using a set of quartz curvettes against the extract of the blank paper (the actual optical density of the blank is

0.001 to 0.002 per cm² against distilled water). The ratio of the optical density at 390 mμ to that at 360 mμ ranges from 0.55 to 0.65 with the common DNP-amino acids while those of DNP-proline and dinitrophenol are 1.4 and 0.8, respectively, as mentioned above. When the ratio of the optical densities is largely different from the above, complete absorption curves should be measured. Millimolar extinction coefficients of usual DNP-amino acids are 17 to 18. Therefore the approximate micromoles of the end group of the protein in question is calculated by multiplying by the optical density reading 17 and by the recovery factor and by dividing by 4 (volume of the extracting solvent). In the cases of bis-DNP-cystine and di-DNP-lysine, it is necessary to multiply further by the factor 0.6 due to the presence of two substituents. The recovery factor of each amino acid consists of the chromatographic recovery (about 90%) and the hydrolytic recovery (varies depending on the amino acid and on the hydrolysis time). It is recommended that recovery factors of all DNP-amino acids after chromatography and after hydrolysis-chromatography, respectively, be individually obtained.

Another convenient characterization system for DNP-amino acids is the *tert*-amyl alcohol-phthalate method of BLACKBURN and LOWTHER [101]. *tert*-Amyl alcohol saturated with the buffer is used for development either in an ascending or descending manner and the buffer saturated with *tert*-amyl alcohol is placed in a chromatographic chamber. DNP-glutamic acid moves ahead of DNP-aspartic acid which moves only slightly ahead of unmovable DNP-carboxymethylcysteine in this system in contrast to the "toluene" system. Mutual separation is enhanced considerably by developing with pH 5 phthalate buffer (50 ml of 0.1 M potassium hydrogen phthalate + 23.9 ml of 0.1 N NaOH made up to 100 ml) for long periods by the descending technique.

B. Dansyl Chloride of Hartley

HARTLEY and MASSEY [427] reported in 1956 that non-fluorescent dansyl chloride (1-dimethylaminonaphthalene-5-sulfonyl chloride, sometimes abbreviated as DNS-Cl) gives a strong yellow fluorescence upon sulfonamide formation with the amino group of amino acids and of peptides as shown in reaction (4). This reaction has been extended to the end group determination of peptides and proteins along similar line to the FDNB method of SANGER [855]. Its main advantage over the SANGER method are the 100-fold increase in sensitivity making it possible to handle less than 1 mμmole of sample, and the increased stability to acid hydrolysis of the DNS-amino acids over the corresponding DNP-amino acids.

The reactivity of functional groups of amino acids and peptides toward dansyl chloride in aqueous solution depends mainly upon their strength as bases. Optimum labeling conditions for most amino acids require the use of a several-fold excess of the reagent in 5 mM concentration at pH 9.5 to 10.5 at room temperature for 2 to 3 h, or at 37 °C for 1 h [368]. Since the reagent is hydrolyzed by water and by hydroxyl ion, amino acids are labeled as a function of their ability to compete for the limited amount of the reagent under the above conditions. Above pH 10.5 the reagent is hydrolyzed too rapidly, and at a pH lower than 9.5 the unreactive protonated form of amino acids is present in too high a concentration.

During the reaction of dansyl chloride with amino acids, some side reactions are known to occur. When excess reagent is present in the reaction mixture, DNS-NH_2 (1-dimethylaminonaphthalene-5-sulfonamide) is produced probably by reaction (5) [722].

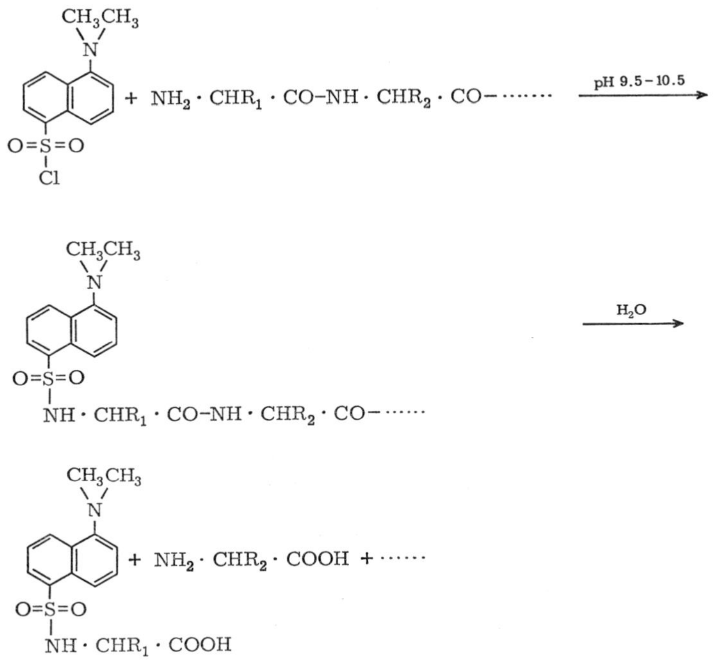

(4)

The extent of the side reaction is dependent upon the nature of the side chains of amino acids as shown in Table 5 [722]. Furthermore, when glutamyl and glutaminyl peptides are subjected to the reaction, especially in highly dilute solutions, they are partly converted into DNS-pyrrolidonyl peptides, the orange fluorescence of which is very weak compared with that of α-DNS-amino acids [1034].

$$\text{DNS–NH·CHR·COO}^- + \text{DNS–Cl} \rightarrow \text{DNS–NH·CHR·COO–DNS} \xrightarrow{OH^-}$$

$$\text{DNS–NH}_2 + \text{RCHO} + CO_2 + \text{DNS–OH} \qquad (5)$$

Dansyl chloride reacts not only with amino groups, but also with imino, phenolic hydroxyl, thiol, and imidazole groups, but the stabilities of the products differ. The DNS-derivatives of thiol and imidazole compounds are very labile in both acid and alkali and are broken down completely under the conditions used to hydrolyze labeled peptides. The degradation products with acid are DNS-OH (1-dimethyl-aminonaphthalene-5-sulfonic acid) and the original group. In mild alkali the dansyl group may be transferred to other bases. DNS-Proline, having a secondary sulfon-amide linkage, is the least stable of the α-DNS-amino acids toward acid, and DNS-tryptophan is extensively degraded under the standard conditions of hydrolysis of

DNS-peptides (5.7 N HCl at 105 °C for 18 h). The recoveries of various DNS-amino acids under these conditions are listed in Table 6 [368]. Upon prolonged hydrolysis with acid of di-DNS-lysine, both the α- and ε-DNS-lysine monoderivatives are obtained in addition to the di-DNS-derivative.

Table 5. *Percent formation of DNS–NH_2 during the reaction of dansyl chloride with amino acids*[a]

Amino acids	% Formation of DNS–NH_2
Glycine	2
Alanine	11
Valine	12
Leucine	15
Proline	1

[a] The reaction mixture consists of 0.1 mmole of amino acid, 0.11 mmole of dansyl chloride, 0.5 mmole of $KHCO_3$, 4.5 ml of acetone and 3.5 ml of water. Data from Ref. [722].

Table 6. *Stability of DNS-amino acids to acid hydrolysis*[a]

Derivate	% Remaining	Fluorescent decomposition products
DNS-Proline	23	DNS–OH[b]
DNS-Serine	65	DNS–NH_2[c] (25%), DNS–OH (10%)
DNS-Glycine	82	DNS–OH
DNS-Alanine	93	DNS–OH
DNS-Valine	100	No detectable breakdown
DNS-Leucine	100	No detectable breakdown
DNS-Isoleucine	100	No detectable breakdown
DNS-Tryptophan	0	DNS–NH_2, DNS–OH (traces of DNS-Gly, DNS-Ser, DNS-Ala)

[a] Percentage remaining as unchanged DNS-amino acid after 18 h of hydrolysis with constant boiling 5.7 N HCl at 105 °C. Data from Ref. [368] by courtesy of Academic Press.
[b] 1-Dimethylaminonaphthalene-5-sulfonic acid.
[c] 1-Dimethylaminonaphthalene-5-sulfonamide.

When much milder conditions are used, recoveries of the acid labile DNS-amino acids increase, but DNS-valyl or DNS-isoleucyl peptides fail to undergo complete hydrolysis [368, 1034], since under the above standard conditions, certain peptide linkages adjacent to DNS-valyl or DNS-isoleucyl residues are resistant to acid hydrolysis. Such DNS-dipeptides can sometimes be identified in addition to DNS-valine or DNS-isoleucine. The DNS-derivatives of glutamic acid and glutamine are partly converted into DNS-pyrrolidone carboxylic acid under acid hydrolysis conditions [1034]. DNS-Carboxymethylcysteine partly changes similarly to the corresponding 6-member ring compound. Upon ammonia treatment, these ring compounds yield their corresponding amides.

As mentioned, it is necessary to use the reagent in a concentration of 5 mM for the effective labeling of amino groups. When very small amounts of a sample are

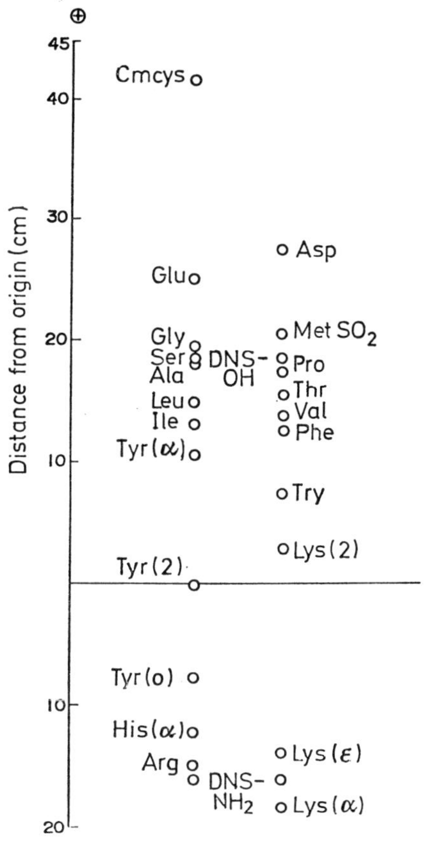

Fig. 3-7. Electrophoretic mobilities of various DNS-amino acids at pH 4.38 (80 V/cm, 2.5 h, 15 °C). Spots are drawn approximately to scale, to give an indication of the resolution attained. Abbreviations: Tyr(α), α-DNS-Tyr; Tyr(2), N,O-di-DNS-Tyr, etc. Reproduced from Ref. [368], by courtesy of Academic Press

treated, the volume of the reaction mixture can be correspondingly reduced. At the end of the reaction, the mixture therefore contains several times more DNS-OH produced by the hydrolysis than the DNS-peptide formed. It is best to remove the by-product, DNS-OH, prior to the acid hydrolysis of DNS-peptides and the subsequent characterization of the DNS-amino acids in the hydrolysate by paper chromatography, thin layer chromatography, or paper electrophoresis.

For the removal of DNS-OH and DNS-NH_2 from the reaction mixture, TAMURA and NAKAJIMA [1034] recommend the use of a small column of DEAE-Sephadex (acetate form, 1 to 2 ml) prior to hydrolysis of DNS-peptide. The reaction mixture is placed on the small DEAE-Sephadex column and unadsorbed DNS-NH_2 is first washed out with 50% methanol. The adsorbed DNS-peptide is then eluted with 50% methanol containing 1 to 5% acetic acid and the yellow fluorescent bands eluted

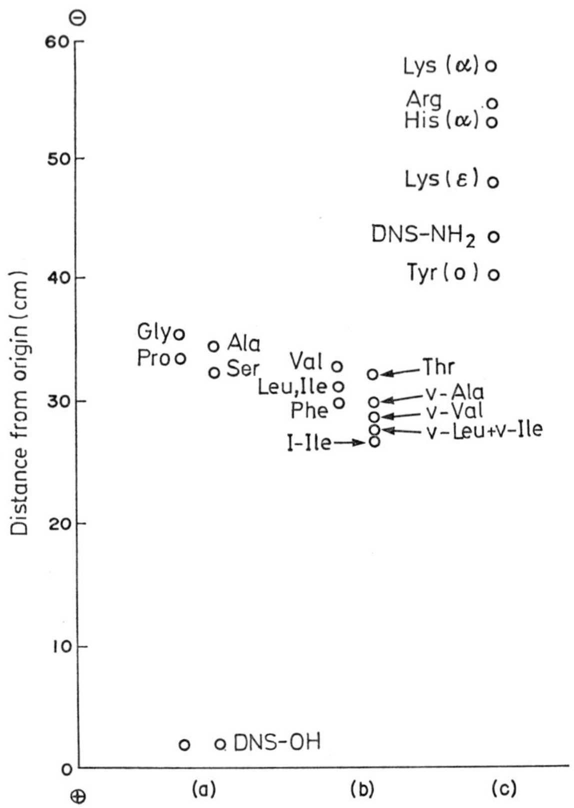

Fig. 3-8. Electrophoretic mobilities of DNS-amino acids at pH 1.9 (50V/cm, 2 h, 20°C). (a) DNS-amino acids which run near DNS-OH at pH 4.4 (Fig. 3-7). (b) Certain other DNS-amino acids and peptides. Abbreviations: I-Ile, Ile-Ile; V-Ala, Val-Ala; etc. (c) Basic DNS-amino acids. Reproduced from Ref. [368], by courtesy of Academic Press

are collected. Under these conditions, DNS-OH is retarded and moves more slowly than the DNS-peptides. However, GRAY [368] suggested that the removal of the by-products of the reaction, mainly DNS-OH, was not necessary, before attempting to identity the terminal group. He recommended the use of electrophoresis on paper for the separation and identification of DNS-amino acids. For paper electrophoresis of DNS-amino acids, an apparatus of the flat-bed type [376] is necessary, since most DNS-amino acids are extracted into the coolant of the MICHL type [683] of apparatus.

With electrophoresis at about pH 4.4 (80 v/cm, 2.5 h at 15 °C), DNS-derivatives of cysteic acid, carboxymethylcysteine, aspartic acid, glutamic acid, methionine sulfone, tyrosine, lysine, histidine and arginine can be identified with certainty, and others with a reasonable degree of confidence (DNS-alanine, -threonine, -leucine, -valine, -isoleucine, and -phenylalanine) as shown in Fig. 3-7. DNS-Derivatives of glycine, serine, and proline and sometimes alanine are obscured in most cases by the strong blue fluorescent band of DNS-OH. These DNS-amino acids can clearly be

Table 7. *R values of DNS-amino acids on silica gel G* [a]

	Benzene-pyridine-acetic acid (80:20:2)	Chloroform-*tert*-amyl alcohol-acetic acid (70:30:0.5)	Chloroform-*tert*-amylalcohol acetic acid (70:30:3)
DNS-Isoleucine	0.92	0.89	1.0
DNS-Leucine	0.89	0.63	1.0
DNS-Valine	0.85	0.75	1.0
DNS-Proline	0.79	0.50	0.93
DNS-Phenylalanine	0.69	0.42	0.94
DNS-Methionine	0.65	0.33	0.92
DNS-Alanine	0.61	0.29	0.89
DNS-Tyrosine	0.55	0.23	0.82
Di-DNS-Lysine	0.48	0.33	0.93
DNS-Glycine	0.38	0.12	0.65
DNS-Tryptophan	0.34	0.25	0.87
DNS-Histidine	0.27	0.09	0.59
DNS-Glutamic acid	0.18	0.08	0.42
DNS-Threonine	0.18	0.02	0.25
DNS-Serine	0.15	0.02	0.20
DNS-Aspartic acid	0.06	0	0.02
DNS-Carboxymethylcysteine	0.05	0	0.02
DNS-Glutamine	0.03	0	0.07
DNS-Asparagine	0.02	0.03	0.02
ε-DNS-Lysine	0.02	0.04	0.08
DNS-Arginine	0	0	0

[a] *R* represents the migration relative to DNS–NH_2. From Ref. [705], by courtesy of Academic Press.

identified by electrophoresis at pH 1.9 as shown in Fig. 3-8. Sometimes pH 12.7 electrophoresis is also used to identify α-DNS-tyrosine, O-DNS-tyrosine, DNS-proline, DNS-threonine, DNS-leucine, and DNS-arginine.

Boulton and Bush [137] reported several useful systems for the paper chromatography of DNS-amino acids. The DNS-derivatives of glycine, alanine, phenylalanine, proline, valine, leucine and isoleucine can be separated by paper chromatography with a mixture of light petroleum-acetic acid-water (100:90:10, v/v) and DNS-derivatives of asparagine, methionine sulfone, serine, aspartic acid, glutamic acid, tyrosine, and glutamine can be separated with benzene-diisopropyl ether-acetic acid-water (1:1:1:1:1, v/v).

The identification of DNS-amino acids also can be made by thin layer chromatography on a plate of silica gel G (or silica gel H, without binders). The extent of migration relative to $DNS\text{-}NH_2$ of the standard DNS-amino acids with several solvent systems is listed in Table 7. In Figs. 3-9 to 3-11 representative patterns of

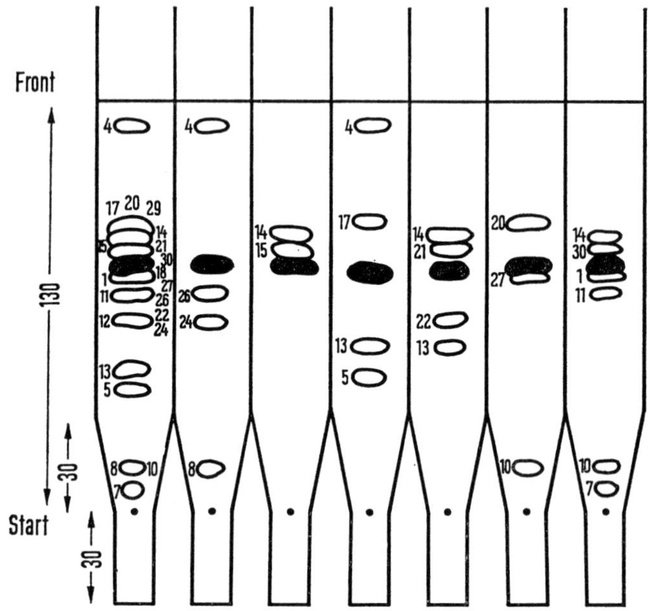

1 Alanine
2 β-Alanine
3 γ-Aminobutyric acid
4 NH_3 ($DNS\text{-}NH_2$)
5 Arginine
6 Asparagine
7 Aspartic acid
8 Cysteic acid
9 Glutamine
10 Glutamic acid
11 Glycine
12 Histidine (di-DNS-His)
13 Hydroxyproline
14 Isoleucine
15 Leucine
16 Lysine (ε-DNS-Lys)
17 Lysine (α,ε-di-DNS-Lys)
18 Methionine sulfone
19 Ornithine (δ-DNS-Orn)
20 Ornithine (α,δ-di-DNS-Orn)
21 Phenylalanine
22 Proline
23 Sarcosine
24 Serine
25 Taurine
26 Threonine
27 Tryptophan
28 Tyrosine (O-DNS-Tyr)
29 Tyrosine (N,O-di-DNS-Tyr)
30 Valine

Fig. 3-9. One-dimensional thin layer chromatogram of DNS-amino acids on a plate of silica gel G. Solvent: methyl acetate-isopropanol-conc. NH_4OH (45:35:20, v/v). Black spot is DNS-OH. Reprinted from Ref. [931], by courtesy of Birkhäuser Verlag

one-dimensional separation of DNS-amino acids on thin layer plates of silica gel G developed with several solvent systems are shown [242, 937]. In most solvent systems $DNS\text{-}NH_2$ runs near the solvent front. The presence of DNS-OH sometimes interferes with the separation of DNS-amino acids, but the blue fluorescent band of this substance is useful as an internal standard due to its easy differentiation from the yellow fluorescence of DNS-amino acids.

During the development of DNS-amino acids, the addition of β-mercapto-ethanol (0.05%) is recommended to eliminate destruction of DNS-amino acids on

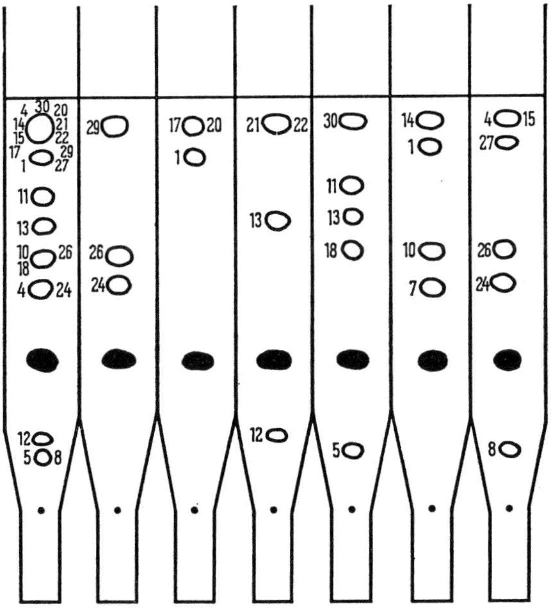

Fig. 3-10. One-dimensional thin layer chromatogram of DNS-amino acids on a plate of silica gel G. Solvent: chloroform-methanol-acetic acid (75:20:5, v/v). Amino acid numbers as in Fig. 3-9. Reprinted from Ref. [931], by courtesy of Birkhäuser Verlag

silica, and particularly to prevent the oxidation of DNS-methionine to its sulfone [1034]. In thin layer chromatography of DNS-amino acids, the *Rf* values are quite variable depending upon the concentrations of salts present in the sample. The unknown sample should be mixed with a series of standard DNS-amino acids and these applied at the origin as shown in Fig. 3-12 [1034]. Although the *Rf* values of the sample and standard DNS-amino acid may be influenced by the presence of salt and appear to be the same, separation of the two can be effected and the relative order maintained when the two compounds are chemically different.

DNS-Amino acids should be viewed under an ultraviolet lamp. Both long-wave (265 mμ) and short-wave (254 mμ) mercury lamps (Chromato-vue, Ultraviolet Products, Inc., San Gabriel, Calif.) are effective in exciting fluorescence. However, the former is preferred because short-wave radiation promotes photo-oxidation of DNS-amino acids to some extent though this is not serious for short exposures [368]. The fluorescence can be visually detected with as little as 10^{-5} μmole of material per spot, but 10^{-4} μmole is about the practical lower limit. Quantitation of DNS-amino acids has been attempted but reproducible results have not so far been obtained.

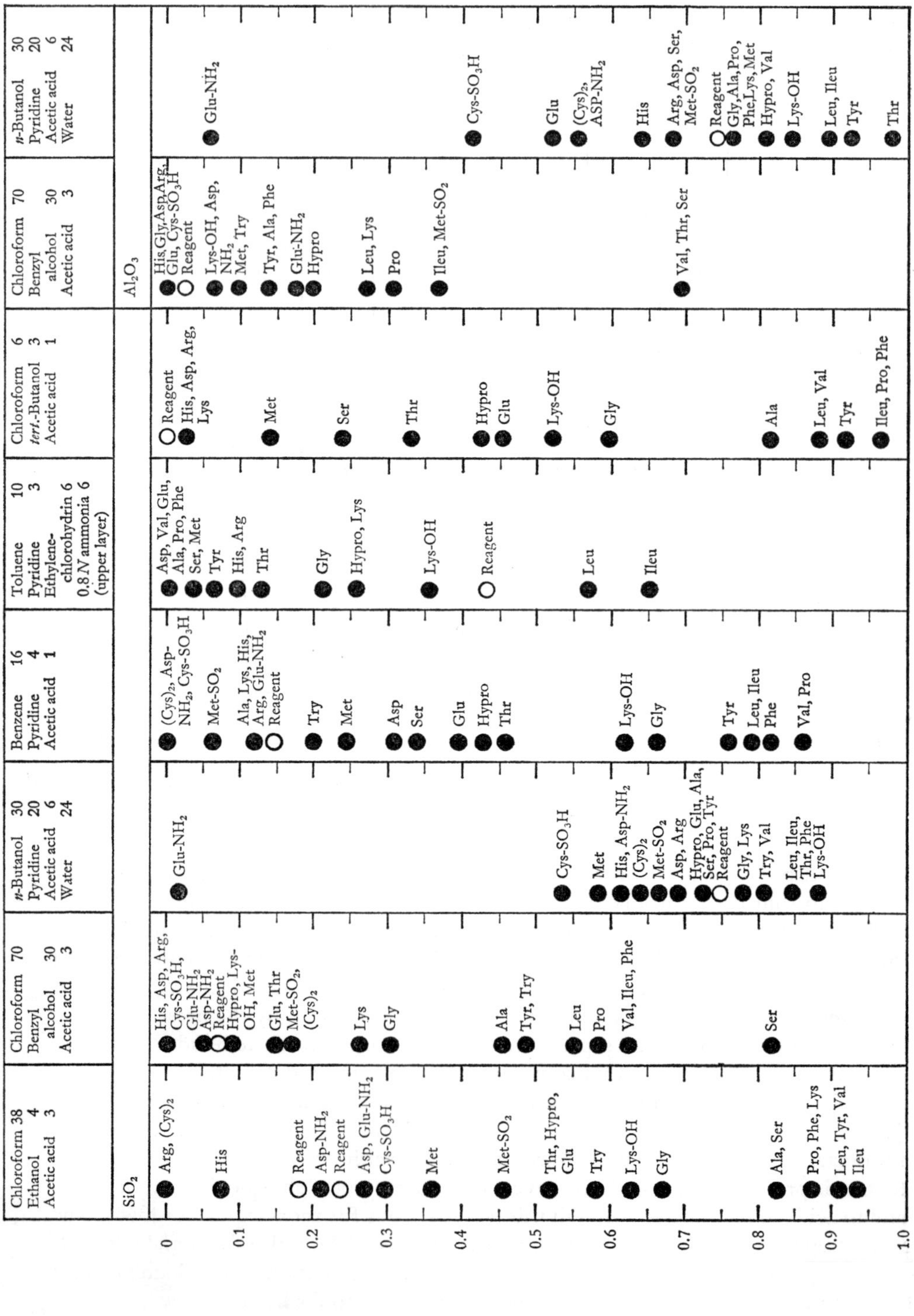

Fig. 3-11. *Rf* values of DNS-amino acids in thin layer chromatograms with different solvent systems and on different carriers. Reprinted

Due to its high sensitivity, the present dansyl chloride method, in combination with Edman's stepwise degradation method, furnishes a powerful tool for the sequence analysis of peptides. After removal of the N-terminal residue of the peptide by the Edman method, the newly released terminal amino acid may be characterized by the present method using only very small amount of the sample. Using this technique, GRAY and HARTLEY [369] elucidated the sequence of a hexapeptide derived from the chymotryptic digestion of *Pseudomonas* cytochrome c_{551} as well as other peptides from different proteins [426, 961]. Other workers also utilized the

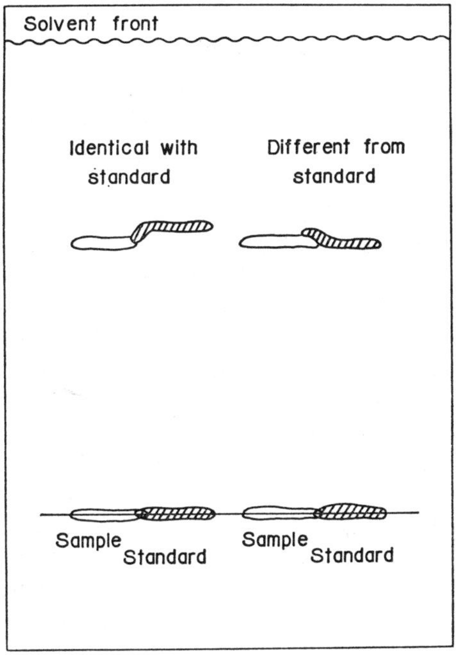

Fig. 3-12. Application of samples and standards of DNS-derivatives on a thin layer plate. Reproduced from Ref. [1034]

above combination of procedures for the structural analysis of proteins and peptides [143, 685, 780]. Standard DNS-amino acids are now available commercially. Their preparations are not described here.

1. Reaction of Dansyl Chloride with Peptides and Proteins [368]

A solution of peptide (0.5 to 5 mμmoles) is transferred into a small test tube. After drying *in vacuo* the peptide is redissolved in 10 to 15 μl of 0.2 M $NaHCO_3$ and the solution is again evaporated *in vacuo* to remove traces of ammonia. The dried residue is redissolved in 15 μl of deionized water and an equal volume of dansyl chloride solution (2.5 mg/ml in acetone) is added. The pH of the solution must be 8.5 to 9.8 at this time. If the pH is below the above range, more base must be added.

The final concentration of the reagent is about 5 mM. The tube is then covered with Parafilm and the reaction is allowed to proceed for 1 h at 37 °C or 2 to 3 h at room temperature. The reaction mixture is then dried *in vacuo*.

For the labeling of proteins, about 10 mμmoles of a protein is dissolved in 0.5 ml of 8 M urea containing $NaHCO_3$ in 0.5 M concentration. To this is added 0.5 ml of dansyl chloride solution (20 mg/ml in acetone) and the mixture is allowed to react for several hours at 37 °C or overnight at room temperature. The mixture is then dialyzed against water to remove salts, urea and DNS-OH, or passed through a small column of Sephadex G-25, if the labeled protein is soluble. Insoluble DNS-protein are collected by centrifugation.

2. Hydrolysis

In most cases it is not necessary to remove DNS-OH from the reaction mixture. To the dried residue, 50 μl (0.5 ml for DNS-proteins) of redistilled constant-boiling HCl (5.7 N) is added and the tube is evacuated and sealed. Hydrolysis is carried out at 105 °C for 16 to 18 h [368] or at 90 °C for 16 to 20 h [1034]. When DNS-valyl-, DNS-isoleucyl- or DNS-leucyldipeptides are found, the hydrolysis time must be longer. The hydrolysate is then dried, and the wall of the hydrolysis tube is washed with one drop of methanol or 50% pyridine to collect the DNS-amino acid in the bottom of the tube. Washing with a drop of methanol or 50% pyridine and drying are repeated until no more fluorescence is visible on the wall. The DNS-amino acids are dissolved in 10 μl of 50% pyridine (dissolves very insoluble di-DNS-tyrosine and -lysine) for application to the electrophoresis paper, the chromatographic paper, or to the silica gel plate for chromatography.

When the removal of the bulk of DNS-OH is required, the DNS-amino acids in the dried acid hydrolysate are extracted twice with 50 μl of ethyl acetate (saturated with water). The pooled extracts are dried in a stream of nitrogen and then redissolved in 50% pyridine for characterization. By this extraction procedure, part of the DNS-OH is also extracted in amounts just sufficient to act as an internal standard. Several DNS-amino acids are extracted only partially, or not at all, and it is usually necessary to examine the residues to identify the DNS-derivatives of cysteic acid, arginine, and α-substituted histidine and lysine.

3. Identification of DNS-Amino Acid

a) Paper Electrophoresis [368]

Gray has shown that most DNS-amino acids can be identified by high voltage paper electrophoresis at pH 4.4, pH 1.9 and sometimes at pH 12.7, using an apparatus of the flat-bed type [376]. Apparatus of the Michl type [683] cannot be used since DNS-amino acids are extracted into the coolant.

Buffer of pH 4.4 is prepared by mixing pyridine, acetic acid and water (10:20:2500, v/v). Buffer of pH 1.9 is made from 8% formic acid (v/v) and pH 12.7 buffer is prepared by dissolving 76 g $Na_3PO_4 \cdot 12\ H_2O$ in 2,000 ml of 0.1 N NaOH. The pH 4.4 buffer is used first for electrophoresis for 2.5 h at 80 v/cm (15 °C). Use of a thick paper such as Whatman 3 mm yields best results. In this case Whatman 3 mm is wetted with about 25 ml buffer per 1,000 cm^2 and the current is adjusted to about 4 to 4.5 ma per cm width. The identification of DNS-derivatives of glycine, serine, alanine and proline is rather difficult at pH 4.4 because strongly contaminating fluorescence of DNS-OH occurs in the same region as these DNS-amino acids. In such instances, a part of the paper where DNS-OH is located is cut from the electrophoretogram and the cut paper strip is stitched across a second sheet with a sewing machine. The second thickness of paper is then cut away from the back of the strip, and the paper is wetted with pH 1.9 buffer. In pH 1.9 buffer no DNS-amino acids migrate toward the anode and only DNS-OH stays near the origin. The potential gradient applied should be 50 v/cm on Whatman 3 mm with a current at about 5 ma/cm width. Under these conditions of electrophoresis (2 h), DNS-glycine, DNS-alanine, DNS-proline and DNS-serine are mutually separated, in this order, from the cathodic side. When pH 12.7 buffer is used for unequivocal identification of α-DNS-tyrosine, DNS-threonine, DNS-leucine, DNS-proline, O-DNS-tyrosine, and DNS-arginine, electrophoresis is carried out at 20 v/cm for 3 h on Whatman 3 mm paper (5 ma/cm). These DNS-amino acids travel in the indicated order from the origin toward the anode at pH 12.7.

b) Thin Layer Chromatography on Silica Gel G Plates

Glass plates are coated with a 1:2 (w/v) slurry of silica gel G in water. Plates are dried for 16 h at room temperature and then heated for 0.5 h at 110 °C.

The DNS-derivatives of the protein-constituent amino acids cannot mutually be separated by the use of one solvent system. Two chromatograms, at least, must be prepared using different solvents. For this purpose, a combination of methyl acetate-isopropanol-concentrated NH_4OH (45:35:20, v/v) and chloroform-methanol-acetic acid (75:20:5, v/v) reported by Seiler and Wiechmann [937] is suitable. Mutual separation of pairs of DNS-valine and DNS-phenylalanine and di-DNS-lysine and di-DNS-tyrosine are insufficient and thus proper solvent systems must be selected from the reported solvents shown in Fig. 3-11 and Table 7.

Paper chromatography also may be used for the separation of DNS-valine and DNS-phenylalanine on Whatman 4 paper with light petroleum ether-acetic acid-water (100:90:10, v/v) [137] as described previously.

C. Cyanate Method of Stark and Smyth

Stark *et al.* [998] observed that contaminating cyanate ions in urea solution react with α- and ε-amino groups in proteins to yield carbamyl derivatives. On the basis of this observation, Stark and Smyth [999] extended the carbamylation reaction with potassium cyanate to the N-terminal analysis of proteins. This method is suitable only for the determination of the N-terminal residue of proteins and large peptides. Potassium cyanate reacts quantitatively with amino groups of proteins at neutral or slightly alkaline pH in the presence of denaturants to yield carbamylated derivatives. Upon heating the carbamyl protein in strong acid, the N-terminal carbamyl residue cyclizes to the corresponding hydantoin, with accompanying cleavage of the peptide bond according to reaction (6).

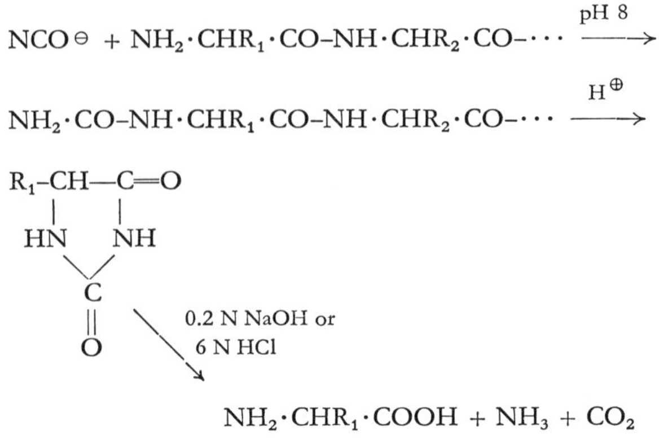

(6)

In contrast to the phenylthiocarbamyl group, the carbamyl moiety requires more drastic conditions for cyclization to the hydantoin of the terminal residue resulting in partial hydrolysis of the residual protein fragment during the cyclization reaction. Thus the cyanate method cannot be used repeatedly as is Edman's stepwise degradation method. The ε-amino group of lysine residues also undergoes carbamylation to yield homocitrulline residues with cyanate at a slow rate (half life is 2 to 3 h) in comparison with α-amino groups (half life is 3 to 4 min) [1000], but the homocitrulline residues do not change under the cyclization conditions. An N-terminal lysine is converted into a carbamyl homocitrulline residue which is then cyclized to the hydantoin of homocitrulline upon heating with strong acid. Sulfhydryl groups also react rapidly with cyanate [998]. In general, reduced and S-alkylated proteins or performic acid-oxidized samples are used in end group determinations of cystine- or cysteine-containing proteins. S-Carbamylcysteine is labile to acid hydrolysis; the products formed are predominantly ammonia and cysteine, plus a small quantity of cystine [998]. Hydantoins are separated from the cyclization mixture by chromatography on Dowex 50-H^+ as will be described later. When an oxidized peptide containing cysteine or cystine is hydrolyzed during the cyclization, cysteic acid is produced. The cysteic acid overlaps with hydantoins derived from the N-terminal carbamylated residues upon elution with water. For this reason, reduction of sulfhydryl groups is preferred to oxidation when the present cyanate methods are employed. However, the N-terminal carboxymethylcysteine does not yield its hydantoin by strong acid treatment and the carbamyl group attached to the terminal nitrogen atom seems to split off (private communication by P. E. Wilcox).

Cyclization of carbamyl amino acid residues into their hydantoins is carried out in a solution of equal volumes of 50% acetic acid and concentrated hydrochloric acid because of the enhanced solubility of carbamyl proteins in 50% acetic acid. Most hydantoins formed from the corresponding carbamyl amino acid residues are quite stable in the mixed acid solvent at 100 °C for 1 h; thus, with the exception of proline (21% yield), less than 3% of any amino acids are formed from the corresponding hydantoins. If only hydrochloric acid is used, approximately twice as much free proline is released. During the cyclization reaction, the thioether sulfur atom in methionine and S-alkylcysteine is partly oxidized in the presence of air. Therefore cyclization must be performed in an evacuated sealed tube.

Hydantoins of acidic and neutral amino acids are respectively acidic and neutral. These hydantoins can easily be separated from other amino acids and peptides in the cyclization mixture by passing the mixture through a column of the hydrogen form of a strongly acidic ion exchanger such as Dowex 50. By this treatment, the hydantoins appear at the breakthrough point, but amino acids, peptides and hydantoins of basic amino acids (histidine and arginine) are retained on the resin column. The basic hydantoins are eluted with 1 N NH_4OH together with amino acids and peptides. The evaporated eluate is repeatedly applied to a Dowex 50 column which has been equilibrated with 0.8 N HCl, and the basic hydantoins are separated by eluting with

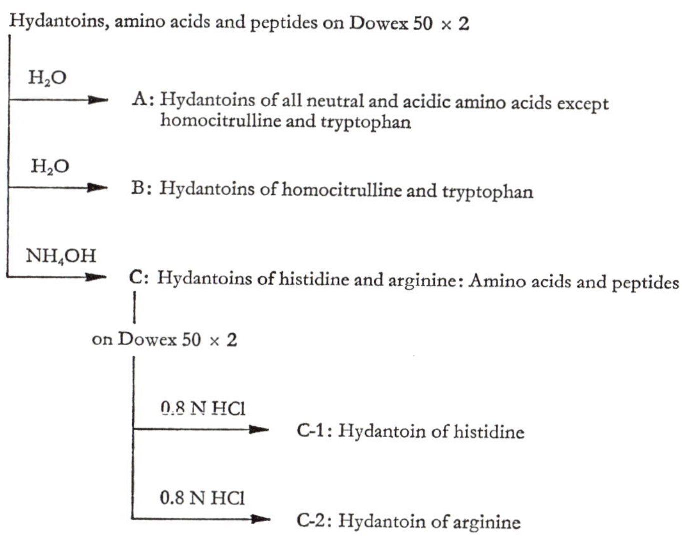

Fig. 3-13. Separation scheme of hydantoins from the products of partial acid hydrolysis according to Stark and Smyth [999], by courtesy of the American Society of Biological Chemists

the same acid as used for the equilibration. Isolation of various hydantoins described above is summarized in Fig. 3-13 [999]. When performic acid-oxidized proteins are used, cysteic acid and peptides containing cysteic acid are contained in the water eluate from the Dowex 50 column together with the hydantoins of neutral and acidic amino acids. In order to separate cysteic acid and its peptides from the hydantoins, the water eluate from Dowex 50 column used previously must be subjected to further chromatography on a Dowex 1-X8 (acetate form) column (0.9 × 6 cm). The hydantoins of neutral and basic amino acids (except tyrosine hydantoin) appear at the breakthrough position by washing the Dowex 1 column with 0.5 N acetic acid with a flow rate of 60 ml/h. The hydantoins of glutamic acid (eluted within 50 ml of effluent) and aspartic acid (100 ml) are somewhat retarded, but strongly acidic cysteic acid and cysteic acid peptides are retained on the column. Tyrosine hydantoin overlaps the positions of the glutamic acid and aspartic acid derivatives.

Heating hydantoins with 6 N HCl regenerates the parental amino acids in nearly quantitative yields with the exception of serine, threonine and tryptophan. Recoveries of serine and threonine are very low due to their destruction under acid conditions and tryptophan is completely decomposed. In the case of alkaline hydrolysis with 0.2 N NaOH, tryptophan is recovered and the recoveries of the two β-hydroxy-α-amino acids increase, but arginine is converted into ornithine, which

Table 8. *Recoveries of amino acids after end group determination*[a]

Amino acid	Recovery (%)
Lysine	84
Histidine	89[b]
Arginine	76[b]
Tryptophan	77[c]
Aspartic acid	99
Threonine	30
Serine	20
Glutamic acid	60 (97[b])
Proline	73
Glycine	98[d]
Alanine	98[d]
Valine	96
Methionine	99[e]
Isoleucine	93
Leucine	97
Tyrosine	96
Phenylalanine	99

[a] All hydrolyses were performed for 20 h in 0.2 N NaOH at 110 °C, except as otherwise indicated. Data from Ref. [999], by courtesy of the American Society of Biological Chemists.
[b] Hydrolysis for 96 h in 6 N HCl at 110 °C.
[c] Determination was performed in the presence of serotonin (see text).
[d] Glycine and alanine were determined separately, not in the presence of other amino acids.
[e] Methionine plus methionine sulfoxides.

overlaps with lysine on the chart of an amino acid analyzer under the standard conditions. With alkaline hydrolysis, recovery of histidine decreases and isoleucine partly epimerizes to alloisoleucine. Therefore the preferred hydrolytic agent for hydantoins changes according to the nature of the hydantoins fractionated. In Table 8 are listed the recoveries of amino acids after the end group determinations. Except for serine and threonine, these recovery factors may be used routinely. Serine and threonine hydantoins decompose somewhat during the cyclization, and the extents of their decompositions are variable; each produces some glycine upon hydrolysis with 0.2 N NaOH, especially threonine hydantoin. For these amino acids it is therefore advisable that each investigator determine the recoveries under his own conditions rather than to use the recoveries listed in the table.

The regenerated terminal amino acids are usually characterized and estimated by an amino acid analyzer. The sensitivity of this method is therefore dependent upon the sensitivity of the analyzer to be used. Procedures of this cyanate method is described in detail by STARK [1000].

1. Carbamylation

Denaturants such as 8 M urea or 4 M guanidine chloride must be employed to obtain carbamylation in homogeneous solutions. Two ml each of N-ethylmorpholine (colorless redistilled preparation, stored in the cold) and water are mixed and the pH of the mixture is adjusted to 8 with glacial acetic acid. To the above buffer is added 2.4 g of solid urea and the final volume is adjusted to 5 ml with water. To a solution of about 50 mg of protein in 2.5 ml of 8 M urea, 2.5 ml of the above buffer is added followed by 250 mg of KCNO. The carbamylation reaction is allowed to proceed overnight at 50 °C.

5 ml of glacial acetic acid is slowly added with stirring to the reaction mixture to decompose the excess cyanate into carbon dioxide and ammonia. If the protein precipitates at this time, it is collected by centrifugation. The precipitate, often a clear gel, should be washed thoroughly, first, with distilled water, then with acetone, and finally dried in the air. If no precipitate was formed on addition of acetic acid, the carbamyl protein is dialyzed against water. In many cases the precipitate forms during dialysis. The contents of the dialysis sac are lyophilized.

2. Cyclization to Hydantoins

A half μmole of carbamyl protein in 1 to 3 ml of 50% acetic acid is placed in a heavy-walled test tube (Pyrex, 1.8 × 15 cm) and an equal volume of concentrated HCl is added. The tube is then evacuated and sealed. After cyclization at 100 °C for 1 h, the contents are evaporated to dryness and 1 ml of water is added to the residue. If the protein contains tryptophan, 50 mg of serotonin creatine sulfate should be present during the cyclization, to prevent destruction of N-terminal tryptophan.

3. Separation of Hydantoins from the Mixture

A column of Dowex 50-X2 (H^+ form, 200 to 400 mesh) is prepared in a 0.9 × 20 cm tube, fitted with a coarse sintered glass disk, such that the column height is 10 cm when subjected to 10 lb/in^2 pressure. The solution of hydantoins is placed onto the column with the aid of three 1 ml rinses of water. The column is developed with water at 20 to 25 °C at up to 100 ml/h. The first 30 ml, including the water displaced by the sample, is collected as fraction A (contains acidic and neutral hydantoins) and the next 40 ml as fraction B (homocitrulline and tryptophan hydantoins). These solutions are evaporated to dryness. The column is washed with 40 ml of 1 N NH_4OH, retained as fraction C, and this solution is evaporated to dryness. The residues from fraction C are dissolved in 2.0 ml of water and 1.8 ml of the solution is placed on another column of Dowex 50–X2 (H^+ form, 0.9 × 10 cm), which has been equilibrated with 0.8 N HCl. The effluent from the column with 0.8 N HCl at a flow rate of about 60 ml/h at room temperature is collected in two fractions, 20 through 35 ml (fraction C–1, contains histidine hydantoin), and 35 through 70 ml (fraction C–2, contains arginine hydantoin). These solutions are evaporated to dryness.

Fraction A contains pyrrolidone carboxylic acid and pyrrolidone carboxyl peptides, if present, derived from glutamine and glutaminyl peptides by cyclization during the cyclization of carbamyl amino acid residues. Hydrolysis of the pyrrolidone groups to glutamic acid or glutamyl peptides is effected with 5 ml of 3 N HCl as the solution is heated at 100 °C for 30 min. After evaporation of the acid, the residues are dissolved in about 2 ml of water and transferred quantitatively onto a Dowex 50–X2 column (H^+ form, 0.4 × 5 cm) with an additional 1 ml of water. The column is washed with 6 ml of water and the effluent is concentrated to dryness. In this author's experience, small amount of non-terminal aspartic and glutamic acids were always found in fraction A. Therefore, the partial acid hydrolysis of this fraction and re-treatment of the hydrolysate with a small Dowex 50 column described above are recommendable.

4. Hydrolysis of Hydantoins

a) Fraction A

2 ml of freshly prepared 0.2 N NaOH is added and 1.8 ml is transferred to a hydrolysis tube (1.8 × 5 cm). The tube is evacuated and sealed, then heated at 110 °C for 20 h. After hydrolysis, the NaOH is neutralized with 1 ml of 1 N HCl and the solution is evaporated to dryness in the tube. 2 ml of citrate buffer of pH 2.2 is added and the solution is centrifuged to remove any precipitate. The supernatant (1.8 ml) is subjected to amino acid analysis for determination of the neutral and acidic amino acids. If the analysis revealed N-terminal glutamic acid, it may be desirable to perform a 96 h acid hydrolysis of the A fraction, since an excellent yield of this amino acid is obtained as shown in Table 8.

b) Fraction B

The residues from fraction B are dissolved in 2.0 ml of 0.2 N NaOH and hydrolyzed exactly as fraction A. The hydrolysate is analyzed with the short column (usually 15 cm) of an amino acid analyzer through the position of lysine.

c) Fraction C

The residues from fractions C–1 and C–2 are dissolved separately in 2.0 ml of 6 N HCl, and 1.8 ml of each solution is hydrolyzed in an evacuated sealed tube at 110 °C for 96 h. The HCl is removed by evaporation and the residues are dissolved in 1 ml of 0.2 N NaOH. The solution is evaporated nearly to dryness to remove excess ammonia. After addition of 1 ml of 1 N HCl, the solution is evaporated to dryness. The residues are dissolved in 2.0 ml of pH 2.2 buffer and 1.8 ml is chromatographed on the 15 cm column, following the positions of histidine for C-1 and arginine for C–2. It is essential that the first 10 ml of the effluent not be mixed with ninhydrin. Large quantities of amino acids derived from neutral peptides are present in this volume which could, if allowed to react with ninhydrin, precipitate in the reaction coil of the analyzer.

D. Other Methods

1. Chemical Methods

Three chemical methods being widely used for the analysis of the N-terminal group have so far been described. In addition, a number of other chemical methods have been proposed but most of them seem to be inadequate for general application to proteins and are seldom used at present. Some of these methods are briefly outlined below. A discussion of the phenylisothiocyanate method, undoubtedly one of the most invaluable methods is reserved for the description by EDMAN in another Chapter in this book.

Arylsulfonyl groups attached to amino groups, in general, are more resistant toward acid hydrolysis than peptide bonds or acylamino linkages. If proper hydrolysis conditions, under which peptide bonds are cleaved but an arylsulfonylamino group is intact, can be found, arylsulfonyl halides can be used as reagents for N-terminal analysis. Dansyl chloride, described in the previous section is a typical example. In addition to dansyl chloride, other arylsulfonyl chlorides such as benzenesulfonyl chloride [385], β-naphthalenesulfonyl chloride [1028] and *p*-iodobenzene-sulfonyl (pipsyl) chloride [1072, 1088] have been examined. The arylsulfonyl amino acids derived from proteins after the reaction and acid hydrolysis are difficult to identify in small amounts, since no characteristic marker groups are involved in these reagents. Pipsyl chloride containing a radioactive I^{131} or S^{35} atom, however, has been shown to be of value for terminal analysis. UDENFRIEND and VELICK [1072, 1088] used this method for the N-terminal analysis of several proteins. Detailed procedures of pipsyl chloride method are described by LEVY [612] and FLETCHER [305]. The half life of the radioactive isotope of iodine or sulfur is relatively short, however, compared with the time necessary for terminal determination.

2,4-Dinitrobenzene sulfonate [277] reacts only with amino groups and releases bisulfate in alkaline media (pH 10.5 to 11.5) yielding the same DNP-derivatives as with Sanger's reagent. Dinitrobenzene sulfonate is sometimes used for the modification of amino groups of proteins, but there seems to exist no advantage as an end group analysis method over the Sanger method. 2,4,6-Trinitrobenzene sulfonate [757, 868, 871] gives the trinitrophenylamino group which is quite labile to hydrolysis and cannot be used for end group analysis, although the reagent is of value for the modification of proteins.

CHIBNALL and SPAHR [201] found that methoxycarbonyl amino acids were readily decomposed in hot strong acid, but upon incubation in 8.6 N hydrochloric acid at 37 °C the extent of hydrolysis of methoxycarbonyl amino acids was relatively low (14 to 20% for 10 to 21 days). On the basis of the above results, they reacted methoxycarbonyl chloride with proteins in an alkaline medium and incubated the methoxycarbonyl proteins in 8.6 N hydrochloric acid at 37 °C for 21 days. After chromatographic separation, they estimated expected amounts of the terminal residues of insulin as the carbomethoxy amino acids. The results indicate that the introduction of methoxycarbonyl group into the terminal amino group labilizes the adjacent peptide bond. This effect was confirmed by using the dipeptide leucylglycine as a substrate for this reaction. The dipeptide was very stable but the corresponding methoxycarbonyl derivative was completely hydrolyzed in 10 days under the above conditions.

Holley and Holley [459] introduced the 2-nitro-4-carbomethoxyphenyl group onto the amino group of peptides, reduced the nitro group catalytically on platinum black to an amino group, and split the terminal peptide bond by heating in neutral aqueous solution with concomitant formation of a lactam ring as shown in the reactions of scheme (7).

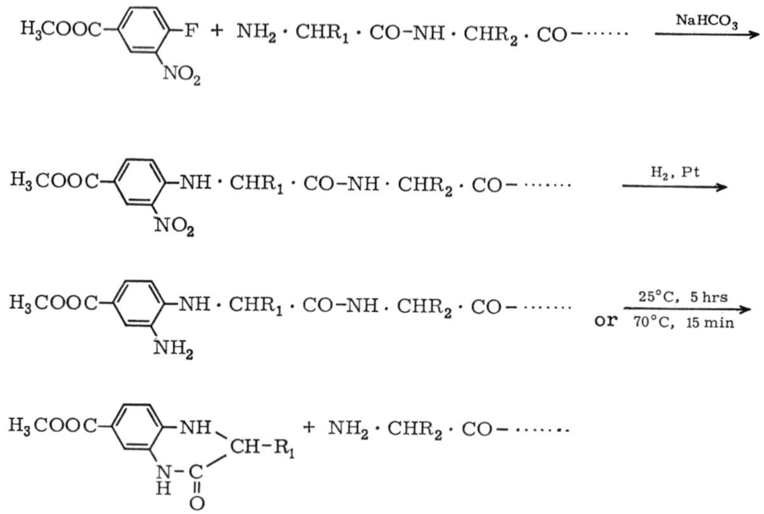

(7)

Ingram [474b] and Jutisz and Ritschard [507] applied the principle of this reaction to DNP-peptides. 2,4-Diaminophenyl peptides produced by catalytic reduction were unstable to atmospheric oxidation and the lactam formation was carried out under a hydrogen atmosphere. Scoffone *et al.* [936] carried out selective reduction of DNP-peptides with ammonium sulfide at 45 to 50 °C in an ethanolic solution to yield 2-amino-4-nitrophenyl peptides, avoiding the formation of the sensitive 4-amino compound. The reduced product was cyclized at pH 2 to 3 by heating at 70 °C for 15 min. The same principle was previously used by Akabori and Narita [11]. They reduced DNP-Taka-amylase with hydrogen sulfide in 70% pyridine and heated the resulting mononitroderivative at 100 °C for 1 h to liberate the terminal residue as the corresponding quinoxaline derivative. This reaction was repeatedly used as a stepwise degradation method. They characterized the newly released end group as a DNP-amino acid according to the Sanger method and succeeded in determining the terminal dipeptide sequence of the amylase as H-Ala-Gly-. However, they observed several DNP-amino acids at the third step of the reaction.

Alkylation reactions of the terminal amino group are also used for the end group determination. N-Dimethylamino acid residues are formed by the action of formaldehyde on the free amino groups followed by catalytic reduction on palladium charcoal [474a]. On hydrolysis of the N-dimethylated peptides terminal residues are recovered as dimethylamino acids which can be identified on a paper chromatogram using the pH indicator, thymol blue.

KORMAN and CLARKE [572] carboxymethylated free amino groups in proteins with bromoacetic acid in a suspension of magnesium hydroxide. They observed that most α-amino groups were quantitatively dicarboxymethylated except glutamic acid, phenylalanine and aspartic acid. Proline was exclusively converted into the monosubstituted derivative. Dicarboxymethylated amino acids liberated, upon acid hydrolysis of the substituted peptides, were precipitated as the mercury salt with mercuric nitrate (or mercuric acetate in 0.1 M sodium chloride). The free amino acid derivatives resulting from hydrogen sulfide treatment are characterized by paper chromatography using the reduced ninhydrin reagent. KORMAN and CLARKE [572] determined the N-terminal aspartic acid of bovine serum albumin by this method.

LEVY [607] demonstrated that carbon disulfide reacts reversibly with amino groups at pH 8.0 to yield substituted dithiocarbamates. Because of the extreme sensitivity of dithiocarbamates to oxidation, the reaction must be carried out under strict anaerobic conditions. CHERVENKA and WILCOX [193] showed that at neutral pH carbon disulfide reacts with the α-amino group only, but at an alkaline pH the ε-amino group of lysine is also reacts with carbon disulfide. Upon acidification of the dithiocarbamates to pH 1 in aqueous solution at room temperature, they cyclize to 2-thio-5-thiazolidones simaltaneous with cleavage of the terminal peptide bond. This is illustrated in scheme (8).

The thiothiazolidone derivatives of amino acid absorb at 280 mμ ($\varepsilon = 13{,}500$), and the parent amino acids can be regenerated from the thiothiazolidone by heating with a mixture of 5 N hydrochloric acid-dioxane. With peptides, the overall yield of thiothiazolidone derivatives is 65% but only 3 to 5% with proteins [601]. Carbon disulfide is used for the reversible masking of amino groups of proteins and peptides [613] rather than as a reagent for the end group analysis.

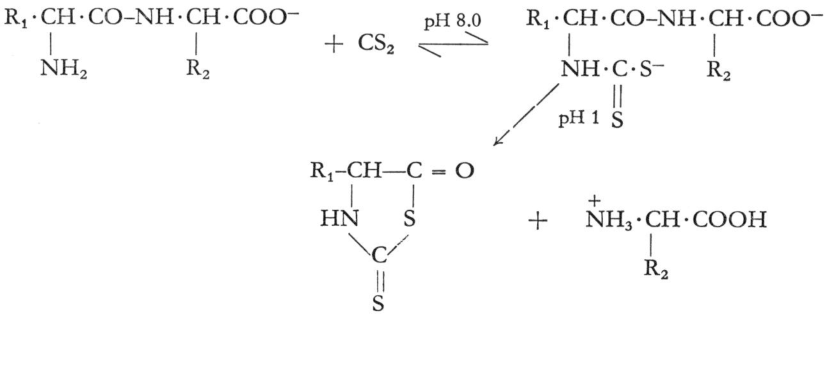

(8)

N-Thiocarboethoxy-peptides prepared from free peptides and methylethylxanthate at alkaline pH give 2-ethoxythiazol-5-one derivatives as intermdiates with fission of the terminal linkage in anhydrous nitromethane saturated with dry hydrogen chloride [536]. This is shown in scheme (9). The intermediate is converted into the N-thiocarboxy anhydride of the amino acid, which yields the parent amino acid by successive treatment with alkali and acid.

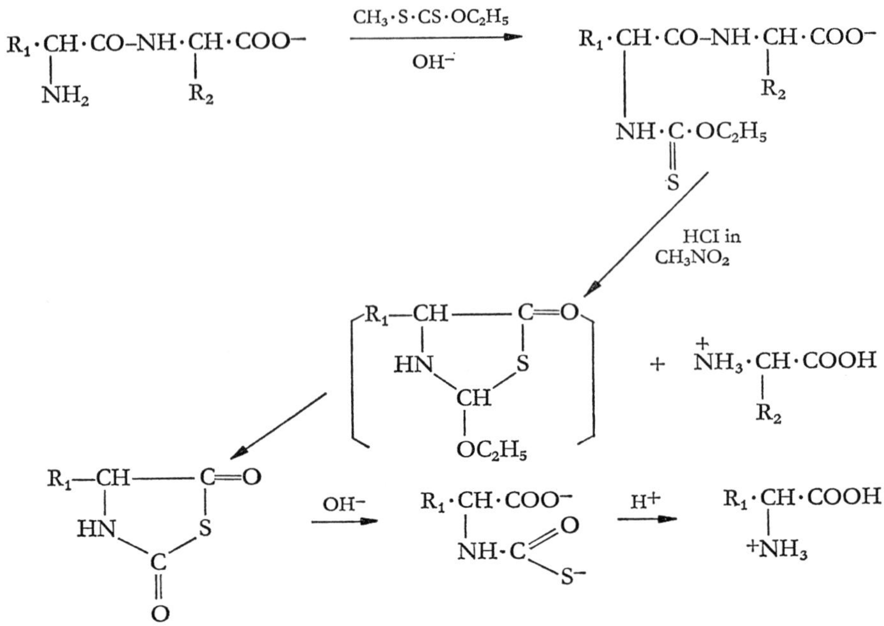

(9)

Elmore and Toseland [281] reported another route to cleavage of the terminal residue. N-Acyldithiocarbamates react at pH 8 with amino group to yield N-acylthio-carbamyl derivatives which are converted into the corresponding thiohydantoin with cleavage of the terminal peptide bond as shown in scheme (10).

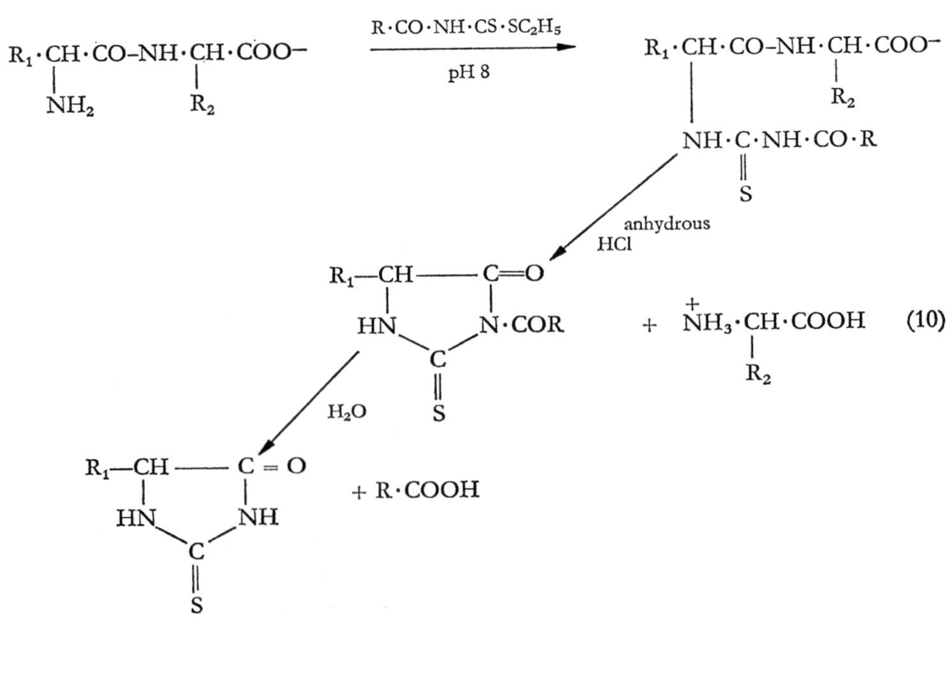

(10)

2. Enzymatic Methods

Leucine aminopeptidase prepared from swine kidney removes amino acids sequentially from the unblocked terminal amino end of peptides and of proteins [965]. Rates of hydrolysis of the terminal amino acids differ, however, depending on the nature of side chain groups. Relative hydrolysis rates of amino acid amides by leucine aminopeptidase are listed in Table 9, assuming the hydrolysis rate of leucineamide to be 100. Rates of hydrolysis for amino acid amides containing bulky aliphatic side chains are high, while that for glycine amide is extremely low. The enzyme catalyzes the hydrolysis of all amino acid of terminal peptide bonds with the exception of the terminal prolyl peptide linkage. When the released amino acids from peptides or proteins are estimated as a function of incubation time with leucine aminopeptidase, the amino acid sequence of small oligopeptides can be deduced. However, when proteins are hydrolyzed with the enzyme for the purpose of sequence analysis, the results are equivocal. Ambiguity derives from difficulty in interpreting the results and from impure enzyme preparations. Leucine aminopeptidase preparations often are contaminated with uncharacterized proteolytic enzymes. The enzyme preparation is usually treated with diisopropyl phosphorofluoridate to inactivate contaminating "active serine" proteolytic enzymes, and sometimes with iodoacetate to inactivate "SH" proteinases, before use. Some preparations of leucine aminopeptidase exhibit carboxypeptidase action and it is recommended to add a carboxypeptidase inhibitor, such as β-phenylpropionate, during the hydrolysis to minimize unfavorable carboxypeptidase action. The procedures are described in detail by Light [619].

Prolonged incubation of proline-free peptides with leucine aminopeptidase yields complete hydrolysis to a mixture of amino acids. The enzyme is exclusively specific for L-amino acid residues and has been extensively used recently for the analysis of the optical purity of synthetic peptides. It is less frequently applied to the sequence analysis of peptides and proteins. Preparation of the enzyme from swine kidney is described by Hill *et al.* [437b] and Light [619].

The enzyme is activated and stabilized by Mg^{2+} and Mn^{2+} and is irreversibly inactivated by chelating agents which remove these divalent cations. It is therefore necessary to dialyze the enzyme against 0.005 M $MgCl_2$–0.005 M Tris buffer, pH 8, at 5 °C before use. Leucine aminopeptidase is labile at pH values above 9 or below 7 and generally is employed at pH 8.5. The hydrolysis of a peptide is performed with 1 mM peptide in 0.0025 M $MgCl_2$–0.05 M barbital buffer of pH 8.5 and with 0.05 to 0.5 mg per ml of enzyme [619].

Wachsmuth *et al.* [1092, 1093] recently isolated another aminopeptidase from swine kidney in a homogeneous state by biochemical and physical criteria. This aminopeptidase is different from leucine aminopeptidase in various respects: the enzyme exists as a bound form with microsomal particle, it is solubilized by tryptic digestion, does not require any divalent metal ion, stable against temperature and extreme pH's, and has almost the same affinities for synthetic substrates of leucine-, alanine- and glycine-*p*-nitroanilide, in contrast to leucine aminopeptidase, which hydrolyzes leucineamide at about 1000-fold the rate of glycineamide as shown in Table 9. Wachsmuth *et al.* reported that their aminopeptidase is a major aminopeptidase present in swine kidney. This enzyme hydrolyzes all the N-terminal amino acids with the exception of proline, at almost equal rates, and is used for test of optical purity of synthetic peptides [189b, 1092, 1093). The enzyme preparation is commercially available as aminopeptidase M.

Table 9. *Relative hydrolysis rates of amino acid amides by leucine aminopeptidase* [a]

Substrate	Mn^{2+} activation	Mg^{2+} activation
L-Leucineamide	100	100
L-Isoleucineamide	20	17
L-Valineamide	17	
L-Alanineamide	3.4	4.9
Glycineamide	0.13	0.10
L-Tryptophanamide	24	
L-Phenylalanineamide	26	17
L-Tyrosineamide	16	
L-Histidineamide	19.1	10
L-Arginineamide	7.1	
L-Lysineamide	7.1	12
L-Isoglutamine	2.2	
L-Aspartic acid diamide	2.9	
L-Serineamide	0.76	
L-Prolineamide	0.71	

[a] Data from Ref. [965].

III. C-Terminal Group Determination

In contrast to the N-terminal determination, only a few chemical methods are available for the determination of the C-terminal group of proteins. Because of this situation, the exopeptidases, carboxypeptidases A and B, are widely used. In carboxypeptidase experiments, it is difficult to draw definite conclusions concerning the C-terminal residue of a large protein unless the relative specificity of the enzyme for the terminal residues is carefully considered. The enzyme method can, however, effectively be used for oligopeptides. Only two chemical methods for the determination of the C-terminal end group are described here; one is the hydrazinolysis method and the other is a qualitative method using the selective labeling of the C-terminal amino acid with tritium.

A. Hydrazinolysis Method of Akabori

Akabori *et al.* [14] found in 1952 that heating proteins with anhydrous hydrazine at 100 °C liberates amino acid hydrazides from the amino acid residues the carboxyl groups of which are involved in peptide linkages. The C-terminal amino acid residue is released as the free amino acid as shown in scheme (11).

$$\cdots\cdots -NH\cdot CHR_m\cdot CO{-}NH\cdot CHR_n\cdot COOH \xrightarrow[100\ ^\circ C]{\text{anhydrous } N_2H_4} \qquad (11)$$

$$\cdots\cdots + NH_2\cdot CHR_m\cdot CO\cdot NH\cdot NH_2 + NH_2\cdot CHR_n\cdot COOH$$

During hydrazinolysis at 100 °C, arginine is decomposed and partly converted into ornithine and triaminoguanidine [710, 741], and cystine is decomposed into unknown product(s). Cysteic acid, S-carboxymethylcysteine and S-aminoethyl-cysteine are stable to the same extent as the other common amino acids, and any of the three residues as the C-terminus can be recovered as free amino acids after hydrazinolysis. It is therefore necessary to carry out the oxidation or reduction of disulfide bonds followed by alkylation prior to hydrazinolysis. The situation with aspartic acid, glutamic acid, asparagine and glutamine residues is complicated [750] as shown in Table 10.

Table 10. *Hydrazinolysis product of acidic amino acids and their amides*

	Original residue	Product after hydrazinolysis
C-Terminal aspartic acid and glutamic acid	–NH–CH–COOH | $(CH_2)_{1-2}$ | COOH	NH_2–CH—COOH | $(CH_2)_{1-2}$ | COOH
C-Terminal asparagine and glutamine	–NH–CH–COOH | $(CH_2)_{1-2}$ | $CONH_2$	NH_2–CH–COOH | $(CH_2)_{1-2}$ | $CO \cdot NH \cdot NH_2$
Non-terminal aspartic acid and glutamic acid	–NH–CH–CO– | $(CH_2)_{1-2}$ | COOH	NH_2–CH–$CO \cdot NH \cdot NH_2$ | $(CH_2)_{1-2}$ | COOH
Non-terminal asparagine and glutamine	–NH–CH–CO– | $(CH_2)_{1-2}$ | $CONH_2$	NH_2–CH–$CO \cdot NH \cdot NH_2$ | $(CH_2)_{1-2}$ | $CO \cdot NH \cdot NH_2$

Separation of the free amino acid derived from the C-terminal position from the bulk of amino acid hydrazides in the protein hydrazinolysate was originally carried out by benzaldehyde treatment [14]. Benzaldehyde reacts with the amino acid hydrazides in an aqueous solution to yield insoluble oil condensation products; thus free amino acids can be recovered from the supernatant. Aspartic β-hydrazide and glutamic γ-hydrazide derived, respectively, from the terminal asparagine and glutamine are also removed with benzaldehyde treatment and thus escape detection.

Amino acid hydrazides cannot be completely removed by a single treatment with benzaldehyde but the hydrazides remaining in the free amino acid fraction are adsorbed tightly on the top of a resin column during amino acid analysis. These amino acid hydrazides are eluted together with the basic amino acids from the short column, hence their separation and the analysis of C-terminal basic residues is rather difficult.

In place of benzaldehyde, isovaleraldehyde [13], enanthoaldehyde [591], and *p*-nitrobenzaldehyde [150] are also used for the same purpose, but there exists no appreciable merit over benzaldehyde.

Akabori *et al.* [12, 13] modified the separation method by introducing the DNP-method to the hydrazinolysate mixture. Hydrazides of amino acids yield neutral di- or tri-DNP-derivatives (substituted at amino group and hydrazide group). However, free amino acid and α- and ω-hydrazides of aspartic acid and glutamic acid give acidic DNP-derivatives, and thus neutral and acidic DNP-derivatives can be separated by fractional extraction with ether or ethyl acetate. Acidic DNP-derivatives are easily separated by two-dimensional paper chromatography with Levy's solvent system [609]. Kawanishi *et al.* [530] further modified the fractional extraction method of DNP-derivatives suitable for the characterization of di-DNP-aspartic and -glutamic monohydrazides.

Niu and Fraenkel-Conrat [741] reported an improved procedure for removing hydrazides by successive benzaldehyde treatment and dinitrophenylation, it being difficult to remove quantitatively amino acid hydrazides by a single aldehyde treatment of the hydrazinolysate. They determined C-terminal amino acid as their DNP-derivatives after characterization by paper chromatography with Levy's solvent system [609]. The recoveries of amino acids after heating at 100 °C for 10 h with anhydrous hydrazine, benzaldehyde treatment, dinitrophenylation, paper chromatography and extraction of DNP-amino acids from the chromatogram with 1% $NaHCO_3$ were estimated by Niu and Fraenkel-Conrat and are listed in Table 11 (2nd column). This author (unpublished experiments) has confirmed their reported recoveries (3rd column in Table 11).

Alternatively, for the mutual separation of free amino acids from the bulk of the hydrazides, relatively short ion exchange columns, such as Amberlite CG-50 (H^+ form, 1 × 3 cm), Amberlite IR-120 (equilibrated with 0.2 M pyridine-acetate buffer of pH 3.1, 1 × 18 cm) or phosphocellulose (equilibrated with 0.4 M pyridine-formate buffer, pH 3.2, 1 × 30 cm) are used and the free amino acid fraction is subjected to amino acid analysis under the usual analysis conditions. Fraenkel-Conrat and Tsung [325] used an aqueous solution of a hydrazinolysate on a column of Amberlite CG-50 in the H^+ form. They recovered acidic and neutral amino acids in 15 ml of the water eluate together with the monohydrazides of aspartic and glutamic acids. The basic amino acids were recovered in 15 ml of the subsequent eluate with 0.1 M ammonium acetate (pH 7). Braun and Schroeder [149] used an Amberlite IR-120 column which had been equilibrated with 0.2 M pyridine acetate, pH 3.1, and a phosphocellulose column which had been equilibrated with 0.4 M pyridine formate, pH 3.2. With the former column, acidic and neutral amino acids are eluted in the pH 3.1 buffer and lysine in 0.2 M pyridine acetate, pH 5.2. From the latter column, all free amino acids and aspartic and glutamic monohydrazides are eluted with the equilibrated buffer. The monohydrazides of the acidic amino acids give no clear peak on the chart by an automatic amino acid analyzer.

Hydrazinolysis of proteins is complicated by the fact that the amino acid hydrazides are not very stable, in particular the dihydrazides of aspartic and glutamic acids, and may undergo hydrolytic and oxidative decomposition. Further, free amino acids with small side chains slowly undergo partial conversion to their hydrazides. It is necessary therefore to find conditions which minimize the conversion of free amino acids derived from the C-terminus into their hydrazides and the regeneration of amino acid from the hydrazides on the one hand, and to maximize the extent of hydrazinolytic cleavage of the C-terminal residues on the other. The longer the time of hydrazinolysis and the higher the hydrazinolysis temperature, the greater

Table 11. *Recoveries of amino acids after heating with hydrazine*[a]

	100 °C, 10 h without catalyst		60 °C, 16 h 1M hydrazine sulfate		80 °C, Amberlite CG–50[f]					
	[c]	[d]	[e]	[d]	11 h	23 h	40 h	60 h	82 h	112 h
Lysine	47	57	90	97	41	36	37	29	31	29
Histidine	43	—	53	—	88	83	85	54	73	60
Arginine[b]	25	32	60	46	—	—	—	—	—	—
Aspartic acid	19	8	99	78	77	63	40	26	24	20
Threonine	52	45	105	92	92	95	80	72	79	68
Serine	39	35	96	90	84	89	67	62	67	51
Glutamic acid	16	8	78	59	81	74	44	41	39	35
Proline	52	68	97	94	100	102	96	80	88	75
Glycine	37	30	—	77	92	88	69	58	60	57
Alanine	45	48	—	79	96	95	85	76	79	74
Half cystine	—	—	78	—	77	52	52	41	25	0
Valine	76	70	—	94	101	108	100	104	108	96
Methionine	56	54	66	72	90	90	86	59	87	79
Isoleucine	54	—	—	86	101	103	104	108	107	103
Leucine	54	60	—	91	99	99	94	94	90	93
Tyrosine	44	—	—	—	95	97	91	89	92	83
Phenylalanine	59	65	—	83	96	97	91	74	93	85
Tryptophan	39	47	46	74	—	—	—	—	—	—

[a] Recoveries of amino acids after hydrazine treatment without catalyst and with hydrazine sulfate are estimated by DNP-technique, and with CG-50 resin by the automatic amino acid analyzer. No value in the Table means that estimation has not been carried out.
[b] Arginine was characterized and estimated as ornithine.
[c] Data from Ref. [741].
[d] Unpublished experiments.
[e] Data from Ref. [144].
[f] Data from Ref. [149].

the extent of the release of the C-terminal amino acids. However, under these conditions, the unfavorable side reactions proceed more extensively. BRADBURY [144] and KAWANISHI *et al.* [530] introduced the catalytic hydrazinolytic technique to include hydrazine sulfate and reported the increased yield of the C-terminal amino acids when proteins were heated with anhydrous hydrazine at 60 to 80 °C for 16 h in the presence of 1 to 2 M hydrazine sulfate. In this author's experience, the recoveries of amino acids after heating the amino acid mixture under the above conditions increased appreciably, but yields of the C-terminal amino acids from several oligopeptides have not been significantly improved suggesting that hydrazinolytic cleavage of the C-terminal residues was not sufficient.

Braun and Schroeder [149] recently reported the effectiveness of the H^+ form of Amberlite CG-50 resin as a catalyst for hydrazinolysis. They heated several proteins with anhydrous hydrazine at 80 °C for 60 to 100 h in the presence of the resin and estimated the C-terminal residues in yields of 60 to 100% without any corrections as shown in Table 12. Recoveries of amino acids after heating them at 80 °C under the same conditions as the resin-catalyzed hydrazinolysis of proteins are listed also in Table 11. The resin-catalyzed technique seems to be superior to the sulfate-catalyzed method since the reported yields of C-terminal residues of several proteins increased significantly in comparison with those of uncatalyzed hydrazinolysis. Furthermore, the catalyst can easily be removed by centrifugation. Braun and Schroeder [149] described difficulties in determining C-terminal asparagine and glutamine residues with confidence in contrast to the results of Kawanishi *et al.* [530]. The situation of the C-terminal asparagine and glutamine is complicated because acidic amino acids and their amides seem to follow the reactions (12) as suggested by Narita and Ohta [720].

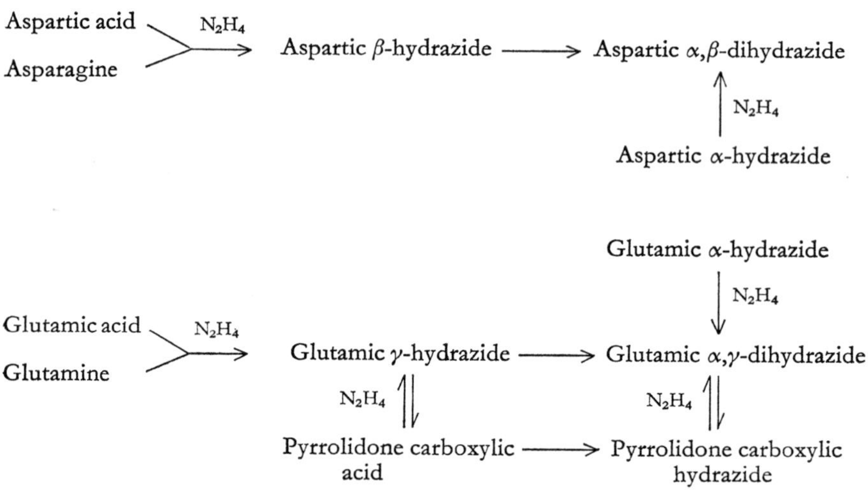

(12)

Furthermore, both aspartic monohydrazides seem to be interconvertible through an assumed intermediate of succinohydrazide as shown in scheme (13).

$$\mathrm{HOOC\cdot CH_2\cdot \underset{\displaystyle NH_2}{\underset{|}{CH}}\cdot CO\cdot NH\cdot NH_2} \rightleftharpoons \begin{array}{l}\mathrm{CH_2\cdot CO}\diagdown \\ | \qquad\qquad \mathrm{N\cdot NH_2} \\ \mathrm{CH_2\cdot CO}\diagup\end{array} \rightleftharpoons \mathrm{H_2N\cdot NH\cdot CO\cdot CH_2\cdot \underset{\displaystyle NH_2}{\underset{|}{CH}}\cdot COOH} \qquad (13)$$

Another important point toward minimizing the unfavorable hydrolytic cleavage of amino acid hydrazides which result from hydrazinolysis of proteins is to avoid prolonged contact with water. Amino acid hydrazides, especially hydrazides of glycine, serine and alanine, are labile in aqueous solution and yield the corresponding free amino acids by hydrolytic cleavage at room temperature. Typical examples are shown in Fig. 3-14 and Fig. 3-15 [719]. Fig. 3-14 shows the released amino acids

Table 12. *Yields of C-terminal amino acids obtained from proteins with Amberlite CG-50 catalyzed hydrazinolysis at 80 °C*[a]

Protein	C-Terminal amino acid and bond to be cleaved	Molecular weight	Hydrazinolysis time (hours)	Yield
Lysozyme	–Arg-Leu-OH	14,600	20	95%
Ribonuclease	–Ser-Val-OH	13,600	95	100%
Myoglobin	–Gln-Gly-OH	17,800	88	100%
Cytochrome *c*	–Asn-Glu-OH	12,400	16	90%
Tobacco mosaic virus protein	–Ala-Thr-OH	17,400	46	72%
Hemoglobin β-chain	–Tyr-His-OH	16,000	95	68%
Insulin	–Cys-Asn-OH		45	52%[b]
	–Lys-Ala-OH	5,700	22	64%
Chironomus hemoglobin	–Met-OH	16,500	72	78%
Bence Jones protein	–Glu-Cys-OH	23,000	45	67%[c]
Glucagon	–Asn-Thr-OH	3,483	75	92%

[a] Data from Ref. [149].
[b] Aspartic monohydrazide fraction was hydrolyzed and estimated.
[c] Determined as cysteic acid after performic acid oxidation of the protein.

from Taka-amylase A by the uncatalyzed hydrazinolysis at 100 °C for various periods. Their amounts are estimated as the DNP-derivatives as mentioned above. The results suggest that serine and glycine apparently are the C-terminal residues. In

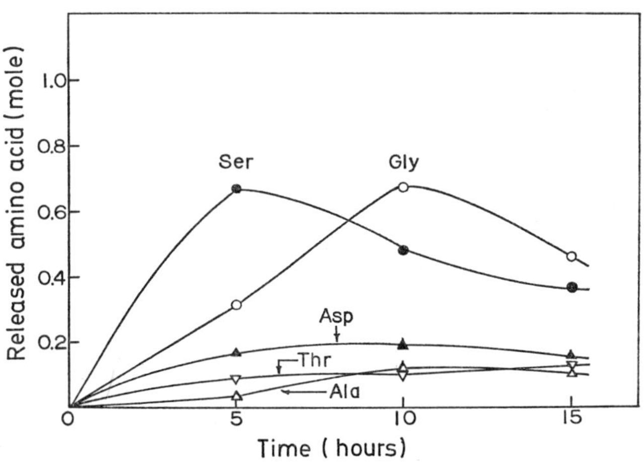

Fig. 3-14. Liberated amino acids from Taka-amylase by uncatalyzed hydrazinolysis at 100 °C as a function of time. Released amino acids were estimated as their DNP-derivatives. Compare difference from Fig. 3-15. Reprinted from Ref. [19], by courtesy of the Japanese Biochemical Society

Fig. 3-15, the released amino acids are estimated by an automatic amino acid analyzer and only serine seems to be the C-terminal group. C-Terminal serine was confirmed by the use of carboxypeptidase A. Hydrazinolysis conditions are the same for the above two experiments, but procedures after benzaldehyde treatment are different from another. In the experiment shown in Fig. 3-14, the supernatant after the aldehyde treatment was reacted with FDNB at room temperature for 1.5 h in $NaHCO_3$ solution, but for the experiment shown in Fig. 3-15, the supernatant was quickly lyophilized to minimize the hydrolysis of amino acid hydrazides and was subjected to amino acid analysis immediately after dissolving the lyophilisate in citrate buffer of pH 2.2. These results clearly suggest that prolonged contact of the hydrazinolysate with an aqueous solution, especially in an alkaline medium, is not favorable.

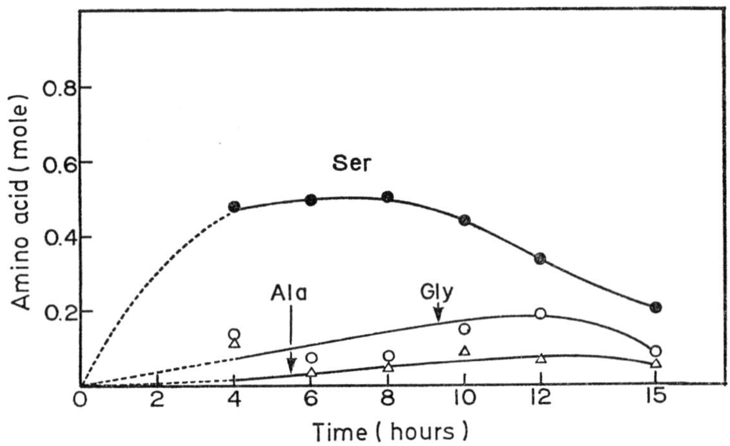

Fig. 3-15. Liberated C-terminal amino acid from Taka-amylase by uncatalyzed hydrazinolysis at 100 °C as a function of time. Amino acids were estimated by an automatic amino acid analyzer. Compare difference from Fig. 3-14. Reprinted from Ref. [19], by courtesy of the Japanese Biochemical Society

When a protein contains small amounts of glycine, serine or alanine, the dinitrophenylation technique can be used without difficulties, especially for small oligopeptides. Hydrazinolysis can therefore be performed even if an amino acid analyzer is not available.

1. Preparation of Anhydrous Hydrazine

A conventional method is to utilize the characteristic of hydrazine to form an azeotropic mixture with toluene. After 50 g of hydrazine (95% +), 300 g toluene, and 50 g of CaO (or 50 g of 80% hydrazine hydrate, 500 g of toluene and 500 g of CaO) are mixed in a 500 ml round-bottom flask and kept overnight, the mixture is refluxed for 3 h while excluding the uptake of moisture with a $CaCl_2$ tube. The azeotropic mixture is distilled at 93 to 94 °C [591]. The lower phase of the mixture is anhydrous hydrazine, of about 98.5% purity, which can be used for the hydrazinolysis of proteins and peptides. It is recommended that distillation be discontinued while the

CaO is still wet, otherwise an explosion is likely. Hydrazine absorbs moisture quickly and the condensed hydrazine must be protected from moisture during distillation. The hydrazine thus prepared is best sealed immediately in a number of ampules of 1 to 2 ml and stored in a refrigerator.

Braun and Schroeder [149] dehydrated hydrazine with NaOH. A mixture of 100 ml of hydrazine (95% +) and 40 g of NaOH pellets in a 250 ml round-bottom flask is heated to the boiling point under nitrogen atmosphere and then cooled to room temperature. During vacuum distillation with an oil pump, nitrogen that has passed through concentrated H_2SO_4 is introduced through a nonflexible capillary. The hydrazine is distilled between 16 to 18 °C and the distillate is condensed in a 100 ml round-bottom flask at the temperature of dry ice.

2. Hydrazinolysis

a) Uncatalyzed Method

To 0.1 to 0.5 μmole of a dried protein or peptide in a small (1 to 1.5 cm × 10 to 15 cm) constricted test tube is placed 0.5 ml of hydrazine by drawing the reagent through the constricted neck under cooling in an alcohol-dry ice bath. Hydrazinolysis is then carried out at 100 °C for 6 to 10 h in a sealed tube. During heating, the sample tube should be shaken gently three or four times. The reaction tube is then opened and the contents are lyophilized. When the reaction mixture cannot be examined immediately, the reaction tube is best stored unopened in a freezer until analysis is begun [149].

b) Resin-Catalyzed Method [149]

About 50 mg of dry Amberlite CG–50 (H^+ form, less than 200 mesh, stored under vacuum in a desiccator over P_2O_5 after drying in an oven at 80 °C) is placed in a heavy-walled Pyrex tube (1.5 × 12.5 cm). A weighed amount of the dried protein (usually around 0.25 μmole) and 2 ml of anhydrous hydrazine are added into the tube, as described in the uncatalyzed method. After hydrazinolysis at 80 °C for 60 to 100 h, the reaction mixture is transferred into a 25 ml round-bottom flask, the reaction tube is rinsed twice with 1 ml of undistilled hydrazine (95% +), and the entire material is then lyophilized. Between rinses, the flask is stoppered to minimize the absorption of moisture. After the totally dry mixture, which contains CG-50 resin, the hydrazides, and the released amino acids, is suspended in 3 ml of water, the resin is centrifuged and washed twice with 1 ml of water. A solution of the hydrazinolytic products in water has a pH between 8.5 and 9, which is lowered to a pH below 2 with a few drops of 2 N HCl before the reaction mixture is chromatographed for the removal of the hydrazides.

3. Removal of Amino Acid Hydrazides from the Released Free Amino Acid

a) Column Chromatography

According to the procedures of Braun and Schroeder [149], Amberlite IR–120 is equilibrated with 0.2 M pyridine acetate buffer of pH 3.1 and a 1 × 18 cm column is prepared. The reaction mixture derived from the above hydrazinolysis is placed on the column at pH 1.5 to 2.0 and the free amino acid are then developed with 0.2 M pyridine acetate buffer at 30 °C with a flow rate of 30 ml/h. 2 ml fractions are collected.

With the pH 3.1 buffer, all amino acid hydrazides are completely retained by the resin, whereas the acidic and neutral amino acids appear between fractions 8 and 30, except tyrosine, which emerges later, between fractions 54 and 60. The buffer is then changed to 0.2 M pyridine acetate, pH 5.2. Lysine emerges between fractions 110 and 120, but histidine cannot be separated from the amino acid hydrazides by the IR–120 column. Since the hydrazides are strongly fixed on the resin, after chromatography, the resin must be regenerated by boiling with 0.5 N NaOH for 1 h, washed successively with water and acid, and, finally equilibrated with the pH 3.1 buffer.

For the separation of histidine from the hydrazides, a phosphocellulose column is employed. A hydrogen form of phosphocellulose (Whatman cellulose phosphate powder P–70, unsifted) is equilibrated with 0.4 M pyridine formate buffer of pH 3.2. The sample at pH 1.5 to 2.0 is placed on a 1 × 30 cm column of the phosphocellulose and free amino acids are eluted with the 0.4 M buffer at room temperature with a flow rate of 30 ml/h. Acidic and neutral amino acids appear between fractions 12 and 22 (1.7 ml fraction), lysine and ammonia between fractions 33 and 43, and histidine between fractions 47 and 53. The monohydrazides of the acidic amino acids appear shortly before the lysine peak and overlap with the lysine position. The efluents are lyophilized and subjected to amino acid analysis by the standardized procedures.

Alternatively, Fraenkel-Conrat and Tsung [325] used the hydrogen form of an Amberlite CG–50 column (1 × 3 cm). Neutral and acidic amino acids appear in 15 ml of the water eluate together with the monohydrazides of the acidic amino acids. Subsequent elution of the CG–50 column with 15 ml of 0.1 M ammonium acetate (pH 7) yields the basic amino acids. They recommend the use of a column of 0.9 × 29 cm (pH 4.26, 0.38 N sodium citrate, 0 to 185 min at 33 °C, followed by pH 5.28 citrate, 185 to 395 min at 55 °C) for the analysis of the basic amino acid fraction including ornithine by the accelerated system of an automatic amino acid analyzer.

b) *Aldehyde Treatment*

The lyophilized hydrazinolysate is dissolved in 2 ml of water, and 0.5 ml of freshly distilled benzaldehyde is added. The mixture is immediately shaken vigorously for 1 h at room temperature and then centrifuged. The supernatant is transferred to a lyophilization flask and the oily residue in the centrifuge tube is washed twice with 1 ml of water. The supernatant and washings are immediately frozen and dried. Just before analysis, the lyophilized material is dissolved in 1 ml of citrate buffer, pH 2.2, and the solution is subjected to amino acid analysis. In this case, the analysis of basic amino acids is rather difficult, because the some of the hydrazides are eluted with basic amino acids by the usual procedures.

c) *Dinitrophenylation Technique* [530]

The lyophilized hydrazinolysate (or effluent of the above chromatography) is dissolved in 3 ml of water, shaken with 0.3 g of $NaHCO_3$ and 6 ml of 2.5% DNFB in ethanol for 2 h. 20 ml of water is added and the pH of the solution is adjusted with 2 N HCl to 1. The solution is extracted with ethyl acetate (10 ml and 3 × 5 ml) and the extract is diluted with 2 volumes of ether. The organic solution is back-extracted with 2% aqueous $NaHCO_3$ (3 × 15 ml) and the aqueous extract is washed with ether (2 × 10 ml). The aqueous phase is acidified with 4 N HCl to pH 1 and again extracted with ethyl acetate (10 ml and 3 × 5 ml). The yellow residue, freed of dinitrophenol by sublimation [689], is chromatographed two-dimensionally according to Levy's procedures [609] or *n*-butanol saturated with 2 N NH_4OH [557] and 1.5 M phosphate buffer of pH 6, as described in the DNP-method. Locations of the DNP-derivatives of the monohydrazides of aspartic and glutamic acids together with usual DNP-amino acids are shown in Fig. 3-16. Spots A and B are artifacts and can be differentiated from the typical DNP-amino acids ($OD_{390}/OD_{360} = 0.55$ to 0.65 in 1% $NaHCO_3$) by their different ultraviolet absorption characteristics (OD_{390}/OD_{360} in 1% $NaHCO_3$ are 1.2 and 0.9, respectively). Millimolar extinction coefficients of

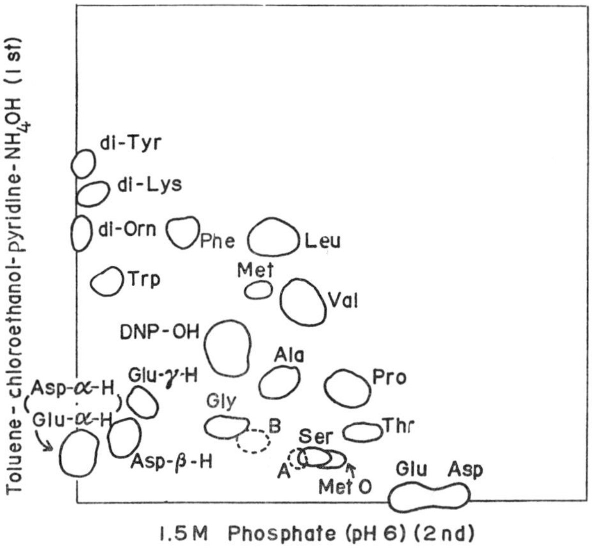

Fig. 3-16. Two-dimensional paper chromatogram of acidic DNP-derivatives possibly present in the dinitrophenylation mixture of hydrazinolyzates of proteins. Spots A and B shown in dotted circles are artifacts. Abbreviations: Asp-α-H, aspartic acid α-hydrazide; Asp-β-H, aspartic acid β-hydrazide; etc. Reproduced from Ref. [530], by courtesy of *the Japanese Biochemical Society*

typical DNP-amino acids at 360 mμ in 1% $NaHCO_3$ average 17 and those of di-DNP-derivatives of aspartic β- and glutamic γ-hydrazides are 26 and 24 at their absorption maximum, 345mμ, in 1 N HCl ($OD_{390}/OD_{360} = 0.55$ to 0.60 in 1 N HCl). The absorption maximum of the di-DNP-derivatives of the above two hydrazides shift to 415 mμ in 1 to 2% $NaHCO_3$ solution, but their millimolar extinction coefficients at the maximum wave length remain unchanged, giving a broader band and OD_{390}/OD_{360} of 1.1.

B. Tritium-Labeling Method of Matsuo

It is well-known that acyl amino acids with the exception of acyl proline and acyl aspartic acid are easily racemized during the condensation reaction with amine components during the synthesis of polypeptides. Neuberger [728], in 1948, suggested as a possible mechanism that racemization of the acyl amino acid might take place through the intermediate compounds, oxazolone and oxazole as shown in scheme (14).

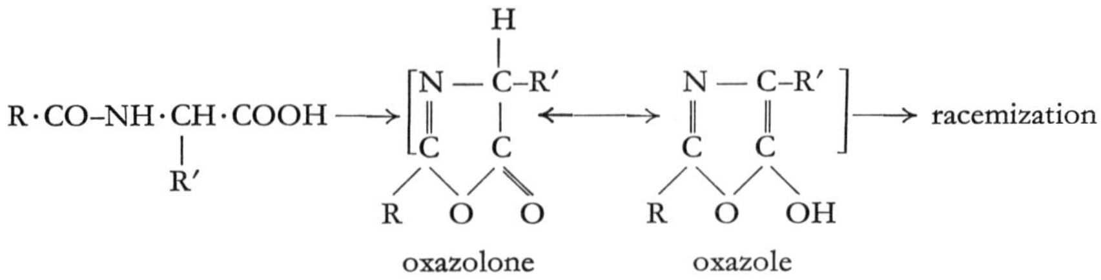

(14)

This racemization mechanism has recently been confirmed by YOUNG [1164] and GOODMAN and LEVINE [357]. MATSUO, FUJIMOTO and TATSUNO [655] also tried to confirm the above mechanism by using heavy water containing isotopic hydrogen (deuterium or tritium) for the ring-opening reaction of the oxazolone derivative.

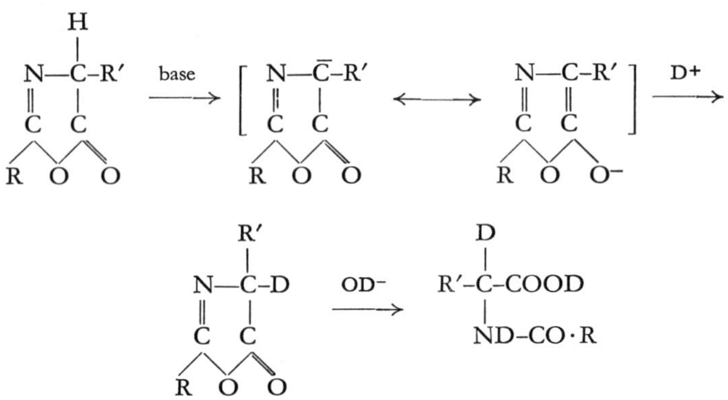

(15)

They originally studied the ring-opening reaction at room temperature with D_2O and the oxazolone derivatives derived from acetyl amino acids and acetic anhydride and found that acetyl derivatives of all of the protein-constituting amino acids, except proline, hydroxyproline and aspartic acid, had undergone racemization via oxazolone derivatives to yield the deuterated acetyl amino acids as shown in scheme (15). Further, they demonstrated by means of NMR that the release of the hydrogen at the 4 position of the oxazolone was a base-catalyzed reaction and that the addition of one drop of pyridine or triethylamine made racemization quantitative regardless of the acyl amino acids employed except for the three acyl amino acids mentioned above.

Under the same conditions of base-catalyzed deuteration of other acetyl amino acids, acetyl aspartic acid gave the intramolecular anhydride, but acetyl asparagine gave the corresponding oxazolone. Unlike acetyl aspartic acid, acetyl glutamic acid was converted into the corresponding oxazolone in a manner similar to acetyl glutamine. Acetyl proline gave the mixed anhydride with acetic acid and escaped the base-catalyzed deuteration reaction under the above conditions. When acetyl aspartic acid and acetyl proline were heated in a mixture of deuterated acetic acid and acetic anhydride, however, they underwent deuteration. Although the mechanism of the isotope labeling reaction of the latter two acetyl amino acids is not clear, the above results seem to provide a novel method of C-terminal characterization of peptides and proteins. The reaction mechanism is illustrated in scheme (16).

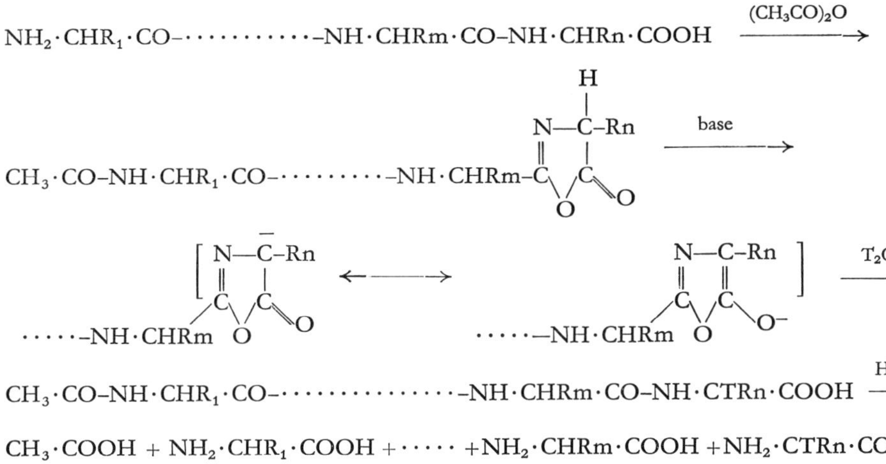

(16)

In order to simplify characterization of the isotopically labeled C-terminal amino acid residue, Matsuo *et al.* [656] replaced D_2O with T_2O (0.05 ml, 50 mc) as the medium, in the presence of one drop of pyridine, for the ring-opening reaction of the oxazolone derivative formed by the reaction of the C-terminal residue with acetic anhydride (2 ml) in dioxane (2 ml) under gentle boiling for 30 min. The tritium labeled C-terminal amino acid was characterized by paper chromatography or similar methods after acid hydrolysis of the tritiated peptide. This method has proved to be useful as a C-terminal characterization with several synthetic oligopeptides and angiotensin II. Due to the insolubility of proteins in dioxane, Matsuo *et al.* modified the reaction conditions for proteins by dissolving them (ca. 2 mg) in T_2O-pyridine (0.1 ml:0.2 ml), under warming, followed by the addition of acetic anhydride (0.05 ml) under ice-cooling and maintaining at room temperature for a few hours. Thus they succeeded in labeling selectively the C-terminal residues of bovine insulin and egg white lysozyme [656]. As one typical example, the radioautogram (by Radiochromatogram Scanner) of the paper chromatogram of the hydrolysate of the tritiated performic acid-oxidized insulin is shown in Fig. 3-17.

When the DNP-protein is used for selective tritium-labeling, both N- and C-terminal groups can be characterized with a single sample [657]. C-terminal aspartic acid and proline can also be selectively labeled by heating peptides (ca. 2 mg) in T_2O-acetic anhydride (0.05 ml, 50 mc/3 ml) for 45 min [656]. The mechanism for this labeling reaction is evidently different from the base-catalyzed racemization as mentioned above.

The reaction conditions used for the tritiation of proteins afforded the tritium incorporation into C-terminal aspartic acid [458], in contrast to the aforementioned conditions; heating peptide with acetic anhydride in dioxane and subsequent addition of T_2O and pyridine. Co-existence of acetic anhydride, pyridine and T_2O may prevent possible formation of the intramolecular anhydride of acyl aspartic acid and may lead to the oxazolone formation followed by the tritiation of the regenerated aspartic acid residue. Although C-terminal proline cannot be tritiated, such conditions are very mild and unfavorable side reactions can be avoided. The C-terminal residues of *Scenedesmus* ferredoxin [658], pig heart aspartate aminotransferases [509] and sheep luteinizing hormone [458] were determined under these conditions.

Isoaspartyl and isoglutamyl residues in peptide chain are also tritiated by the above base-catalyzed reaction. For example, glutathione gave radioactive glutamic acid and glycine with the standard base-catalyzed tritiation reaction (acetic anhydride-pyridine-T_2O) and subsequent acid hydrolysis [657]. When DNP-glutathione (oxidized

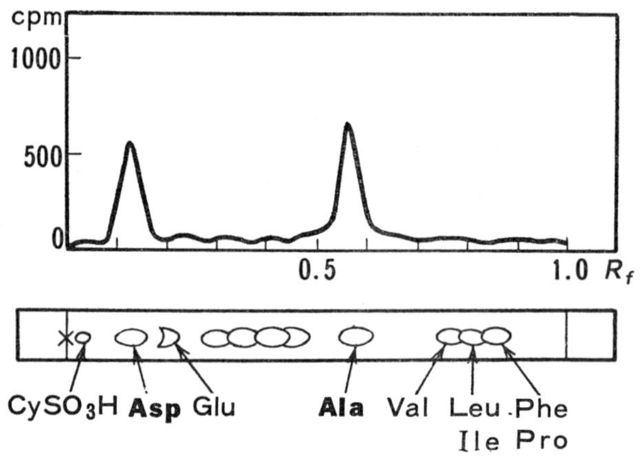

Fig. 3-17. Radio-chromatogram of the hydrolyzate of the tritiated insulin. Solvent used for paper chromatography is phenol saturated with water. Chromatogram was scanned by a Packard radio-chromatogram scanner (Model 7200). Reproduced from a review by MATSUO [KAGAKU, **36**, 244 (1966), in Japanese]

form) was used, however, only C-terminal glycine was tritiated, since the DNP-group is not an acyl function. The present selective tritium-labeling method, therefore, seems to be applicable to the identification of isoaspartyl and isoglutamyl residues in proteins.

The extent of tritium incorporation into C-terminal residues of peptides and proteins has not so far been quantitatively investigated. Nevertheless, this method is useful for qualitative C-terminal analysis as an adjunct to other methods, and it could become the second most powerful C-terminal determination method once the quantitation procedures are established.

Procedures

Tritium water of high specific activity such as 1 c/ml is diluted with distilled water to 10 ml and it is stored in a freezer (10 mc/0.1 ml). Pyridine and acetic anhydride are stored in a cold room after redistillation of analytical grade reagents.

In order to handle the tritiation reaction and subsequent hydrolysis in one-vessel, use of an apparatus shown in Fig. 3-18 is convenient [654]. The apparatus consists of a small test tube (A, 1 × 20 cm) for the reaction and hydrolysis, a connecting glass adaptor with an arm tube (B) leading to a vacuum pump and a glass trap (C).

Protein or peptide (0.05 to 0.5 μmole) is placed in a test tube (A) and 0.1 ml of T_2O (10 μc to 100 μc, specific activity to be used depends upon sensitivity of method to count radioactivity of amino acids) and 0.2 ml of pyridine are added. It is important to make solution clear by gentle shaking or slight warming; otherwise tritium incorporation sometimes becomes unsatisfactory. When sample protein is sparingly soluble, addition of urea crystals is effective. To the above protein solution 0.05 ml of acetic anhydride is added under ice-cooling. The tube is tightly stoppered and the mixture is shaken for a while. The reaction mixture is allowed to stand for 30 min in an ice-bath and then several hours or overnight at room temperature.

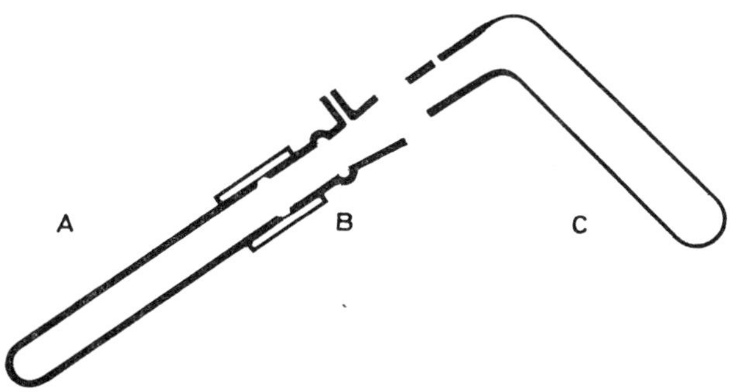

Fig. 3-18 A—C. Apparatus for C-terminal tritiation of proteins (see the text). Reproduced from Ref. [654]

Prolonged standing may increase the extent of tritium incorporation into proteins. When the solution gels during the reaction, addition of pyridine (0.2 ml) and acetic anhydride (0.05 ml) and standing for additional 3 h appreciably improves the extent of tritiation.

After completion of the tritiation reaction, the test tube is connected by the aid of thick rubber tube with B which has been connected with C with silicone grease shown in Fig. 3-18. The solution in A is frozen in a dry ice-acetone bath and whole apparatus is evacuated by a vacuum pump through the arm of B. Then C is rotated to close the vacuum line and the dry ice-acetone bath is removed. Trap C is inserted in the dry ice bath and, T_2O, pyridine and acetic acid in A are distilled into C. During this distillation process, care should be taken to avoid foam formation. The highly radioactive distillate in C is discarded carefully. To the residue in A, 0.5 ml of water is added to remove exchangeable tritium introduced into carboxyl, amide, etc. and above distillation procedure is repeated. Repeating 5 times of this washing process is enough to remove washable tritium.

Tritiated protein at the C-terminus in the test tube A was directly hydrolyzed with 6 N HCl as usual. Several methods such as paper chromatography, paper electrophoresis and thin layer chromatography on cellulose powder plates can be used for characterization of tritiated amino acids. After staining of all constituent amino acids on a chromatogram with very dilute ninhydrin reagent (0.02%), radioactivity of each cut filter paper, where amino acid is located, may be counted by a liquid scintillation counter. This method gives rather low counts because tritiated amino acids are partly decomposed by the ninhydrin reaction. When a radiochromatogram scanner is available, it is convenient to know the location of radioactivity on a one-dimensional chromatogram or electrophoretogram prepared at neutral pH. Of course, standard amino acids are also developed in the same sheet of paper with the tritiated protein hydrolysate. Thus approximate identification of the tritiated amino acids can be made by comparing the stained chromatogram with the recorded chart of radiochromatogram scanner, as shown in Fig. 3-17. The use of other solvent systems for the extracted tritiated amino acid with water from the paper make possible exact characterization of the tritiated amino acid. When accurate counts of radioactivity are required, the filter paper is cut where the radioactive amino acid is located, and this is subjected to a liquid scintillation counter. Radioactivity of the same size of the chromatogram, where no tritiated amino acids are contained, as the sample cut paper is also measured as a blank and this value is subtracted from the counts for sample.

When aspartic or glutamic acid was found by the base-catalyzed tritiation method, it cannot be concluded that this had been arisen from the C-terminal acidic amino acid. Discrimination of C-terminal asparagine or glutamine from aspartic or glutamic acid can be made by digesting tritiated protein with carboxypeptidase A as usual. The C-terminal radioactive asparagine or glutamine is rather quickly released from the tritiated protein but aspartic or glutamic acid slowly under the usual conditions.

C. Carboxypeptidases

Carboxypeptidases remove amino acids sequentially from the C-terminus of a polypeptide chain. Carboxypeptidases are classified at present into three groups, A, B and C, depending on their substrate specificities. Carboxypeptidase A (Cpase A), which was originally isolated by Anson from bovine pancreas, releases most rapidly the C-terminal residues with an aromatic or large aliphatic side chain (tyrosine, phenylalanine, tryptophan, leucine, isoleucine, methionine, threonine, glutamine, histidine, alanine, valine and homoserine), at moderate rates, asparagine, serine and methionine sulfone, and very slowly, glycine, aspartic acid, glutamic acid, S-carboxymethylcysteine and cysteic acid. The enzyme does not release C-terminal proline, arginine and lysine. Recently Ambler [23] suggested that C-terminal lysine can be released by Cpase A at a slow rate. The end products after digestion of oligopeptides composed of amino acid residues having strong affinity to Cpase A are the N-terminal dipeptide and the constituent amino acids of the parent peptide, unless extremely high concentrations of the enzyme are used. If high concentrations of the enzyme are used, Cpase A completely hydrolyzes oligopeptides into the constituent amino acids. Relative rates of hydrolysis of some synthetic substrates by Cpases A and B are listed in Tables 13 and 14. Cpase B, which was isolated by Folk and Gladner [309] also from bovine pancreas, releases C-terminal arginine and lysine rapidly, but it also releases neutral amino acids. All of the protein-constituent amino acids at the C-terminus of polypeptide chain can therefore be removed by the action of either Cpase A or Cpase B with the exception of the C-terminal proline. ZUBER [1184] recently isolated Cpase C from the peel of citrus fruit. Cpase C also has been isolated from plant sap [819, 1111]. C-terminal proline as well as other amino acids was released by Cpase C at roughly similar rates. Citrus Cpase C releases, optimally at

Table 13. *Relative hydrolysis rates of synthetic substrates by carboxypeptidase A*[a]

Substrate	Relative rate	Reference	Substrate	Relative rate	Reference
Cbz-Gly-L-Phe-OH	100	[964]	Cbz-L-Ala-L-Phe-OH	85.9	[75]
Cbz-Gly-L-Tyr-OH	47.9	[964]	Cbz-L-Ala-L-Tyr-OH	49.5	[75]
Cbz-Gly-L-Trp-OH	36.2	[964]	Cbz-L-Glu-L-Phe-OH	6.11	[75]
Cbz-Gly-L-Leu-OH	20.0	[964]	Cbz-L-Glu-L-Tyr-OH	3.76	[75]
Cbz-Gly-L-Met-OH	9.22	[234]	Cbz-L-Trp-L-Ala-OH	9.25	[964]
Cbz-Gly-L-Ile-OH	4.15	[964]	Cbz-L-Trp-L-Glu-OH	0.052	[964]
Cbz-Gly-L-Ala-OH	0.29	[964]	Cbz-L-Trp-L-Pro-OH	0.019	[964]
Cbz-Gly-Gly-OH	0.018	[964]	Cbz-L-Arg-L-Leu-OH	46.2	[39]
Cbz-Gly-L-Arg-OH	0	[39]	Cbz-L-Phe-L-Arg-OH	0	[39]
			Bz-Gly-L-Lys-OH	0	[309]

[a] Relative values of the first order proteolytic coefficients of the peptides are listed with that of Cbz-Gly-L-Phe-OH selected to be 100. Abbreviations of Cbz and Bz denote carbobenzoxy and benzoyl, respectively.

Table 14. *Relative hydrolysis rates of synthetic substrates by carboxypeptidase* B[a]

Substrate	Relative rate
Bz-Gly-L-Lys-OH	100
Bz-Gly-L-Arg-OH	71.5
Bz-Gly-L-Orn-OH	29.5
Bz-Gly-L-Homoarginine	1.19
Bz-Gly-L-Pro-L-Lys-OH	0.040
Bz-Gly-D-Lys-OH	0
Bz-Gly-L-Lys-NH_2	0

[a] Relative values of the zero order proteolytic coefficients of the peptides are shown with that of Bz-Gly-L-Lys-OH selected to be 100. Bz and Cbz are abbreviations of benzoyl and carbobenzoxy, respectively. Data from Ref. [309].

pH 3.5, amino acids at the following relative rates: most rapidly, leucine, proline, glutamic acid, aspartic acid, lysine, histidine, alanine, less rapidly, serine and, glycine, appreciably more slowly. A penultimate glycine residue greatly diminishes the hydrolysis rate of the terminal residue, and peptides with C-terminal sequence of -Gly-Pro-OH or -Pro-Pro-Gly-OH are practically not hydrolyzed [1185].

Preparations of Cpases A and B are commercially available. Cpase A is practically insoluble in water and is usually available as a crystalline suspension in water. Crystalline Cpase A is contaminated with other pancreatic proteolytic enzymes and treatment with diisopropyl phosphorofluoridate (DFP) is required to inactivate contaminating trypsin and chymotrypsin prior to use. The DFP-treated enzyme is commercially available and this preparation can be used after washing the crystals with water to remove contaminating amino acids [321].

A crystalline Cpase B has been reported [1142] but the commercial preparation is an aqueous solution. The B enzyme is rather unstable and must be kept frozen since lyophilization leads to inactivation. The enzyme preparation is generally

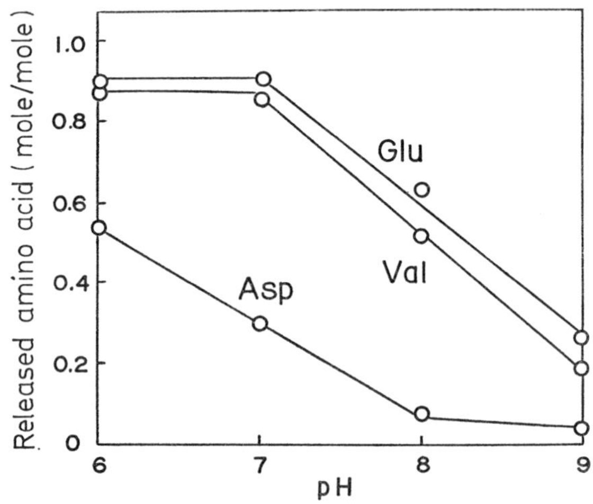

Fig. 3-19. The effect of pH in digestion of N-acetyl-Gly-Asp-Val-Glu-OH (N-terminal peptide of bovine heart cytochrome *c*) with carboxypeptidase A. The digestion was carried out at room temperature and the three pH's indicated in the figure with the enzyme-substrate ratio of 1:1 by weight. The released amino acids after 15 h incubation were plotted. Reproduced from Ref. [1060], by courtesy of the Japanese Biochemical Society

contaminated with proteolytic enzymes and free amino acids; thus DFP-treatment is essential. The DFP-treated Cpase B is also available. The enzyme is best used after purification by gel filtration on a Sephadex G–25 column in 0.01 M Tris-hydrochloric acid buffer of pH 8.0 to remove free amino acids [23].

The two Cpases are stable between pH 7 and 10 and the enzyme digestion is normally carried out at pH 7.5 to 8.5. However, when a peptide terminates with glutamic or aspartic acid, digestion with Cpase A at lower pH's (pH 5 to 6) increases the rate of release of the acidic amino acid from the C-terminus [371, 1060] as shown in Fig. 3-19.

When the released amino acids in the incubate of a protein or peptide with Cpase A or B are characterized and estimated as a function of time by a suitable method, for example by an automatic amino acid analyzer, the amino acid sequence of the C-terminus can be deduced. When only one amino acid is released from the peptide chain in a stoichiometric amount, the interpretation of the results is quite simple. Tobacco mosaic virus and egg white lysozyme are such typical examples. Stoichiometric amounts of threonine and leucine are released from tobacco mosaic virus [419] and lysozyme [1050], respectively, by the action of Cpase A. The C-terminal sequences of the two proteins are respectively -Pro-Ala-Thr-OH- and Gly-Cys-Arg-Leu-OH. In the former case, the alanine, which is the second residue from the terminus, should be liberated after the terminal threonine is removed, but no alanine is released. This fact suggests that the release rate is affected predominantly by the C-terminal amino acid and influenced also by the penultimate residue. Such a tendency is evident with synthetic substrates as shown in Table 13. In contrast to tobacco mosaic virus, even when the penultimate residue is proline, the terminal amino acid can sometimes be released. Adrenocorticotropin, possessing the sequence, -Pro-Leu-Glu-Phe-OH, is such a case [421]; C-terminal phenylalanine is released very quickly, glutamic acid at a moderate rate and leucine at a slightly slower rate. Isoleucine, on the carboxyl side of a proline residue, was reported to be partially removed from Ac-Ala-Tyr-Asn-Pro-Ile-Thr-Ser-OH after liberation of serine and threonine [716]. This peptide was isolated from the protein of cucumber virus 4. Generally the C-terminal amino acid which is linked to a penultimate proline residue is difficult to remove quantitatively with these enzymes.

When several amino acids are released at similar rates from relatively large molecular peptides or proteins, the interpretation of the results is quite difficult even when the relative specificity of Cpase is carefully considered. In such a case, it is dangerous to draw definite conclusions regarding the C-terminal amino acid of a protein based upon only Cpase experiments. Conclusion should be carefully made considering the results of other chemical methods for the C-terminal analysis.

An excellent review of Cpases A and B has been written by Neurath [737], and procedures for the end group and sequence analyses with Cpases have been presented by Fraenkel-Conrat, Harris and Levy [321], and by Ambler [23].

1. Preparation of the Enzyme Solution

Commercial preparations of Cpase A are normally aqueous suspensions of crystals at a concentration of 50 mg /ml.The suspension is stable at 4 °C for years, but lyophilisation will cause inactivation. The enzyme is active in crystalline suspension but usually the crystals are dissolved immediately before use. The enzyme solution cannot be stored for a long time prior to use. A suspension containing 1.25 mg protein (normally 25 μl) is diluted with 1 ml water in a centrifuge tube. The crystals sediment very quickly and the supernatant containing free amino acids is discarded. Into the sedimented crystals is added 0.1 ml of 1% $NaHCO_3$ with ice-cooling and 0.1 N NaOH (2 to 3 drops) is added with shaking until the crystals dissolve. The pH of the solution is then brought back to 8.0 with 0.1 N HCl (2 to 3 drops), and the volume is made up to 1.25 ml such that the enzyme concentration becomes 1 mg/ml with 0.2 M buffer of pH 8.0. When the digestion is followed by the ninhydrin reaction, buffers containing ammonia, amines or phosphate must be avoided. Phosphate buffers result in precipitation during the usual ninhydrin analysis. Ambler [23] recommended the use of an N-ethylmorpholine acetate buffer. Care must be taken to prevent over-addition of hydrochloric acid during neutralization of the enzyme alkaline solution since the enzyme inactivates rapidly below pH 6 and precipitates. The enzyme (molecular weight, 34,300) solution is, if necessary, treated with 50-fold molar excess of a solution of DFP (15 μl of 0.1 M DFP in dry isopropanol per mg of the enzyme) for 1 h at 25 °C. The concentration of the enzyme solution is measured spectrophotometrically, assuming that the enzyme has a molar absorbancy of 8.6×10^4 at 278 mμ [736].

Commercial Cpase B solution normally contains free amino acids and other proteolytic enzymes of pancreatic origin as discussed. If the preparation has not been treated with DFP, DFP-treatment should be carried out as described for Cpase A. In order to remove free amino acids, gel filtration on a Sephadex G–25 column or dialysis must be performed before use. The enzyme concentration may be measured spectrophotometrically by using a molar absorbancy of 7.4×10^4 at 278 mμ [311].

2. Reaction of Carboxypeptidases with Protein

Native proteins generally are resistant to the action of Cpases and prior denaturation of proteins is necessary. Proteins containing disulfide bonds or cysteine residues usually require prior reduction and subsequent alkylation or performic acid oxidation. Proteins thus treated are of course denatured. Cpase A is not denatured in a solution of 6 M urea [402] or 0.056 M sodium dodecyl sulfate [381] and such a denaturant solution can be used during the reaction. Denaturant solutions can be used effectively to solubilize insoluble protein substrates at pH 8.0.

The amount of denatured protein to be digested depends upon the sensitivity of the analytical method for the amino acids released and upon the number of aliquots to be taken as a function of suitable time intervals. Usually 0.1 to 2 μmoles of denatured protein are dissolved in a volatile 0.2 M N-ethylmorpholine acetate buffer of pH 8.0 to a concentration of 0.2 to 0.5%. The enzyme solution is then added in a substrate: enzyme ratio of 1/20 to 1/200 (molar basis) and the mixture, in a screw-capped or stoppered tube, is incubated at 25 to 37 °C. At scheduled time intervals (for example, 0, 0.5, 1, 2, 4, 8, and 16 h) 0.05 to 0.25 μmole portions of the mixture are removed, and acidified (trichloroacetic, acetic or hydrochloric acid) to pH 2 to stop the reaction. Any resulting precipitate is removed by centrifugation. The supernatant is subjected to amino acid analysis.

Under standard analysis conditions, the peaks of asparagine and glutamine overlap that of serine and the three amino acids cannot be estimated separately. For the quantitation of the three amino acids, asparagine and glutamine must be hydrolyzed to aspartic and glutamic acids and then estimated. For this purpose the released amino acids must be separated from the protein and the enzyme prior to the hydrolysis. Utilization of the molecular sieve effect of ion exchange resins is applicable to such separations [1050]. The hydrogen-form of sulfonated polystyrene beads (Dowex 50–X10–X8, 20 to 60 mesh) should be added until the pH becomes 2 to 3 [23]. Fine beads should not be used to minimize adsorption of proteins. The mixture is then shaken for 20 min, the supernatant is removed and the resin beads are washed twice with 2 resin volumes of water. The adsorbed amino acids on the resin are extracted three times with 2 resin volumes of 5 M NH_4OH and the combined extracts are evaporated to dryness. The residue is dissolved in an appropriate volume of pH 2.2 citrate buffer and the solution is divided into three portions. One portion is dried, hydrolyzed with acid and analyzed for neutral and acidic amino acids. The other two portions are analyzed without hydrolysis for acidic and neutral amino acids, and for basic amino acids, respectively. The increase in aspartic and glutamic acid values after acid hydrolysis gives the amide content. For mutual separation of asparagine, glutamine and serine by an automatic amino acid analyzer, use of the lithium form of the resin and lithium citrate buffer is recently proposed.

D. Other Methods

The first attempt to characterize the C-terminal group of a protein based on an organic chemical reaction was made by FROMAGEOT *et al.* [330, 501] in 1950. They reduced the carboxyl groups of insulin to the corresponding alcohol groups with $LiAlH_4$ in N-ethylmorpholine at 55 °C for 8 h. The amino alcohols derived from the C-termini of the reduced insulin in the ether extracts of the acid hydrolysate were characterized by paper chromatography. CHIBNALL and REES [198] reduced insulin methyl ester with $LiBH_4$, a much milder reducing agent, and identified the amino alcohols derived from the C-terminal alanine and asparagine in the expected amounts. In addition to the two amino alcohols, they obtained ethanolamine derived from one of the glycyl peptide bonds. CHIBNALL and REES extensively studied the reduction method, but this method has not been widely used for end group determination because of two side reactions: reductive cleavage of peptide bonds and reduction of peptide carbonyls to methylene groups. Despite the problems of the reduction method, BAILEY [48] proposed an interesting procedure for the sequential removal of the C-terminal residue as the amino alcohol based on the coupling of reduction with an N,O-acyl shift reaction as illustrated in scheme (17).

$$\begin{array}{ll}
R_1\cdot CO\text{–}NH\cdot CHR_2\cdot CO\text{–}NH\cdot CHR_3\cdot COOH & \xrightarrow{CH_3OH+HCl} \\
R_1\cdot CO\text{–}NH\cdot CHR_2\cdot CO\text{–}NH\cdot CHR_3\cdot COOCH_3 & \xrightarrow{AlH_3} \\
R_1\cdot CO\text{–}NH\cdot CHR_2\cdot CO\text{–}NH\cdot CHR_3\cdot CH_2\cdot OH & \xrightarrow{H^+} \\
R_1\cdot CO\text{–}NH\cdot CHR_2\cdot CO\text{–}OCH_2\cdot CHR_3\cdot NH_3^+ & \xrightarrow{LiBH_4} \\
R_1\cdot CO\text{–}NH\cdot CHR_2\cdot CH_2OH + HO\cdot CH_2\cdot CHR_3\cdot NH_2 &
\end{array} \qquad (17)$$

He reduced the *p*-toluenesulfonyl peptide ester in tetrahydrofuran with AlH_3 at −40 °C and carried out the N,O-acyl shift with phosphorous oxychloride at room temperature. Reductive cleavage of the β-amino esters with $LiBH_4$ gave the terminal amino alcohol in 85 to 90%. Although the principle of Bailey's method is of interest, side reactions have obviated its use with proteins.

Another sequential determination method for the C-terminus is the thiohydantoin method originally proposed by SCHLACK and KUMPF [882] in 1926. Peptides substituted at the terminal amino group are converted to the peptidyl thiohydantoin by heating for 30 min at 100 °C with ammonium thiocyanate in acetic anhydride. The thiohydantoin corresponding to the C-terminal residue is released by treating with 0.1 N NaOH at room temperature for a few minutes. The reaction is shown in scheme (18).

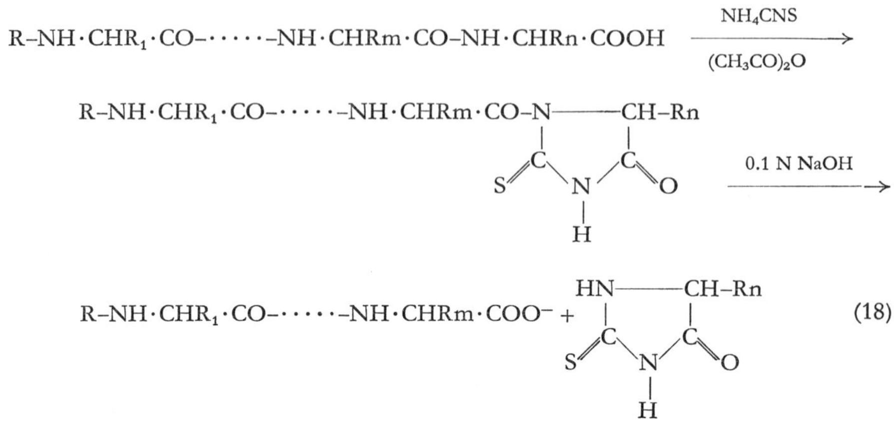

(18)

The released thiohydantoin is extracted with organic solvents and the parental amino acid is regenerated by hydrolyzing with hydrogen bromide. Thiohydantoins of amino acids can be characterized directly by paper chromatography [225, 275]. This method has been successfully applied to smaller oligopeptides but has not always been successful with proteins [320, 1091] without modification of the procedures. Recently, STARK [996] reinvestigated this sequential degradation method and applied his modified procedures to a number of peptides. 100 mg of recrystallized ammonium thiocyanate were dissolved in a mixture of 4.0 ml of redistilled acetic anhydride and 1.0 ml of glacial acetic acid. To this solution was added 0.5 ml or less of 50% acetic acid solution of a peptide and the mixture was kept for 6 h at 50 °C. After incubation for 18 h with an additional 100 mg of ammonium thiocyanate at 50 °C, the peptidyl-thiohydantoin was isolated by the aid of a Sephadex G–25 column. For the release of the thiohydantoin derivative of the C-terminal amino acid from the peptidyl-thiohydantoin, he recommended the use of 0.1 M acetohydroxamate in 50% pyridine solution at 50 °C for 2 h. A portion of 50% acetic acid extract of the dried mixture was subjected to amino acid analysis and the amino acid removed was characterized and estimated in comparison with the amino acid composition of the parent peptide, without attempting direct characterization of the released thiohydantoin of the C-terminal amino acid. Even with these modified procedures, the reaction was not quantitative. Only two to three residues from the C-terminus could be determined with confidence. Aspartic acid and proline at the C-terminal end could not be determined by this thiohydantoin method in similar to the tritium labeling method described above.

A similar method has been proposed by KENNER *et al.* [534] which was diphenyl phosphoroisothiocyanatidate. An acylated peptide is reacted with the reagent in anhydrous acetonitrile in the presence of triethylamine to yield the peptidyl thiohydantoin, from which thiohydantoin is released on alkaline treatment as shown in scheme (19).

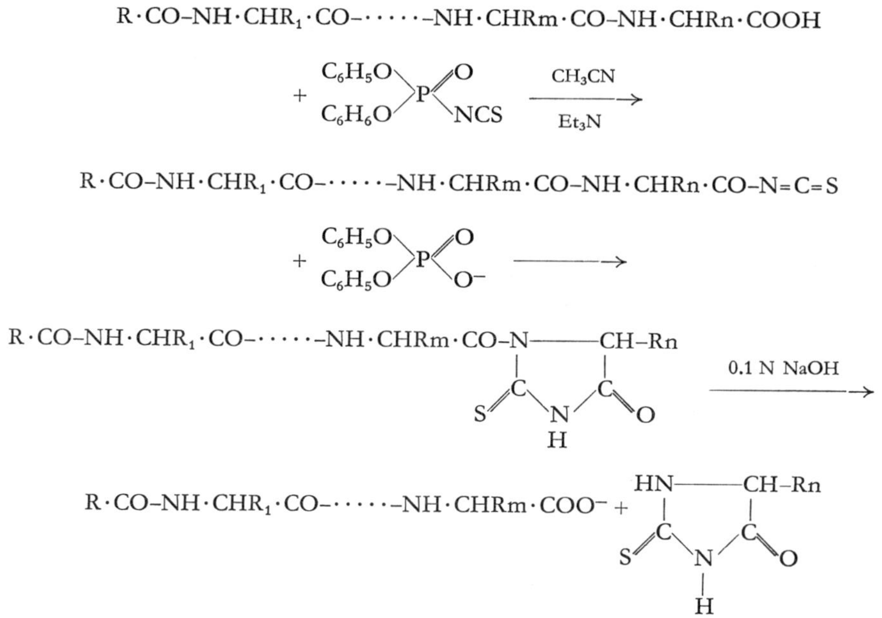

(19)

In 1936 BERGMANN and ZERVAS [78] proposed a sequential degradation of peptides beginning at the C-terminus using the Curtius rearrangement. The benzoyl peptide ester is converted into the corresponding hydrazide and then into the peptide azide with nitrous acid. The peptide azide is then converted with benzyl alcohol, into the benzyl urethane derivative. Catalytic reduction yields the benzoyl peptide amide with one residue less than the parent peptide, an aldehyde derived from the C-terminal residue, ammonia, carbon dioxide and toluene as shown in scheme (20).

$$C_6H_5\cdot CO\text{–}NH\cdot CHR_1\cdot CO\text{–}\cdots\cdots\text{–}NH\cdot CHRm\cdot CO\text{–}NH\cdot CHRn\cdot COOC_2H_5 \xrightarrow{N_2H_4}$$

$$C_6H_5\cdot CO\text{–}NH\cdot CHR_1\cdot CO\text{–}\cdots\cdots\text{–}NH\cdot CHRm\cdot CO\text{–}NH\cdot CHRn\cdot CO\cdot NH\cdot NH_2 \xrightarrow{HNO_2}$$

$$C_6H_5\cdot CO\text{–}NH\cdot CHR_1\cdot CO\text{–}\cdots\cdots\text{–}NH\cdot CHRm\cdot CO\text{–}NH\cdot CHRn\cdot CO\cdot N_3 \xrightarrow{C_6H_5\cdot CH_2\cdot OH}$$

$$C_6H_5\cdot CO\text{–}NH\cdot CHR_1\cdot CO\text{–}\cdots\cdots\text{–}NH\cdot CHRm\cdot CO\text{–}NH\cdot CHRn\cdot NH\cdot COO\cdot CH_2\cdot C_6H_5 \xrightarrow[Pd]{H_2}$$

$$C_6H_5\cdot CO\text{–}NH\cdot CHR_1\cdot CO\text{–}\cdots\cdots\text{–}NH\cdot CHRm\cdot CO\cdot NH_2 + Rn\cdot CHO + NH_3 + CO_2 + C_6H_5\cdot CH_3 \quad (20)$$

Although this reaction can be repeated sequentially by converting the peptide amide into the peptide hydrazide, the procedures are very difficult when applied to proteins.

KHORANA [541] proposed a novel though difficult to apply route for the selective removal of the C-terminal residues. He reacted an acyl peptide with di-(*p*-tolyl)-

or di-(*p*-dimethylaminophenyl)-carbodiimide to yield the corresponding acyl peptidyl urea which undergoes degradative formation of N-*p*-tolyl-carbamyl amino acid *p*-toluidide on treatment with 0.01 N NaOH in 75% ethanol at room temperature as shown in the reaction scheme (21).

$$C_6H_5 \cdot CO\text{–}NH \cdot CHR_1 \cdot CO\text{–} \cdots\cdots \text{–}NH \cdot CHRm \cdot CO\text{–}NH \cdot CHRn \cdot COOH$$

$$\xrightarrow{CH_3 \cdot C_6H_4 \cdot N{=}C{=}N \cdot C_6H_4 \cdot CH_3} \qquad (21)$$

$$C_6H_5 \cdot CO\text{–}NH \cdot CHR_1 \cdot CO\text{–} \cdots\cdots \text{–}NH \cdot CHRm \cdot CO\text{–}NH \cdot CHRn \qquad \xrightarrow[75\%\ C_2H_5OH]{0.01\ N\ NaOH}$$
$$\rangle C{=}O$$
$$H_3C \cdot C_6H_4 \cdot HN \cdot CO \cdot N \cdot C_6H_4 \cdot CH_3$$

$$C_6H_5 \cdot CO\text{–}NH \cdot CHR_1 \cdot CO\text{–} \cdots\cdots \text{–}NH \cdot CHRm \cdot COO^- +$$
$$CH_3 \cdot C_6H_4 \cdot NH \cdot CO \cdot NH \cdot CHRn \cdot CO \cdot NH \cdot C_6H_4 \cdot CH_3$$

The product derived from the terminal residue can be converted into the parent amino acid by hydrolysis with acid or alkali.

IV. Masked Terminal Groups

Gramicidin S was the first oligopeptide for which the amino acid sequence was determined [213]. Unlike the usual oligopeptides derived from proteins by partial hydrolyses, gramicidin S is a cyclic decapeptide devoid of both the terminal groups. Thereafter, the inability to find terminal residues by the usual terminal analysis methods led to the conclusion that the protein or peptide in question contained a cyclic chain or tailed cyclic chain structure devoid of one or both of the free end groups. With the exceptions of some antibiotic and toxic oligopeptides cyclic and tailed cyclic polypeptide chain structures have not been found with a frequency commensurate with such conclusions. The discovery of masked terminal residues has shed light on these ambigious problems.

A. N-Acylated Terminal Group

The many efforts to elucidate the N-terminal group of the protein of tobacco mosaic virus by the usual analytical methods had been unsuccessful leading to the postulation that the peptide structure was that of a tailed cyclic chain having a C-terminal threonine [151, 324]. Narita [714, 715] eventually clearly demonstrated that the terminal α-amino group was blocked with an acetyl group. Since then, the proteins of cucumber virus 4 [716], turnip yellow mosaic virus [418], α-melanocyte stimulating hormone [416], ovalbumin [717, 718] and cytochromes *c* from vertebrates [582, 1060] have been shown to be blocked with an acetyl group. In Table 15 are listed the N-terminal acetyl proteins thus far reported. Proteins with terminal groups once reported to be acetylated (bovine fibrinopeptide B and α_1-acid glycoprotein) but later found to be pyrrolidone carboxyl (pyoglutamyl) residues are omitted from the table. Table 16 lists the terminal pyroglutamyl peptides and proteins.

Some oligopeptides, such as gramicidin A [863] and bee venom, melittin [581], and the newly synthesized proteins in cells of *E. coli* [6, 1106], are found to be masked with a formyl group at their N-terminal α-amino groups. The terminal formyl group of *E. coli* proteins is believed to be split enzymatically immediately after protein synthesis has been completed in the bacterial cells.

No method for end group determination can be directly applied to such terminal acylated proteins and pyroglutamyl proteins. In most cases peptides containing a terminal acyl group or pyroglutamyl residue are isolated from their proteolytic digests and characterized separately as outlined in Fig. 3-20 and in Chapter 6.

Table 15. *N-Terminal acetyl proteins*[a]

Protein	Sequence	Reference
Tobacco mosaic virus	Ac-Ser-Tyr-	[714, 715]
Cucumber virus 4	Ac-Ala-Tyr-Asn-Pro-Ile-	[716]
α-Melanocyte stimulating hormone	Ac-Ser-Tyr-Ser-Met-Glu-	[416]
Turnip yellow mosaic virus	Ac-Met-Gln-Ile-Asp-Lys-	[418]
Cytochromes *c* (vertebrate)	Ac-Gly-Asp-Val-Glu-	[582, 1060]
Ovalbumin	Ac-Gly-Ser-Gly-Ile-Ala-	[717, 718]
Histone IV (calf thymus)	Ac-Ser-Gly-Arg-Gly-Lys-	[235, 785]
Human fetal hemoglobin ($F_1\gamma$)	Ac-Gly-	[903]
Chicken hemoglobin $A_1\beta$	Ac-Val-Thr-Leu-	[869]
Enolase (rabbit muscle)	Ac-Ala-	[1141]
Myosin (rabbit muscle)	Ac-Ser-Ser-Asp-Ala-	[748]
Wool	Ac-Ala-	[746]
Tropocollagen	Ac-	[466]
Carbonic anhydrase (human erythrocyte)	Ac-Ala-Ser-Pro-Asp-	[649, 471]
α-Glycerophosphate dehydrogenase (rabbit muscle)	Ac-	[1075]
Lactic dehydrogenase (H_4)	Ac-	[1003]
Carp hemoglobin α	Ac-Ser-Leu-Ser-Asp-Lys-	[443b]
Actin (rabbit muscle)	Ac-Asp-Glu-[Thr, Ala]-	[20, 337]
Tropomyosin	Ac-[Met, Asp]-Ala-	[21]
Apoferritin (horse spleen)	Ac-Ser-Ser-Gln-Ile-Arg-	[1022]
Hyaluronidase	Ac-[Tyr, Leu, Phe, Ala, Glu]-	[985]
Glyceraldehyde 3-phosphate dehydrogenase (lobster)	Ac-Ser-Lys-Ile-Gly-Ile-	[423]
Frog hemoglobin α	Ac-Ala-Leu-	[1896]

[a] Abbreviation Ac denotes acetyl group.

1. Acetylated Terminus

A protein for which an N-terminal residue cannot be found by the routine terminal analysis methods, is digested with proteolytic enzymes possessing a broad specificity, such as pronase, an extracellular proteinase produced by *Streptomyces griseus*. If the acetyl peptide is large, it probably will contain basic amino acids giving a neutral, or perhaps basic peptide inspite of the blocked N-terminal amino group which would tend to make the peptide more acidic. Trypsin therefore cannot be used as a proteolytic enzyme, since tryptic peptides always contain lysine or arginine at the C-terminal end except the C-terminal peptide of the parent protein. When the terminal acetyl peptide is thought to contain basic residues, the functional groups of the lysine and arginine residues in the protein must be blocked prior to proteolytic digestion. Suitable techniques for this purpose are succinylation of the amino groups with succinic anhydride at alkaline pH [394, 613] and benzylation of the guanido groups with dibenzoyl reagents [487]. The latter reaction is carried out under strongly alkaline conditions and a possibility of partial hydrolysis of peptide

Table 16. *N-Terminal pyroglutamyl peptides and proteins*[a]

Peptide	Structure	Reference
Eisenin	Pyrglu-Gln-Ala-OH	[749]
Fastigiatine	Pyrglu-Gln-Gln-OH	[233]
Eledoisin	Pyrglu-Pro-Ser-Lys-Asp-Ala-Phe-Ile-Gly-Leu-Met-NH_2	[290]
Physalaemin	Pyrglu-Ala-Asp-Pro-Asn-Lys-Phe-Tyr-Glu-Leu-Met-NH_2	[291]
Fibrinopeptide B (human)	Pyrglu-Gly-Val-Asn-Asp-Asn-Glu-Glu-Gly-Phe-Phe-Ser-Ala-Arg-OH	[108]
Fibrinopeptide B (bovine)	Pyrglu-Phe-Pro-Thr-Asp-Tyr(SO_3H)-Asp-Glu-Gly-Gln-Asp-Arg-Pro-Lys-Val-Gly-Leu-Gly-Ala-Arg-OH	[110]
Gastrin II	Pyrglu-Gly-Pro-Trp-Met-Glu-Glu-Glu-Glu-Glu-Ala-Tyr(SO_3H)-Gly-Trp-Met-Asp-Phe-NH_2	[374]
Snake venom peptide	Pyrglu-Gln-Trp-OH	[518]
Snake venom peptide	Pyrglu-Asn-Trp-POH	[518]
α_1-Acid glycoprotein	Pyrglu-Pro-Ile-	[470]
Immunoglobulin (human)		
light chains	Pyrglu-Ser-Val-Leu-	[463]
heavy chains	Pyrglu-Val-Thr	[799]

[a] Abbreviation Pyrglu denotes pyrrolidonecarboxyl group.

bonds exists. This is however of no consequence at the present step. For the modification of arginine residues, reaction with 1,2-cyclohexanedione can also be used under much milder alkaline conditions [1064].

In general, acetyl peptides present in digests of the terminal acetylated protein are acidic. Other peptides in the digest are amphoteric, at least, based on the presence of free terminal amino and carboxyl groups. The hydrogen form of a strong cation exchanger such as Dowex 50 adsorbs these amphoteric peptides but does not retain the acetyl peptides or the pyroglutamyl peptides, if present. The digest is placed on a Dowex 50 column (X2–X16, H^+ form) and the column is washed thoroughly with water until the effluent is no longer acidic. The acidic peptides (acyl peptides, pyroglutamyl peptides, and peptides containing cysteic acid or phosphoserine residue) are contained in the water eluate together with larger peptides not adsorbed on the resin column due to the molecular sieve effect of the resin [714, 718]. The water eluate is then applied to a column of a strongly basic anion exchange resin such as Dowex 1 (chloride or acetate form) having an identical or lower degree of cross-linking compared to that of the cation exchange resin previously used. The larger peptides again are not adsorbed because of the molecular sieve effect. These are removed by washing with water. The adsorbed acidic peptides are then developed chromatographically and eluted with diluted hydrochloric, acetic or formic acid [714, 717, 1060]. A typical example is shown in Fig. 3-21. However, acetyl peptides in the effluent cannot be detected by the usual ninhydrin reaction but can be performed only after alkaline hydrolysis [445a] of each fraction from the column [718, 1060]. The Folin reagent [628] containing cupric ion is also used for detection of

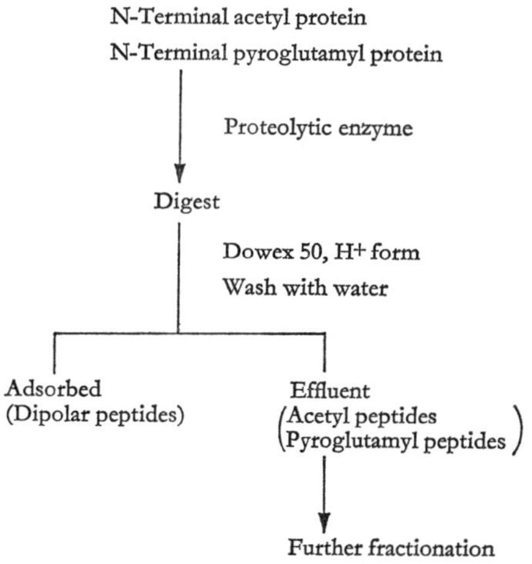

Fig. 3-20. Fractionation scheme for acyl peptides from proteolytic digests of the N-terminal acylated proteins

acetyl peptides [714, 716]. The acetyl peptides in the water eluate can be fractionated directly by paper chromatography or electrophoresis if large peptides are not contained in the digest [716].

When glutaminyl peptides are present in the digest, they are quite labile and easily converted into pyroglutamyl peptides. This reaction is accelerated by contact with the hydrogen form of a sulfonic acid exchanger and also by heating at neutral or slightly acidic conditions [715, 730]. It is quite difficult to distinguish acetyl peptides from pyroglutamyl peptides without structural analysis since both groups of peptides are negative to the ninhydrin reaction and to other reactions based on amino groups. Acetyl and pyroglutamyl peptides can be revealed on paper chromatogram by chlorine-iodide-starch reaction of Rydon and Smith [843] or by a modification in which chlorine gas is replaced with *tert*-butyl hypochlorite in cyclohexane (1%) [260].

If the acetyl peptide is small enough for identification of the acetyl group, several simple methods are available. Narita identified the acetyl hydrazide, amino acid hydrazide and C-terminal amino acid simultaneously by paper chromatography of the hydrazinolysate of acetyl dipeptides isolated from proteolytic digests of the proteins of tobacco mosaic [714, 715] and cucumber [716] viruses and ovalbumin [717, 718], respectively. Acetyl and amino acid hydrazides are positive to the alkaline-silver nitrate reagent used for reducing sugars, and amino acids and their hydrazides are reactive to the ninhydrin reagent. Rf values of some of acyl and amino acid hydrazides are listed in Table 17. It should be noted that pyroglutamyl peptides yield an unknown spot on a paper chromatogram which cannot be distinguished from the spot of acetyl hydrazide even when the peptide does not contain any acetyl substituent.

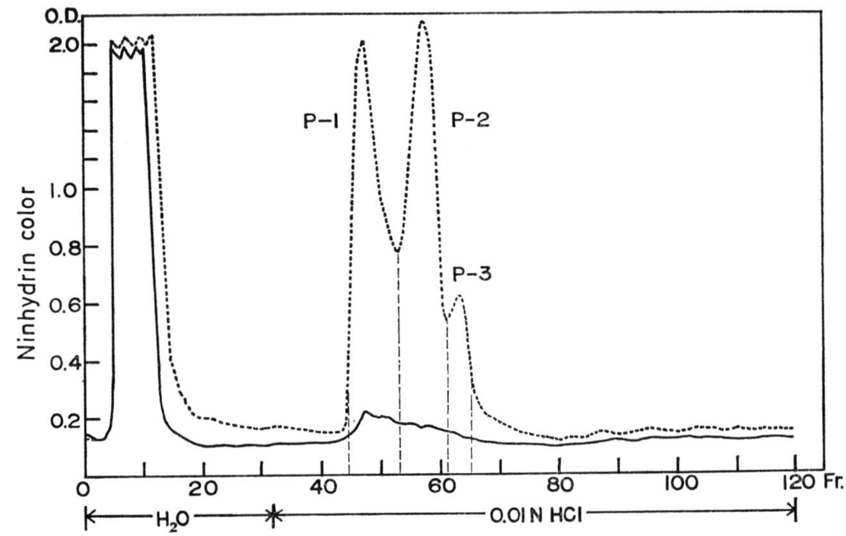

Fig. 3-21. Fractionation of acetyl peptides in the peptic digest of ovalbumin. The water eluate from a Dowex 50 column of the digest (see Fig. 3-20) was chromatographed on a column of Dowex 1-X2 (2.3 × 30 cm). Solid and dotted curves denote the ninhydrin color values before and after alkaline hydrolysis, respectively. Main components in peaks P-1 and P-2 are, respectively, N-acetyl-Gly-Ser-Gly-Ile-OH and N-acetyl-Gly-Ser-Gly-Ile-Ala-OH. Reprinted from Ref. [718], by courtesy of the Japanese Biochemical Society

Therefore, when a ninhydrin negative peptide containing glutamic acid residues is isolated, the possibility of pyroglutamyl peptides must carefully be examined as described later. Acetyl hydrazide is also identified chromatographically as its DNP-derivative (dinitrophenylation at pH 3.0 and 38 °C for 4 h) [786], extractable with ether from aqueous solution at pH 3. Phillips [786] described the formation of propionyl hydrazide from cystine residues on heating with anhydrous hydrazine.

When acetyl peptides are subjected to methanolysis by heating (100 °C, 5 h) with methanol (0.5 ml) and concentrated sulfuric acid (25 μl), methyl acetate is formed. Distilled methyl acetate (condensed in a dry ice-acetone bath) can be identified and estimated by gas chromatography or by chromatography of the acetohydroxamate formed with a neutralized concentrated hydroxylamine solution [630]. Fraction of volatile methyl esters from the methanolysis products of whole proteins, the N-terminal residues of which are known to be unblocked, were heated with hydrazine, and the products (fatty acid hydrazides) were examined by paper chromatography. The paper chromatogram revealed several spots positive to the alkaline-silver nitrate reaction and Rf values of some of them corresponded to the formyl, acetyl and propionyl hydrazides. It is recommended therefore, that the acidic peptide fraction, rather than the whole protein be used for the analysis of acyl groups.

Alternatively, acetate can be detected after partial hydrolysis of acetyl proteins or, preferably, acetyl peptides, in 6 N sulfuric acid at 100 °C for 5 h. It is difficult to distill acetic acid quantitatively from the hydrolysate and Phillips' method [785] is preferable. He extracted acetic acid with ether and subjected the ether extracts to

Table 17. *Rf values of some acyl and amino acid hydrazides*

	Pyridine-aniline-water[a] (9:1:4, v/v)	Collidine-water[b] (10:2, v/v)	Collidine-picoline water[c] (5:3:2, v/v)
Formyl hydrazide	0.56	0.28	
Acetyl hydrazide	0.66	0.35	0.45
Propionyl hydrazide	0.75	0.50	
Glycine hydrazide	0.37	0.06	0.11
Serine hydrazide	0.51	0.07	0.16
Alanine hydrazide	0.46		
Valine hydrazide	0.65		0.43
Aspartic α-hydrazide	0.30		0.02
Glutamic α-hydrazide	0.34		0.02

[a] From Ref. [714 to 718, 1060).
[b] From Ref. [714].
[c] From Ref. [717, 718, 1060].

paper chromatography by the method of Brown and Hall [159]. Ward *et al.* [1101] extracted acetic acid with *tert*-butyl ethyl ether after addition of solid Na_2SO_4 (anhydrous) and estimated the extracted acid by gas chromatography.

The structures of acylated peptides may be determined from their C-terminal ends by the use of carboxypeptidase A. However if the acyl peptides contain residues (proline, aspartic acid, glycine, etc.) resistant to carboxypeptidase, they must be further degraded into fragments by enzymes or by chemical methods, and the structures of the fragments elucidated separately. The acetyl-amino acid linkage fortunately is rather resistant toward concentrated hydrochloric acid at room temperature for 24 to 48 h [582, 1060]. Thus most acetyl amino acids can be released by this treatment from the parental acetyl peptides.

a) Fractionation of Acetyl Peptides

About 10 to 20 μmoles of a protein is digested with a proteolytic enzyme (pronase, chymotrypsin, pepsin, subtilisin, etc.) for 24 h and the digest is placed on the top of a Dowex 50–X2 (H^+ form, 50 to 100 mesh) column, 1 × 15 cm (can handle about 1 g of protein digest). The column is then thoroughly washed with water until the effluent becomes neutral (about 200 ml). Acidic peptides, in general, are eluted at the break-through point. The water eluate is lyophilized and the lyophilizate is redissolved in 1 to 2 ml of water. The solution is chromatographed on a column of Dowex 1–X2 (chloride form, 200 to 400 mesh, 1 × 30 cm). The column is first washed with water and about 100 ml of the effluent is collected. Large peptides which have escaped adsorption on the Dowex 50 column are contained in this fraction. This fraction is ninhydrin positive. Acidic peptides on the column are developed with a gradient of HCl reaching the high concentration of 0.03 N and 2 to 3 ml fractions are collected with a fraction collector. Most acidic peptides are eluted within 100 fractions. Each fraction is then analyzed with the ninhydrin reagent before and after alkaline hydrolysis according to the method of Hirs, Moore and Stein [445a]. Fractions which are ninhydrin negative before alkaline hydrolysis but positive after hydrolysis are pooled and lyophilized for structural analysis.

b) Identification of the Acetyl Group of the Acetyl Peptide

About 1 μmole of the purified acetyl peptide is placed in a small test tube and 50 μl of anhydrous hydrazine is added. The tube is then sealed and heated at 100 °C for 8 to 10 h. The hydrazinolysate is frozen and dried in an evacuated desiccator. The hydrazinolysate is examined by paper chromatography with the following solvent system: pyridine-aniline-water (9:1:4, v/v), collidine-water (10:2, v/v) or collidine-α-picoline-water (5:3:2, v/v). Acyl hydrazides and amino acid hydrazides are characterized by the alkaline silver nitrate reagent, and the C-terminal amino acid and amino acid hydrazides can be stained with ninhydrin reagent. The silver nitrate reagent is prepared by adding 0.1 ml of saturated $AgNO_3$ aqueous solution to 50 ml acetone. The paper is first dipped through the reagent and the acetone is blown off. The dried paper is then dipped through 0.5% KOH in ethanol and the solvent is again blown off. Hydrazides begin to appear at once. The background of the chromatogram prepared by the silver nitrate reagent becomes gradually brown by light. To prevent formation of Ag_2O from the excess $AgNO_3$ by light, the paper chromatogram is quickly washed successively with 5% sodium thiosulfate and water and finally dried. The chromatogram can be stored over a year. The ninhydrin negative and alkaline silver nitrate positive spot is acyl hydrazide and the reverse one is the C-terminal amino acid of the acyl peptide. Positive spots for both reactions are derived from the amino acid residues between the termini.

Standard acyl hydrazides and amino acid hydrazides can be prepared from their esters by heating with hydrazine in ethanol at 60 to 100 °C for 1 h. The dried residue of the mixture can be used as a standard.

For the conversion of acyl hydrazides into their DNP-derivatives [786, 985] the dried hydrazinolysate is dissolved in 1 ml of 0.2 M citrate (pH is adjusted to 3 with 1 N HCl), and 3 volumes of ethanol and 0.5 ml of 5% FDNB in ethanol are added. The reaction is allowed to proceed for 4 h at 38 °C. After evaporation of the mixture and addition of 2 ml of water, DNP-derivatives are extracted 5 to 6 times with 2 to 3 ml of ether. The ether extracts are dried and dinitrophenol is removed by vacuum sublimation as described in the DNP method. The DNP-acyl hydrazides (brown color by exposing to ammonia vapor) are characterized by paper chromatography or thin layer chromatography (silica gel G, hexane-ethyl acetate, 1:1, v/v) and estimated colorimetrically at 420 mμ in comparison with known amounts of the authentic acyl hydrazides which are treated under the same conditions as the hydrazinolysate of the acetyl peptide.

2. Formylated Terminus

The N-formyl group is labile in comparison with N-acetyl and can be removed by incubating with 1 N hydrochloric acid in methanol for 1.5 h at room temperature without serious cleavage of peptide bonds in proteins [948]. Isolation of formyl peptides and characterization of the formyl group are carried out by the same techniques as described for the terminal acetylated proteins.

Recently GUSSIN *et al.* [386] isolated deformylase from *E. coli* and they reported that this enzyme attacks N-formyl peptides and not formyl amino acids. When terminal formyl proteins or peptides are found, this enzyme is probably useful for the removal of the terminal formyl group in order to elucidate amino acid sequences of such proteins and peptides. Further studies demonstrated that this deformylase also hydrolyzes N-acetylated peptides at an appreciably slower rate [56].

3. Pyroglutamyl Terminus

Pyroglutamyl peptides are separated by the same techniques as those for the acyl peptides. However, it is rather difficult to conclude with confidence, that the isolated pyroglutamyl peptides were derived from the N-terminus of the parent protein since glutaminyl peptides derived from the internal part of the molecule by proteolytic enzymes are known to be transformed easily, sometimes quantitatively, into pyroglutamyl peptides during the digestion and fractionation, unless the complete amino acid sequence is established. When negative results for terminal analysis and for the identification of acetyl group are obtained, and when a pyroglutamyl peptide is isolated in quantitative yield, an N-terminal pyroglutamyl residue is then indirectly assumed for the parent protein in question. IKENAKA *et al.* [470] carefully investigated the terminal pyroglutamyl residue of α_1-acid glycoprotein and reported that isolation of acidic peptides with a Dowex 50 column, H^+ form, at lower temperatures such as 4 °C minimized the transformation of N-terminal glutaminyl residue into pyroglutamyl groups. Furthermore, they showed that alkaline incubation of Pyrglu-Pro-Ile-OH, which is the N-terminus of the glycoprotein, at 24 °C for 40 and 72 h achieved ring opening in 36 and 50%, respectively, without appreciable cleavage of the peptide bonds. However, when the same conditions were used for native α_1-acid glycoprotein, cleavage of several peptide bonds was observed.

Recently DOOLITTLE and ARMENTROUT [250] reported the isolation of pyrrolidonyl peptidase from *Pseudomonas fluorescence* which hydrolyzes specifically the N-terminal pyroglutamyl linkage. Incubation of N-terminal pyroglutamyl proteins with this enzyme, therefore, opens blocked proteins to usual terminal analytical methods.

B. C-Amidated Terminal Group

The only known substituent at the C-terminus of naturally occurring polypeptides is the amide group. Glycinamide was first demonstrated as the C-terminus of oxytocin [259] and vasopressin [993] by DU VIGNEAUD and associates. Several amidated C-terminal groups in polypeptides are now known; glutamine amide for melittin [581], valineamide for α-melanocyte stimulating hormone [416] and secretin [712], phenylalanineamide for gastrin [374], and methionineamide for eledoisin [290] and physalaemin [291]. The C-terminal tryptophan of Gramicidin A is known to be amidated with ethanolamine [863]. This constitutes an exception to the general rule. The structures of these polypeptides have been completely or nearly completely elucidated and the nature of these amidated groups is established. It is difficult, however, to characterize the amidated C-terminal residues of large molecular weight proteins. Although this problem has not been extensively explored, one possible method for the characterization of such residues is the isolation of basic peptides (blocked C-terminus) from proteolytic digests of proteins in a manner similar to that described above for N-terminal peptides.

The enzyme digest is passed through a basic anion exchange resin column, e.g., Dowex 1–X2, and the column is thoroughly washed with water. Peptides possessing free carboxyl groups are adsorbed on the column, while the basic peptide amide generally free of acidic residues is eluted from the column with water. The water eluate should contain the peptide amide together with arginine and arginine peptides, the basicities of which are similar to that of free arginine. The basic peptide fraction is then fractionated by suitable chromatographic techniques. A basic peptide, negative to the Sakaguchi reaction, locates the peptide amide. The amino acid sequence of the amidated peptide can be determined by the usual terminal analysis methods without appreciable difficulties.

Chapter 4

Amino Acid Composition Determination

John W. Eveleigh and George D. Winter

I. General

A. Early Methods

The pioneering studies of Emil Fischer [301] at the turn of the century established the role of amino acids and the peptide bond in the composition and chemistry of proteins. In the light of contemporary techniques, his methods of protein degradation and subsequent fractionation of the products appear cumbersome and of historical interest only. Yet it is significant that for nearly 50 years these original publications provided the major methods available to the analytical protein chemist.

The basic amino acids were precipitated with phosphotungstic acid and subsequently fractionated as their silver salts [575]. Neutral and acidic amino acids were esterified with ethanol and hydrochloric acid and the esters separated by fractional distillation [298]. These and allied methods were fraught with practical difficulties and the variable yields made them far from quantitative. The availability of isotopes in the late 1940's enabled the early separation methods to be quantitated using the isotope dilution principle [828]. The isotopes C^{14}, N^{15}, S^{35}, and deuterium were successfully employed. However, the low specific activity of the radioactive isotopes available at that time coupled with poor efficiencies of detection systems necessitated the use of relatively large quantities of protein sample.

An alternative to the purely chemical approach was offered by the development of the microbiological determination technique [977].

This method utilized cultures of *Neurospora* and *Lactobacilli* strains whose metabolic characteristics enabled the determination of specific amino acids at the milligram level. By turbidimetric assay of the growth rate directly, or titration of the lactic acid or carbon dioxide produced, an estimate of the amino acid concentration could be obtained. Non-specificity of the mutant strains and the purely relative nature of the method contributed largely to its replacement by chromatographic methods.

In 1944, Consden, Gordon and Martin [212] introduced the technique of paper chromatography. Its application to complex mixtures of amino acids allowed hitherto impossible separations to be readily achieved with a minimum of expense and experience. The inherent sensitivity and the ease of adaptation to specific applications made the method one of the major analytical aids to protein research.

Concurrent with the development of the paper chromatographic method was the application of column separation techniques. Early attempts used differential absorption of the amino acids on solids such as aluminum oxide, silica gel and activated charcoal. The use of starch columns eluted with butanol, propanol and dilute hydrochloric acid, was described by MOORE and STEIN [700]. The colorimetric reaction of the collected fractions with ninhydrin allowed the first truly quantitative separation of a protein hydrolysate. The amount of bovine serum albumin used in these early column procedures was 2.5 mg.

With the introduction of the polymeric ion-exchange resins the column separation techniques became rapidly more versatile. In 1958, SPACKMAN, STEIN and MOORE [987] published details of an automatic analyzer designed specifically for the separation and quantitation of amino acids. The subsequent development of this original system has resulted in the sophisticated, fully automated instruments currently available and in routine use for the determination of the amino acid composition of proteins and peptides.

B. Preparation of Hydrolysates

The spectacular advances in amino acid separation techniques and associated instrumentation have tended to obscure the importance of sample preparation. Full realization of the potential of contemporary methods, with regard to separation capability and precision of quantitation, depends largely upon the techniques applied in the preparation of the protein or peptide hydrolysate.

Basically, hydrolysis of the peptide bond is simple:

$$\text{R-CO-NH-R}' + H_2O \longrightarrow \text{R-COOH} + \text{R}'\text{NH}_2\,.$$

The rate of hydrolysis in aqueous solutions is, however, extremely slow and hence catalysts such as strong acids and alkalis or enzymes are used for preparative purposes. Unfortunately the hydrolysis rate is dependent upon the nature of the adjacent residues and degradation of some amino acids occurs concurrently with their release. It is for these reasons, and the effects of contamination on the process, that attention to small but important practical details of the technique is essential.

1. Acid Hydrolysis

The major method used to hydrolyze protein and peptide samples to amino acids is that of heating with hydrochloric acid. Although the method results in low recoveries of certain amino acids and the total destruction of tryptophan, in capable hands it can yield good results. Assuming a reasonable quantity of protein sample, the following procedure is recommended to obtain reliable and reproducible composition analysis.

The protein sample is allowed to equilibrate to atmospheric conditions in a weighed open bottle for several days. A sample is removed for total nitrogen analysis by micro Kjeldahl and another for ash content. Further weighed samples are removed for timed hydrolysis and the remainder weighed and dried at 105 °C overnight. After cooling in a dessicator, the dried protein is reweighed and the data is hereafter expressed in terms of the dry, ash-free protein.

There are many cases, however, in which this procedure is unwarranted or impractical. Limitations in sample quantity, the presence of known impurities and derivatives of proteins often preclude this type of manipulation or make it unnecessary. In these cases it is good practice to relate the analytical results of amino acid composition to an independently measurable quantity. Examples are enzymic activity, radioactivity, or protein concentration as measured by U.V. absorption or the biuret and Folin reactions. In the analysis of peptides, in which relative ratios are required rather than absolute values, this type of approach is obviously unnecessary.

As it is essential that the hydrolysis is performed in an inert atmosphere or in vacuum, the sample is usually quantitatively transferred to an ampule. Solutions are preferably rotary-evaporated to dryness unless the fluid volume is small. If manipulative losses are expected an internal sample standard should be added. The recovery of this standard, which should obviously be extremely stable during hydrolysis, will enable suitable correction to be made for subsequent dilution or handling of the sample. It is common practice to use a solution of norleucine for this purpose.

The protein or peptide sample is diluted with as large a volume of 6N hydrochloric acid as is practical. The use of large dilutions (in the order of 500 fold) considerably reduces the decomposition of certain amino acids. The purity of the hydrochloric acid used for hydrolysis is very important. Commercial acid usually contains trace elements as well as ammonium chloride. Distillation from silica vessels to a constant boiling mixture and subsequent careful storage is usually adequate for purification. The composition and cleanliness of the ampule is of similar importance; borosilicate glass pre-cleaned with chromic acid and very will rinsed has proven satisfactory. Elution of peptides directly from paper chromatograms or "finger-prints" with 6N hydrochloric acid invariably leads to contamination with extraneous amino acids, notably serine, glycine and alanine.

Flushing the ampule with nitrogen, or evacuation is essential before sealing or heating. Although evacuation is to be preferred, the practical difficulties can sometimes be exasperating. Protein solutions invariably foam badly under reduced pressure and violent "bumping" can occur causing loss of sample. Provided a good grade of nitrogen is used (less than 0.1% oxygen) a combination of reduced pressure and flushing can be very effective. The acid solution can be carefully cleared under reduced pressure and nitrogen admitted by means of a 3-way stopcock after allowing a period for equilibration the process can be repeated several times before sealing the neck of the ampule. Freezing of the sample and evacuation is not recommended as significant amounts of air are dissolved or occluded by the ice. Some amino acids are particularly susceptible to oxidation; cystine and S-carboxymethyl cysteine are good examples, and their recovery is highly dependent upon technique.

After flame sealing, the ampules are placed in a thermostatically controlled oven devoted to the purpose. This precaution is advisable as the opening and insertion of cold glassware, for example, causes wide temperature variations and increases the possibility of thermal shock and explosion of the ampules. The temperature used for hydrolysis varies widely among workers in the field, 110 °C being the most commonly used. The maintenance of a constant temperature is more important than the choice of an absolute value.

During acid hydrolysis there is decomposition of several amino acids. Tryptophan is invariably totally destroyed and serine, threonine, cystine and tyrosine are partially decomposed. The rates of decomposition are markedly affected by variation of temperature, hence the necessity of strict control of this variable. The amount of decomposition can be ascertained by performing timed hydrolysates and extrapolating to zero time. Fig. 4-1 shows the recovery of serine and threonine falling with time in the hydrolysis of myokinase. Unfortunately the decomposition depends upon the nature of the protein as well as conditions of hydrolysis. Decomposition values of 15% for serine, 6% for threonine, 10% for tyrosine and 4% for cystine

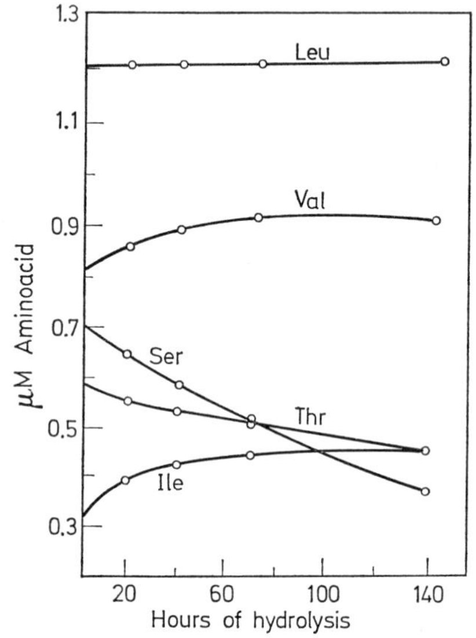

Fig. 4-1. Recovery of representative amino acids from HCl hydrolysis as a function of time. (Reprinted from Techniques in Amino Acid Analysis; D.I. Schmidt, ed., Technicon Instruments Co., Ltd. Chertsey, England)

can be expected during an overnight incubation. The decomposition of tyrosine can be prevented by the addition of thioglycolic acid, phenol or hydrazine to the hydrochloric acid [861].

Concurrent with decomposition of some amino acids, Fig. 4-1 also illustrates an apparent increase in recovery of valine and isoleucine with time. The reason for this lies in the resistance of sequences containing these residues to hydrolysis. Extrapolation to zero time is obviously erroneous in this case and lengthy hydrolysis times must be employed to obtain maximum yields. For protein analyses, a series of timed hydrolysates of 24, 48 and 72 h appears necessary but for small peptides an overnight hydrolysis of 16 h is usually considered adequate. The acid hydrolysis of protein and peptide derivatives raises similar problems of recoveries. Di-nitrophenyl derivatives are notorious for their instability during hydrolysis, despite specific variations of acid concentration and temperature [798]. Similarly, 5-dimethylamino-1-naphthalene sulphonyl ("dansyl") derivatives, although generally more stable, give special problems in hydrolysis.

The final operation after hydrolysis is the removal of the hydrochloric acid before analysis is undertaken. The simple method of placing the sample in an open vessel contained in a dessicator over solid hydroxide is not to be recommended for quantitative studies. The gradual concentration of the sample to a syrup can give rise to losses by esterification and oxidation, particularly at room temperature. Lyophilization is difficult owing to the low melting point of 6N hydrochloric acid. A far more reliable method is to film-evaporate, usually by a rotary-evaporation apparatus, as rapidly as is practical. Storage of hydrolysates can also give problems; aqueous solutions are prone to microbial contaminations, while dry residues being hygroscopic can give rise to losses similar to those occurring in "syrup" conditions. The preferred method is to keep the unopened ampules in a refrigerator until facilities for analysis are available.

2. Alkaline Hydrolysis

As tryptophan is destroyed by acid, the estimation of this amino acid by chromatography requires the use of alkaline conditions. One of the best methods has been provided by NOLTMAN [743] using barium hydroxide. The protein is placed in ampules, a small amount of water and sufficient solid barium hydroxide is added to give a final molarity of about 4. After evacuation and sealing, the ampules are heated at 110 °C for various periods of time to obtain a recovery calibration. Tryptophan release from protein can take up to 50 h for complete recovery.

The decomposition of some amino acids, arginine and cystine in particular, and the occurrence of racemization has limited the method mainly to the determination of tryptophan. Direct methods for determining this amino acid in proteins and peptides are gradually replacing this indirect method.

3. Enzymatic Hydrolysis

The complete degradation of a protein to amino acids by enzymatic means is a difficult undertaking. However, limited application of the method is possible for peptides and for specific sequence studies. In conjunction with the column separation techniques, the use of proteases and peptidases has provided valuable information on partial hydrolysates. Digestion at the optimum pH of the enzyme and direct application to the analytical column allows quantitative studies of released residues.

One distinct advantage this method has over acidic or alkaline hydrolysis is that the amides, asparagine and glutamine, retain their integrity. The disadvantages are that the majority of proteolytic enzymes show considerable specificity toward certain sequences, hydrolyzing some of the peptide linkages rapidly and others not at all.

C. Paper Chromatography and Electrophoresis

Since the publication by CONSDEN, GORDON and MARTIN [212] on the separation of amino acids by paper chromatography, there has been an overwhelming number of publications on the method. Several excellent books have also been written on this subject. For this reason, the method will be summarized only in this contribution.

The original publication described a two dimensional system using phenol-ammonia for up to 72 h, followed by collidine for a further 48 h in the second dimension. The high toxicity and somewhat unpleasant odors associated with the method stimulated the development of several new solvent systems. LEVY and CHUNG [610] introduced butanol-acetic acid-water (4:1:5) for one dimension and borate buffered phenol-cresol in the second elution. A rapid screening method was introduced by ROCKLAND and UNDERWOOD [832] employing tert-butanol-formic acid-water followed by phenol-ammonia. The complete separation took approximately 24 h.

Although the separations obtained are adequate, the two dimensional system has several disadvantages. The sample size is limited by distortion effects and only one sample per paper sheet is possible. More important, quantitation is nearly impossible. During the 1950's a large number of one-dimensional systems were proposed and championed by their originators. Many workers in the field have been perplexed by the many alternatives offered and to some extent, by lack of success in repeating the methods. The careful studies of Hanes and his co-workers did much to characterize the many variables inherent in the technique and to introduce more reliable methodology. Their final paper [413] refers to the wealth of experimental detail and apparatus described in earlier papers of the series.

The methods used by HANES give excellent separations and offer hitherto unattainable precision of quantitation. The time required for chromatography is in the order of 40 h and the technique of dissecting the chromatograms and reacting with ninhydrin is relatively tedious. These disadvantages plus the strict control of solvent and paper purity, and of the chromatographic environment, make this excellent method somewhat tedious.

The introduction of high-voltage electrophoresis by MICHL [683] and the subsequent development of the cooled plate apparatus by GROSS [376] overcame the disadvantage of the long time periods required by previous methods. Adaptations of these early methods have become the major alternatives to column chromatography systems. ATFIELD and MORRIS [44] developed a three buffer system with separations achieved at pH 5.2, 1.85 and 1.81. The amino acids are quantitated using a cadmium acetate-ninhydrin reagent.

Electrophoretic separation methods are extremely useful in peptide composition studies where accurate quantitation is not mandatory. Being capable of producing separations in a very short time, this method is also useful for radioactive or fluorometric (dansyl) derivative studies. Similarly the use of electrophoresis in the separation of peptides has one particular advantage over alternative methods. The direction of migration at neutral pH is indicative of overall ionic charge of the peptide. This makes it possible to distinguish between glutamic acid and glutamine, or aspartic acid and asparagine residues, when electrophoresis is used in combination with hydrolysis and amino acid analysis method.

Thin layer chromatography, with its inherent capability of high resolution in relatively short times, has also found a place in the analytical methods of protein composition. Despite several attempts to quantitate results, however, the method appears to be highly subjective and is mainly used for qualitative studies.

D. Column Chromatography

The introduction of ion-exchange chromatography can be considered as a breakthrough in the technique of amino acid analysis. Earlier work by MOORE and STEIN using polymeric ion-exchange resins laid the foundation of the now classical procedure published in collaboration with SPACKMAN [987]. Using two columns packed with sized, crushed Amberlite IR-120 and an eluting sequence of sodium citrate buffers, they obtained separation of the protogenic[1] amino acids in 22 h. By the continuous addition of ninhydrin reagent to the column effluent and passing the mixture through a heated capillary, concentrations could be continuously monitored photometrically. The quality and originality shown in this early work can be judged by the fact that several present-day commercial systems are basically similar to this unique instrument.

The first major variation of this method was based on the concept of gradient elution introduced by PETERSON and SOBER [783]. A multichamber gradient elution device, called a Varigrad, was used by PIEZ and MORRIS [788] to separate the protogenic amino acids on a single resin column. Spherical resins were employed by HAMILTON [404] in a single column, high resolving, fully automatic procedure. This later contribution can be considered another classic as it contains the first serious studies on the effect of pH, molarity, temperature, etc. on elution.

Continuing research efforts and modifications to these methods during the past few years have culminated in the present day fully automated systems. Resin technology has advanced producing finer resin particles which, when packed in smaller columns, have enabled the analysis time to be reduced to the order of 1 h [292].

II. Practical Aspects of Ion-Exchange Chromatography

In common with many analytical techniques, the successful analysis of amino acids by ion-exchange chromatography is largely dependent on careful attention to multiple, sometimes minor, operational details. The emphasis in the following discussion will therefore be practical rather than theoretical or descriptive.

The original separation procedure of SPACKMAN, STEIN and MOORE [987] is still the most widely used method for the analysis of amino acids. In all the variations of this method, duplicate samples of the hydrolyzed protein are carefully applied to the resin, packed in each of two columns, and then eluted in sequence by appropriate buffers. The more "basic" amino acids are commonly separated on a short resin column by a single buffer. Two or more buffer solutions are passed in succession through a longer column, to separate the "acidic" and "neutral" amino acids. Many investigators prefer to use only one column, however, and a single sample. In this case, the acidic, neutral and basic amino acids are eluted by a progression of buffers that may be passed through the column in succession according to the Hamilton [404] technique, or as a continuously changing buffer gradient as described by PIEZ and MORRIS [788]. Two commercially available systems are used in the following discussion to illustrate salient features of these two major variations. The Technicon

---

[1] The term "protogenic" was proposed by Professor STANFORD KING in 1967 as a generic term for the amino acids derived from proteins.

Model NC-1 Amino Acid Analyzer is an essentially manually operated single column system employing gradient elution, while their Model TSM Amino Acid Analyzer is a sophisticated example of a completely automated dual column system.

A. Single Column Gradient Elution System

The single column Amino Acid Analyzer mentioned above is illustrated in Fig. 4-2 and represented schematically in Fig. 4-3. The principle of operation is relatively simple; buffer solution from a nine chamber gradient forming device is pumped through a thermostatted separation column by a positive displacement (P.D.) pump. The column effluent, containing the separated amino acids, is continuously mixed with ninhydrin reagent and segmented with nitrogen, to maintain resolution, by means of a multiple tube peristaltic proportioning pump, represented by the numbered block in Fig. 4-3. The mixture is then passed through a coil in the heating bath where the amino acids quantitatively react with the ninhydrin, producing a blue colored derivative. The stream is cooled and led through a series of colorimeters in which the relative absorbance is transformed into electrical signals suitable for operating the three-point recorder.

1. Ion-Exchange Resin and Column

The resins almost universally employed for amino acid separations are referred to as nominal 8% cross-linked, sulfonated polystyrene resins. This means that approximately 8% divinylbenzene (DVB) is co-polymerized with a styrene monomer. The polymer is then sulfonated to produce a strongly acidic ion exchange resin. Minor variations in manufacturing reaction conditions or procedures, or in purity of the materials used, can have profound effects on the separating characteristics of the resin. For this reason, it is fortunate that the ion exchange resin is considered an integral part of all commercial amino acid analyzers and the instrument manufacturers supply these resins with the necessary specifications for successful operation in their respective systems.

A newcomer to the method is often perplexed by the variety of resins offered by instrument manufacturers and specialist firms. This abundance of resin types has developed from the many variations of technique that have been published in recent years, many apparently requiring a particular type or size of resin. These innovations, in general, advocate the use of different buffer systems with the ultimate objective of improving resolution and decreasing the time of analysis. Thus the original NC-1 system was designed for a 21 h analysis and employed a nominal 8% cross-linked resin having 22 μ diameter spherical particles. Later modifications reduced the time to $12^1/_2$ h using a 17 μ resin, then to $4^1/_2$ h with a 13 μ particle material.

The major distinguishing feature of these different resins is that of particle size. Over the range of particle diameter that is of interest in amino acid separations (5 to 30 μ), peak height varies inversely with particle diameter in an approximately linear manner. A reduction of 20% in particle diameter will result in approximately a 13% reduction in peak width or a corresponding increase in peak height. Limits to the reduction in particle diameter are, in practice, determined by the hydraulic pressures encountered with tightly packed, very fine resins. For a given column length and diameter, the pressure developed at constant volumetric input varies

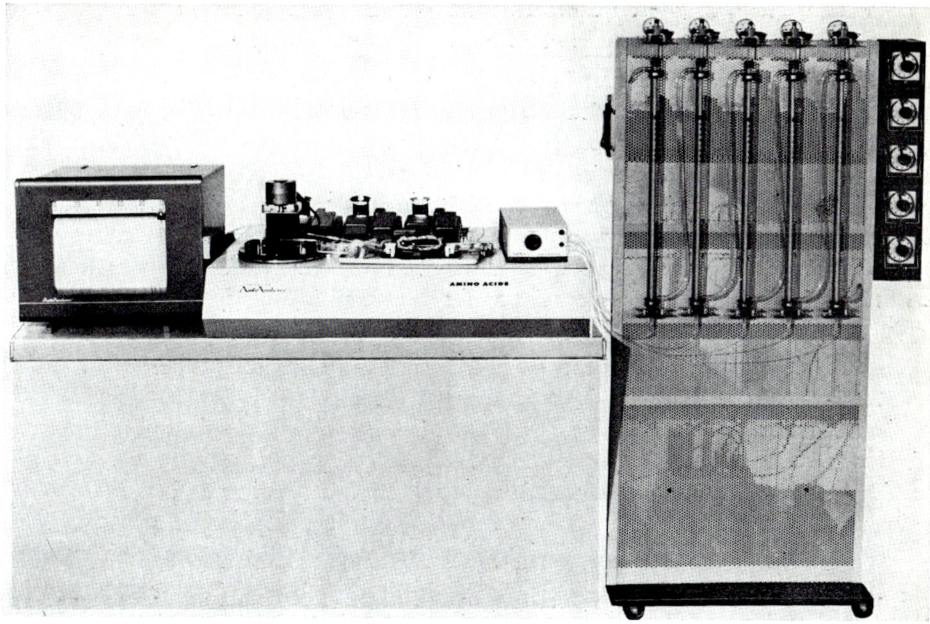

Fig. 4-2. Five column, Model NC-1 Amino Acid Analyzer, showing from left to right: recorder, three colorimeters (in rear), reaction heating bath, proportioning pump with tubing manifold, automatic shut-down programmer, multiple column support with five columns, gradient devices, positive displacement pumps, circulating heating bath and sequential timing mechanism

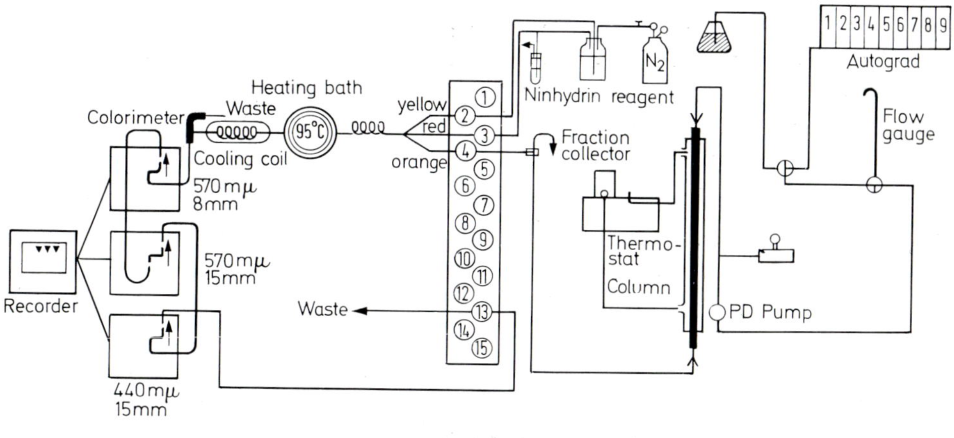

Fig. 4-3. Schematic flow diagram of Model NC-1 Amino Acid Analyzer

with the reciprocal of the square of particle diameter. Since pressure is also proportional to column length, the choice of resin size and column length depends largely upon the pressures that can be tolerated in the system.

For this reason, some standardization in particle size and column dimensions have been necessary in the commercial system. The NC–1 uses a glass column 6.3 mm internal diameter and 140 cm length. This is packed with either 22 μ or 17 μ resins, depending on the type of analysis to be performed. The accelerated system uses a 75 cm long column which is packed with resin having a mean diameter of 13 μ. The sizing of these resins can be a costly and time consuming procedure unless very reliable starting material is available. Articles by Hamilton [403] and Whitaker and Strasdine [1124] should be consulted if resin fractionation is contemplated for exploratory or other research purposes.

Pretreatment or cleaning of resin before packing a column is essential. Quite often, new resin, and almost invariably used resin, becomes contaminated both chemically and by microbes. The citrate buffers used in the analytical procedure are excellent growth media for fungal and bacterial contaminants, which can eventually destroy the function of the column. The procedure for cleaning such contaminated resin requires pouring the resin into a funnel having a fine porosity sintered plate, followed by repeated additions of 6N cold nitric acid until the filtrate solution is quite clear. The resin is washed thoroughly with several volumes of distilled water, then repeatedly with 2N sodium hydroxide solution until the filtrate is again clear. After rewashing with water, the resin is ready for use. Gross proteinaceous contamination may require the use of warm (do not exceed 60 °C) acid and alkali treatment as above. Similarly, resin contaminated with grease, from stopcocks and joints, will require pretreatment with organic solvents. Carbon tetrachloride and acetone, are acceptable solvents if 70 °C is not exceeded. If a non-polar solvent is used, it is important to wash with a polar solvent, e.g. acetone, before the final washing with water. For long term storage, the resin should be left in distilled water or under a solution of about 3% formaldehyde (Normal formalin solution, diluted 10:1).

The attention to detail applied to the quality and cleanliness of resin should also be given to the column itself. Manufacturers are usually very careful in the fabrication of their columns and adhere closely to the strict specifications dictated by theory and experience. Quite often an amateur replacement of a commercial column can result in significant chromatographic differences as well as hitherto unencountered problems. One of the more critical dimensions is the internal diameter; although minor deviations may have little effect upon the chromatographic process, at least in theory, they can be troublesome for mechanical reasons. For example, a glass column that has an internal taper can give rise to channelling or, worse, compaction of the resin with resultant high back pressures. Leakage around the top and bottom seals, due to irregularities in manufacture can be exasperating and structural weakness can result in column breakage under pressure, with the loss of expensive resin.

The top and bottom fittings used in most columns depend upon "O" rings to effect a seal. The reliability of this design is dependent on correct dimensions as well as on cleanliness. A slight increase in the internal dimensions of the column results in seepage under pressure, while a decrease invariably leads to excessive internal stresses and cracking of the column when the fitting is forced in.

The lower column fitting contains a support for the resin, which permits the eluant buffer to flow through while retaining the small resin spheres. The frit, as the support is called, usually is made of sintered plastic such as Teflon. While this material is admirably suitable for the purpose, it is subject to blockage either by contamination or mechanical distortion. The presence of ultra-fine resin particles that may plug the holes in the frit should be suspected if continuous occlusion occurs. If high pressures are encountered during the column packing procedure, distortion of the frit itself may be responsible. The frit should be gently inserted into its recess and never forced in.

These column fittings are held in the column by spring-loaded clamps which provide some factor of safety against pressure rupture of the column. These clamps are also useful for diagnostic purposes. Opening of the lower fitting clamp often indicates a blocked frit; an upper clamp distortion may be the result of excessive resin compaction or contamination.

The column is surrounded by a water jacket through which thermostatically controlled (usually 60 °C) water is circulated. It is recommended that a 20% solution of glycol type anti-freeze be used in the circulating bath, to prevent corrosion of pump parts and to discourage mold growth. The use of glycol also retards evaporation and minimizes maintenance procedures. The control of column temperature is quite important as the mobility of some amino acids, such as aspartic and glutamic acids and tryptophan, is very sensitive to changes in temperature. Viscosity of the buffer solutions is also affected by temperature changes with the result that column back pressures may be increased dramatically by failure of the heating bath.

The method of packing the column with resin is extremely important. After being degreased to avoid later channelling in the resin, the column should be packed as uniformly as possible. If the resin used has been fractionated to a very uniform particle population, the column may be packed by simple sedimentation at room temperature. The column, with a filling bulb connected on top, is filled with distilled water. A slurry of resin is then poured into the bulb and allowed to settle into the column overnight. This technique cannot be used with resins containing a variety of particle sizes; the larger spheres will tend to sediment more rapidly than the smaller, resulting in a graded resin bed and poor resolution.

Resins of this type are packed under pressure with the P.D. pump. A small aliquot of a thick (50% v/v) slurry of resin is poured directly into the column and pressure applied to pack it as rapidly as possible. When one portion is packed, further aliquots are added and the process repeated, until the column is filled to the desired height. The resin should be packed in the sodium form and at the operating temperature of the column. The advantage of this method, in addition to better homogeneity, is that the column is compacted from the start and reduction in bed length does not develop during operation.

2. Gradient Elution

A steep gradient or fast change in the parameters which govern the elution of an amino acid will result in closely spaced peaks in the chromatogram. By a careful selection of the gradient it should be possible to time the appearance of each amino acid and to bring them closer together or separate them further, depending on the number of peaks appearing in a given section of a chromatogram. In order to find a suitable gradient for any pattern of elution it is necessary, of course, to know the exact environmental conditions at which each of the amino acids is eluted, and to relate this environment at the desired time to the contributions from each of the chambers of the gradient device. The PETERSON and SOBER [783] formula can predict the environmental conditions as a function of time, knowing what each chamber contains. It is far more difficult to compute the chamber compositions

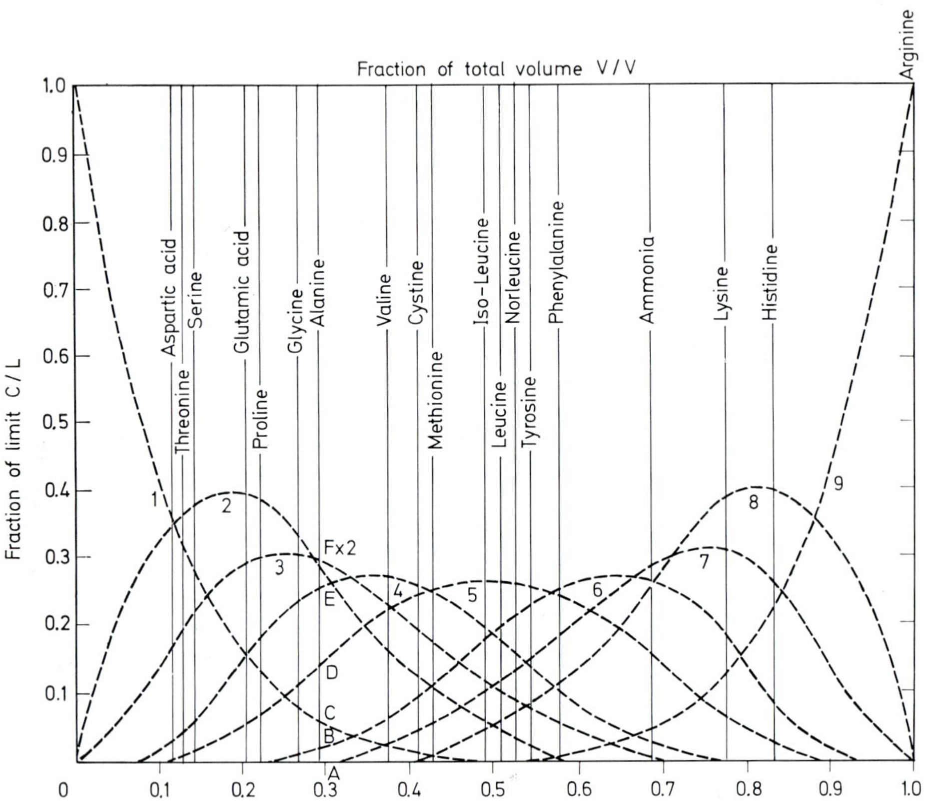

Fig. 4-4. Diagrammatic representation of the contributions of individual chambers of a buffer gradient producing device, adapted from PETERSON and SOBER [784]

from the desired gradient. To cope with this particular computation, BURNS, KACSER and CURTIS [171] have devised a computer program whereby, given any type of gradient, it is possible to determine what should be the composition of each of the 9 chambers. Without a computer program, however, it is still possible to arrive at a satisfactory gradient by a system of trial and error, using a graphic method. Fig. 4-4 illustrates the chart which is used for this type of graphic computation. The abscissa on this chart is the time at which the amino acids are eluted, while the ordinate (y axis) represents that fraction of the total effect which is contributed by each of the chambers. If all chambers have the same composition, then the concentration would not change from unity in going from time zero to time 0.3 and the contribution of each of the chambers to the total is given by adding up the increments AB, AC, AD, AE, AF, the total of which will equal AZ. If, however, one of the chambers contains a smaller quantity of a particular constituent, let us say 20%, then its contribution will only be 20% of the value shown on the chart. Hence if chamber 4, for example, has only 20% salt concentration as compared to the other chambers, then the contribution AD given by chamber 4 would only be a fifth as much and the line AZ would be depressed by just that amount from the 1.0 ordinate.

The application of this graphical method, plus a knowledge of the effects of pH and salt concentration on the elution of amino acids, has permitted simple operating tables of gradient device filling buffers to be prepared. The buffers commonly used in this system are made up from a basic solution containing:

73.55 gm Na_3 Citrate$\cdot 2H_2O$

4500 ml H_2O

125 ml 2 N NaOH

which is titrated to the required pH with 6 N hydrochloric acid. A wetting agent, BRIJ 35 (10 ml/liter) is added to all solutions and thiodiglycol (5 ml/liter) is added to the buffers that are used in the first five chambers of the gradient device. The function of BRIJ 35, a substance which is not ninhydrin positive, is to keep the resin and columns clean so that the eluting front is kept horizontal. The anti-oxidant, thiodiglycol, prevents oxidation of the susceptible amino acids, notably methionine and cystine.

Methyl alcohol may be added to improve the separation of threonine and serine. Its use is based on the different solubility and pK of these acids in alcohol. The concentration of methyl alcohol must be restricted to a minimum because the long term effects of excess alcohol may be troublesome. The use of excessive concentrations of methyl alcohol will tend to delay elution of glutamic acid until it is superimposed on proline and to delay cystine until it is eluted with methionine. Versene (Na_4EDTA) in a concentration of 0.05% may be added to avoid metal ion interference with the ion exchange process. Some examples of gradients used in this system are given in Tables 1, 2 and 3.

The main criticism of the single column elution system has been the common appearance of a dramatic rise in the baseline toward the latter part of the chromatogram. However, this "ammonia plateau", as it is called, can be practically eliminated by careful preparation and storage of the buffer solutions. Ammonia and allied amines occur widely in laboratory atmospheres and as contaminants in reagent grade chemicals. If these compounds are present in the eluting buffers, they tend to be concentrated by the ion exchange process and to be eluted in the form of a very broad peak underlying the basic amino acids.

These volatile basic compounds can be effectively removed by distillation techniques which can be applied to most of the components of the citrate buffers. The water for buffer preparation should be quartz distilled from dilute sulphuric acid. The hydrochloric acid used in pH adjustment should be distilled as a constant boiling mixture and hermetically stored in glass. The stock citrate buffer, which is alkaline when prepared, is readily cleansed by simple boiling before dilution and titration. Thiodiglycol is notoriously difficult to purify, but fortunately a high purity grade is now commercially prepared, specifically for amino acid analysis. If this is

Table 1. *All columns 0.62 cm diameter. Standard 21-h gradient for urines, physiological fluids and hydrolysates Volumes = 75 ml/chamber*

Chamber	pH	Salt, N	Methyl alcohol
1	2.875	0.2	10%
2	2.875	0.2	3%
3, 4	2.875	0.2	—
5	70 ml—3.80	0.2	—
	5 ml—5.00	0.8	
6	6 ml—2.875	0.2	
	9 ml—3.80	0.2	
	60 ml—5.00	0.8	
7, 8, 9	5.00	0.8	

Column 140 cm 60 °C with "A" (22 μ) or "B" (17 μ) chromobeads.
Flow 30 ml/h.
Variation 2 h initially at 45 °C improves serine-threonine separation.
Reduce temperature to 45 °C after ammonia, displaces tryptophan to the far side of histidine.

Table 2. *5$^1/_2$-h chromatograms for protein hydrolysates*

Chamber	pH	Salt, N	Methanol
1	2.75	0.2	10%
2, 3	2.90	0.2	
4	5 ml—2.90	0.2	
5	70 ml—3.80	0.2	
6, 7, 8, 9	3.80	0.2	
	10.0	1.2	

Volume 75 ml/chamber.
Column "A" Resin, 140 cm, 60 °C.
Flow 90 ml/h.

not available passage through a column of cationic resin in the hydrogen form is often required to prepare the reagent grade for use.

Precautions in the storage of buffers is as important as preparation. As previously mentioned, microbial contamination may develop and the metabolic products may not only produce artifacts, but may also affect quantitation of the chromatogram. Storage under anaerobic conditions with the inclusion of a fungicide and bacteriocide is considered mandatory. Phenol, chlorophenate and thymol are frequently used and are usually completely effective. A further source of contamination, particularly by ammonia, is the use of polyethylene storage vessels. This material is somewhat permeable to ammonia so that utilization of high-density polypropylene or glass vessels is preferred. Soda glass and some varieties of boro-silicate glass can effect the pH of their contents because sodium ion may be leached from the glass. Siliconizing or waxing is usually effective in preventing this common problem.

Finally, heavy metals contamination can produce unexpected chromatographic abberations. The contaminants may be present in the buffers or the samples, or derived from metallic components of the system. The addition of sequestering agents such as

Table 3. *3-h Chromatogram for protein hydrolysates*

Chamber	pH	Salt, N	Methanol
1	2.75	0.2	10%
2	27 ml— 2.75	0.2	10%
	13 ml— 2.875	0.2	
3	2.875	0.2	
4, 5	3.80	0.2	
6	17 ml— 5.00	0.2	
	23 ml— 6.10	1.2	
7	6.10	1.2	
8	20 ml— 6.10	1.2	
	20 ml—11.68	1.2	
9	11.68	1.2	

Column 60 cm, C_2U Resin, 60 °C
Volume 40 ml/chamber

EDTA to the buffers can help prevent this problem with many metallic elements. Contamination by any metals, particularly chromium, as indicated by a green color formation at the top of the resin bed and by the oxidation of methionine, requires thorough chemical cleaning of the resin as outlined earlier.

3. Column Operation

The elution buffer is forced through the chromatograph column by a positive displacement pump. One of the major essentials for reproducible elution patterns is constant flow of buffer through the column. Contemporary pumps show great improvements over earlier designs; pistons are made of sapphire and the graphite seals have been replaced with modern polymers, etc. Most designs however, still depend on ball valves for their operation so that exclusion of particulate matter from the buffers is still mandatory. Many otherwise inexplicable variations in elution time, or occasional asymmetric peaks, are caused by an incompletely closing ball valve or a small amount of air in the piston chamber. A sudden drop in operating pressure accompanied by an increase in the base-line of the chromatogram are indicative of this problem and cleanliness is its only solution.

The application of the sample can be another source of error. Ideally the sample should be at or slightly below the pH of the starting buffer and should be of such concentration that it may be in a minimal volume. It should be applied carefully to the exposed surface of the resin bed, forced into the resin under moderate air pressure (a hand pressurized 50 ml syringe with appropriate tubing connection is ideal for this purpose) and finally washed in with a small volume of buffer which is forced into the resin in the same way. The space above the resin is then carefully refilled with buffer to complete the procedure. Before the column fitting is replaced, it is important to flush the buffer supply tubing with the proper starting buffer to avoid contamination by buffer material from a previous chromatogram.

After the gradient buffers have been pumped through the column, the amino acid separation is complete. It is then necessary to wash and regenerate the resin before starting a new analysis. This is accomplished, in some systems automatically, by pumping a 0.2 N NaOH solution followed by starting buffer. The total volume of these regenerants is not usually considered critical but it is good practice to remain consistant in the procedure and it is preferable to continue regeneration until the column effluent has the same pH as the starting buffer.

4. Analytical System

Most amino acid analyzers use separate P.D. pumps for pressure elution of the column and to pump the ninhydrin reagent. The column effluent is mixed with ninhydrin and this reaction mixture is then pumped through a long plastic, capillary tubing immersed in a constant temperature reaction bath. The length of this tubing is adjusted to keep the reaction mixture at the selected temperature for approximately 15 min. To prevent longitudinal diffusion of the reaction products, the tubing bore is very small usually in the order of 0.5 to 0.8 mm internal diameter.

The analyzer used as an example in this discussion is unique because it uses a relatively large diameter glass tube (2 mm I.D. or larger) which is relatively insensitive to blockage by dirt or by precipitates from excessive chemical reaction. In this system, the flowing stream is segmented by nitrogen bubbles that fill the lumen of the tubing. These bubbles divide the stream into separate segments that remain discrete during passage through the reaction bath, etc. and prevent any diffusion from taking place along the tubing.

In the Model NC–1 Analyzer, an adjustable portion of the column effluent is removed from the total stream by the peristaltic pump and the remainder is diverted to waste or to a fraction collector. This sample stream is then mixed with approximately three times its volume of ninhydrin reagent and segmented with nitrogen by the same peristaltic pump. (The flow rates of these streams are determined and maintained by the use of precalibrated tubes in the pump.) The mixture then passes through a glass tubing spiral maintained at 95 °C in a heating bath. During the 15 or so minutes that the mixture is heated, the characteristic blue reaction product is formed.

Few reagents can be as universally disliked as the ninhydrin reagent for amino acid analysis; it is unstable in light and air, is very pH sensitive and is both noxious and poisonous. Yet for 20 years it has remained almost the only reagent for the determination of trace quantities of amino acids. In spite of its disadvantages the sensitivity and relative non-specificity of ninhydrin makes it uniquely suited to this type of analysis.

In spite of the fact that the reagent has been used for something over 55 years, relatively little is known about the mechanism of its color formation with amino acids. One point, however, is clear; hydrindantin, the reduced form of ninhydrin, must also be present in order that the color formed is linear with concentration, stable and of the requisite sensitivity, as shown by MOORE and STEIN [699]. An excellent article on the chemistry of ninhydrin by McCALDIN [662] reviews many of the reaction mechanisms proposed since 1910.

Several reagent formulations have been routinely used in current systems. The ninhydrin-hydrindantin reagent recommended by Technicon Corp. consists of a 2 g-% solution of ninhydrin with 0.15 g-% hydrindantin in methyl cellosolve containing 350 ml/l of a 4 N sodium acetate buffer (pH 5.5). This solution is diluted with four volumes of 50% methyl cellosolve prior to use. Other manufacturers recommend a solution of 20 g ninhydrin and 0.4 g stannous chloride dihydrate in methyl cellosolve containing 250 ml/l of 4 N sodium acetate buffer for use in their systems.

Both of these reagents require preparation and storage under an inert atmosphere and the storage life can be limited, depending upon purity of all the reagent components. The major factor in reagent deterioration is usually peroxide contamination of the methyl cellosolve. These peroxides can be tested for with potassium iodide (yellow color) and removed by re-distillation over ferrous sulfate. Occasionally a batch of cellosolve becomes opaque upon dilution with water, the cause being dissolved varnish from the supply drum. Commercial nitrogen used to provide the inert atmosphere for ninhydrin storage can be readily de-oxygenated by bubbling through acidified vanadyl sulfate solution that is kept over zinc amalgam. Formation of a green color indicates exhaustion of the amalgam.

A recent promising development in our laboratory makes use of hydrazine sulfate as a "reducing" agent for ninhydrin. A 0.6 g-% solution of ninhydrin is prepared in 50% methyl cellosolve containing 10% v/v 4 N sodium acetate buffer, and the solution brought to pH 5.4 to 5.5 with acetic acid. This solution is mixed continuously in the Analyzer and immediately prior to use, with a 2 mM hydrazine sulfate solution in the ratio of about 2:1 (0.7 ml/min ninhydrin reagent, 0.32 ml/min hydrazine and 0.44 ml/min sample). The greatest advantage of this variation is the fact that the solutions are insensitive to oxygen and can be prepared and stored in open vessels. Similar techniques using sodium cyanide with ninhydrin, suffer from the tendency to oxidize and from the volatility of undissociated HCN.

Several factors in addition to reagent purity and composition will affect the reaction of ninhydrin with amino acids. The pH of the final reaction mixture also influences the yield of colored product. The optimal value is approximately pH 5.0 and most reagent systems are designed to achieve this. Alteration of the ratio of buffered reagent to sample can, however, lead to losses in sensitivity. Another important factor is that of reaction rate. Reaction rate constants differ between individual amino acids and all are dependent on the temperature of reaction. A compromise temperature is the customary solution. In some systems, a bath temperature of 95 °C with a residence time of 10 to 15 min is used, while other analyzers use the same residence time in a boiling water bath.

With the exception of proline (an imino acid), amino acids react with ninhydrin to give a product whose absorption maximum is close to 570 mμ. Virtually all amino acids, under optimal conditions, produce significantly different color yields or, more precisely, possess individual, characteristic absorption coefficients. However, as they are separated from each other in the chromatographic method, this presents little practical difficulty. It is common practice to relate the color yields to that of leucine and refer to the ratio values obtained as "color factors" or "leucine equivalents". Although tables have been published, it is preferable to determine these factors, if required, with the particular system and reagent being used.

To obtain quantitative results it is essential that the relationship between optical density and amino acid concentration be linear. In contemporary systems this linearity holds up to about 1.5 optical density (O.D.) units. As the method is essentially a relative one, that is, standardization against synthetic amino acid mixtures of known concentration is obligatory; absolute values of absorbance are not required. For this reason, however, it is essential that any amino acid analyzer should reproduce the results obtained over an extended series of analyses.

5. Colorimeter and Recording Components

After reaction the segmented stream is cooled and passed through a series of colorimeters. Before the first of these, the nitrogen bubbles are removed by a glass "T" fitting referred to as a "de-bubbler". The first colorimeter measures the relative absorbance at 570 mμ in a 15 mm flow cell. The second, colorimeter also monitors at 570 mμ but at half sensitivity, using an 8 mm flow cell. The final colorimeter is used primarily to measure reacted proline concentration at 440 mμ. The monochromatic light is obtained from stabilized tungsten lamps by means of interference filters.

The system described here uses two channels in each colorimeter. One contains the small volume flow cell and the other serves as a reference channel. The electrical outputs of these channels are expressed as a ratio by the potentiometric recorder. Should the light output from the lamp vary for any reason, it will be electrically compensated for by this method of presentation. The recorder is a three point device that prints the absorbance value of each flow cell on the chart, in sequence. Potentiometric controls are provided to set a baseline value and to electrically zero the instrument at infinite optical density. To aid in computation, the chart paper is scaled logarithmically, i.e., in optical density units.

6. Multiple Column System

A variation of the NC–1 analyzer uses up to five accelerated analysis columns to permit a series of analyses to be performed in an overnight operation. The columns are pre-loaded with samples and then operated by a control unit which turns on the five P.D. pumps and gradient devices in sequence, as each chromatogram is completed. The effluent tubing connections from these columns are connected to a common manifold device and then to a single analytical system. This type of analyzer is relatively efficient in terms of analytical through-put, but suffers from some disadvantages. Chief among these are the difficulties associated with matching columns so that all produce the same type of chromatogram, etc. This system has been superseded by the more sophisticated apparatus described below.

B. Two Column Chromatographic Systems

Recently the Technicon Corp. introduced an automated version of an amino acid analyzer that is based on the two column method of SPACKMAN, STEIN and MOORE [987]. This system is designed to operate completely automatically, with a potential through-put of more than 80 hydrolysate samples per week. This instrument can be considered as illustrative of a new generation of automatic analyzers and will, therefore, be discussed as a modern example of the two column system. The design incorporates some novel features, including a new concept of automatic sample loading, which are described in detail.

1. General Principle

The model TSM Amino Acid Analyzer is shown schematically in Fig. 4-5 and photographically in Fig. 4-6. The two chromatographic columns are supported under a sample loading device. The longer, 35 cm column is used for the separation of the

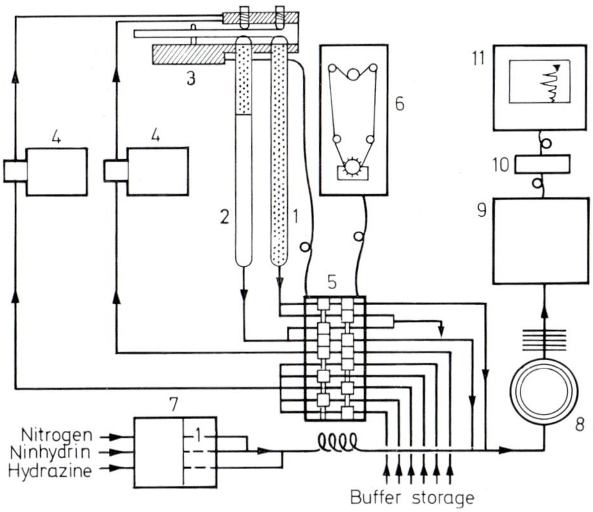

1 Acidic/neutral column	5 Multi way valve	9 Colorimeter
2 Basic column	6 Film loop timer	10 Amplifier
3 Sample loading device	7 Peristaltic pump	11 Recorder
4 Column pumps	8 Heating bath	

Fig. 4-5. Schematic flow diagram of Model TSM Amino Acid Analyzer

acidic and neutral amino acids, while the other column, containing 6 cm of resin, performs the basic amino acids separation. Eluting buffers are delivered to the columns through the sample loading device from two positive displacement pumps. The supply of these buffers from the storage reservoir is controlled by a multiple-channel programmed valve. This device determines the sequence of buffers required in the step-wise elution of the amino acids, and also acts as a control for the column effluents.

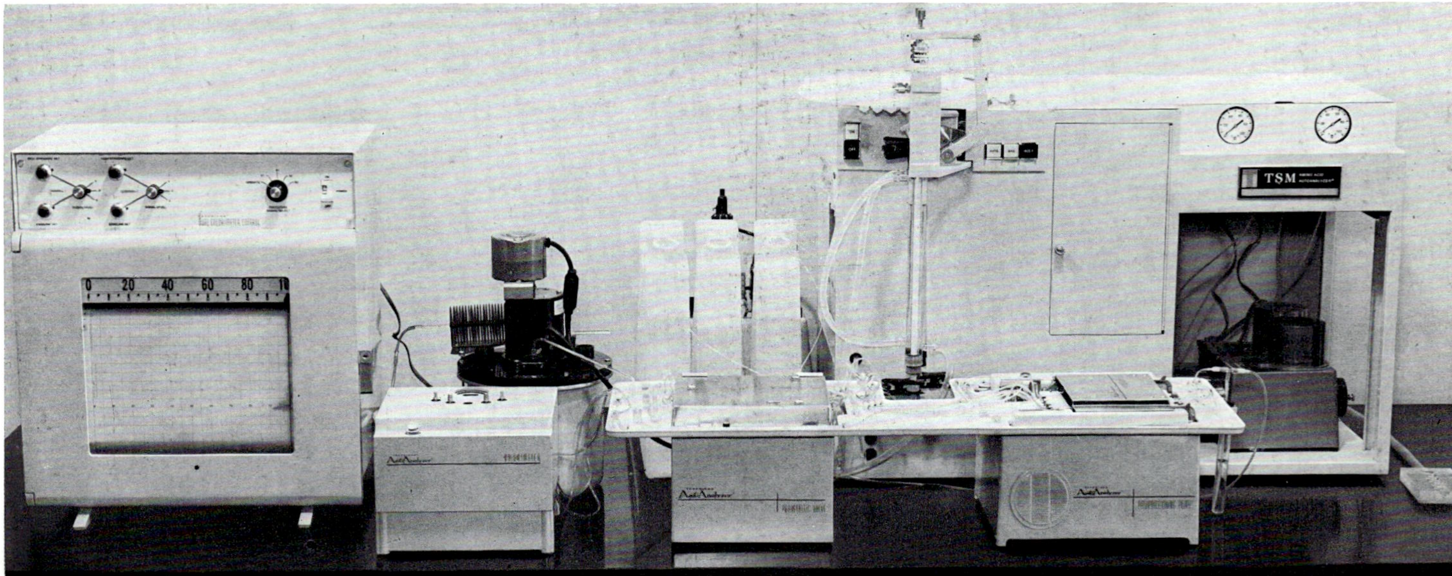

Fig. 4-6. Model TSM Amino Acid Analyzer showing from left to right: 2 pen recorder with colorimeter control mounted above, dual colorimeter, reaction heating bath (rear), multiple channel peristaltic valve, proportioning pump, bench top cabinet supporting sample presentation device, columns and tape timer (behind door) and positive displacement pumps

The valve is in turn controlled by a tape-strip control unit to which it is electrically connected.

The column effluent tubes pass through the control valve and are introduced into the analytical system at the appropriate times. The effluent from one column at a time is mixed with a segmented stream of reagent propelled by a peristaltic pump and passed through a heating bath. The cooled reaction mixture is monitored in a dual flow cell colorimeter which has incorporated new circuitry to amplify and linearize the signals that are presented to a conventional two-pen potentiometric recorder.

2. Sample Introduction Device

Many of the potential advantages of accelerated analysis systems cannot be fully realized unless some form of automatic sample loading is used. Several instruments that would perform this function have been described by Hirs [443a], Murdock

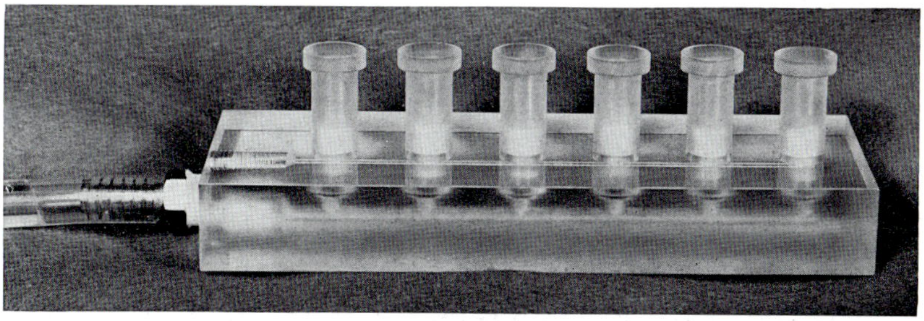

Fig. 4-7. Aspiration block with six sample cartridges in position

et al. [711], Dus *et al.* [252] and Eveleigh and Thomson [293]. The principle used in this instrument is based on the design of the last authors, in which pre-loaded sample cartridges are mechanically incorporated into the column elution system at appropriate times.

The sample cartridge consists of a small plastic cylinder containing a quantity of ion-exchange resin, similar to that used in separation columns, that is supported between two sintered discs. One cartridge (and one aliquot of sample) is used for each column. The reusable cartridges are loaded with samples in a six position aspiration block, shown in Fig. 4-7, or by individual connection to an aspirator. The sample solution is adjusted to approximately pH 1.5 and simply dispensed into a cartridge. During aspiration, the amino acids are adsorbed on the resin while the excess fluid passes through. With the exception of cysteic acid, this amino acid adsorption is quantitative, provided that the sample volume is less than 0.5 ml. Larger samples can be used if they are adjusted to a lower pH and if a slight adjustment of the subsequent elution sequence is made.

The loaded cartridges are placed in a rotatable sample tray capable of holding 80 cartridges which is then positioned in the application unit as shown in Fig. 4-8. On receipt of an activating electrical pulse, this unit indexes one position and the following sequence of operations is performed: the positive displacement pumps are inactivated and the column back pressures are permitted to fall to atmospheric by drainage through the columns. After a 1 min delay, the clamping mechanism is opened, releasing the pair of cartridges from the preceding analysis cycle. The sample tray is rotated to place a new pair of samples in position and the clamp mechanism is closed. The cartridges are then securely sealed between the eluant supply tubing and the top of the columns ready for a new analytical cycle to commence.

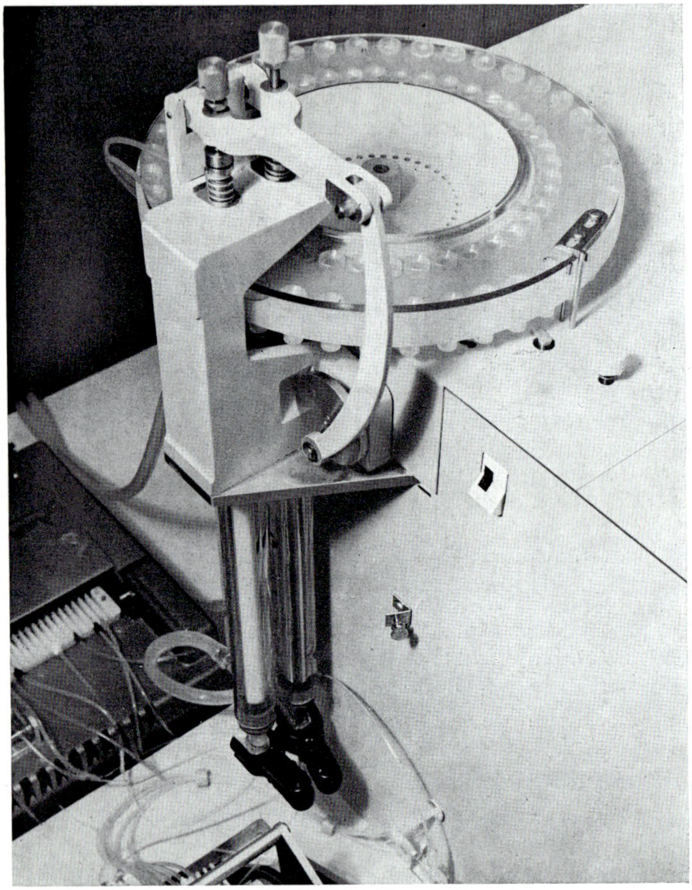

Fig. 4-8. Rotary sample plate and automatic sample presentation device with sample cartridges clamped in position, in contact with columns mounted below

An advantage of the cartridge concept over other systems which employ capillary tubing for sample storage, is that the sample volume is not restricted to a defined amount. Furthermore, as sealing is performed at atmospheric pressure, the difficulties associated with high pressure rotary valves avoided. It will be observed that new cartridges can be added to the storage tray and used ones removed during the elution without interfering with the analysis, i.e., the order of sample application can be readily changed at any time.

3. Multi-Channel Valve

This device is essentially similar to that described by Eveleigh and Thomson [293]. Its main purpose is to control the sequence of elution buffers supplied to the column pumps and to switch the column outlets into the analytical system. As can be seen in Fig. 4-9, a rotating spool, supporting twelve sets of rollers is positioned below a spring-loaded platen, across which are stretched a number of plastic tubes. Each

Fig. 4-9. Multiple channel peristaltic valve shown with platen and cover removed

roller is made up of a number of spacers of two different diameters that are arranged in the required order along a shaft. At the highest position of the drum, these spacers are in contact with the plastic tubes on the platen. The larger diameter spacers occlude their corresponding tubes, while the small spacers allow fluid to flow through the tubes they contact. The roller width and tube positions are adjusted to correspond, so that any desired tube(s) may be opened or closed at any valve position.

The valve is motor driven and activated by electrical impulses from the tape timer. Each pulse turns on the motor to index the valve one roller position, thus permitting liquid in some tubes to flow through to the pump, while other tubes previously open may be closed. The sequence and the operational period of the eluting buffers is easily programmed and reliably controlled by this means, as illustrated in Fig. 4-10. (The twelve position valve is simplified and shown with only six roller positions in this diagram and the time periods are illustrative only.) Minor

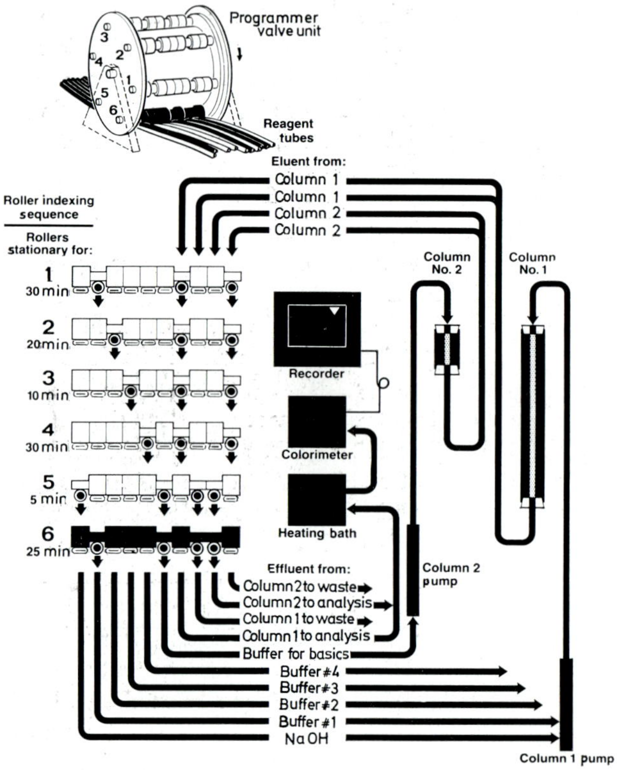

Fig. 4-10. Schematic representation of multiple valve function, showing sequential presentation of buffers

modifications in the elution protocol or a complete change of program for a different type of analysis may be accomplished by simply reprogramming the rollers.

In addition to the spacers used to open and close tubes, the valve has five spacers on the end of each shaft that activate separate microswitches. These switches control the sample introduction device, the two column pumps and the valve itself. The fifth switch activates a "fail-safe" mechanism that will turn off the analyzer in the event of system failure, loss of program synchrony or electrical supply interruption.

4. Tape Loop Timer

This device provides electrical impulses to activate the multi-channel valve. As illustrated in Fig. 4-11, it consists essentially of a loop of perforated 16 mm plastic tape, driven past a microswitch by a synchronous motor. The tape is pre-punched by

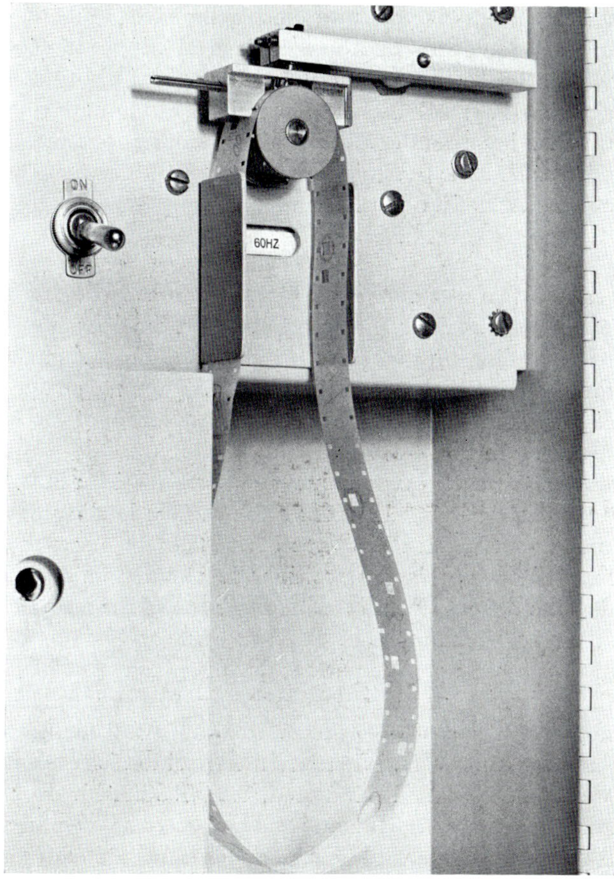

Fig. 4-11. Plastic tape timer with programmed tape loop mounted in operating position

the operator with a series of holes to correspond to the desired operation sequence of the valve. The tape is driven past the switch (which is activated by falling into a hole) at a constant rate so the linear distances between holes represent the desired time intervals. The sequence is repeated indefinitely because the tape is prepared in the form of a loop, one revolution corresponding to one analysis cycle. The advantage of this type of timer is that precise timing is considerably simplified and alternative programs can be readily set up and stored in a convenient form.

5. Analytical System

The analytical system of the TSM Amino Acid Analyzer is similar in principle to the system described earlier. Some modifications and refinements have been made, however, to improve the quality of recording and increase sensitivity. The nitrogen used to segment the reaction stream is introduced by mechanical cam action for greater precision and the temperature of the reaction bath is regulated by a proportional control device instead of a thermoregulator.

The colorimeter contains two separate flow cells having much smaller volume than previous models, and uses photo-tubes instead of photo-voltaic cells. Log amplifiers in the control unit provide linear (with respect to optical density) output to the recorder and signal amplification that permits 0.1 O.D. unit to be expanded to full scale deflection are incorporated. With these modifications and additions to the earlier analyzers, it has become possible to perform quantitative analyses of samples having only one nanomole of a typical amino acid.

6. Operation

It has been shown that the apparatus functions in a cyclic manner; one cycle representing one complete analysis. The total time and the operations within a cycle are readily varied depending on the type of analysis required. As an example of the detailed operation of the instrument, a 2 h elution program for analysis of the amino acids of a protein hydrolysate is described below. The pattern of this analysis cycle is shown diagrammatically in Fig. 4-12.

The start of an analysis cycle occurs when the tape loop timer indexes the valve to position No. 1. An electrical signal activates the sample loading device to place a pair of samples in stream. When these have been sealed in place, the acidic-neutral column pump is turned on. In this first position, the valve supplies Buffer 1, pH 3.25, to the pump and opens the effluent tube of this column to the analytical system. The effluent tube of the second column is diverted to waste and its pump is inoperative.

At the end of the prescribed time period, during which cysteic acid and aspartic acid are eluted, the valve receives a signal from the tape loop and indexes to position 2. The second buffer is now supplied to the pump, and, subsequently, at positions 3 and 4, the third and fourth buffers. When the fifth position is reached and nearly all of the neutral amino acids have been eluted, 0.2 N sodium hydroxide solution is supplied to column 1 pump and column 2 pump becomes operative. During the following 5 min or so, the acidic and neutral amino acids are eluted from the short column (2) and are diverted by the valve to waste while tyrosine and phenylalanine from the first column are flowing into the analytical system. Finally at position 6, regenerating buffer is supplied to column 1, its effluent tubing is diverted to waste, and the basic amino acids from the second column are introduced into the analytical stream, to complete the analysis cycle.

Several additional operating features of the system are omitted from the schematic illustrations to avoid complexity. A shut-down procedure, initiated by positioning a pin on the sample tray relative to the last sample cartridge, permits unattended operation with a limited number of samples. When all samples on the tray have been analyzed, the analytical system is flushed with solvent to remove ninhydrin before operation ceases. A series of safety features have also been incorporated for protection against loss of samples, should an unexpected malfunction occur. A generalized

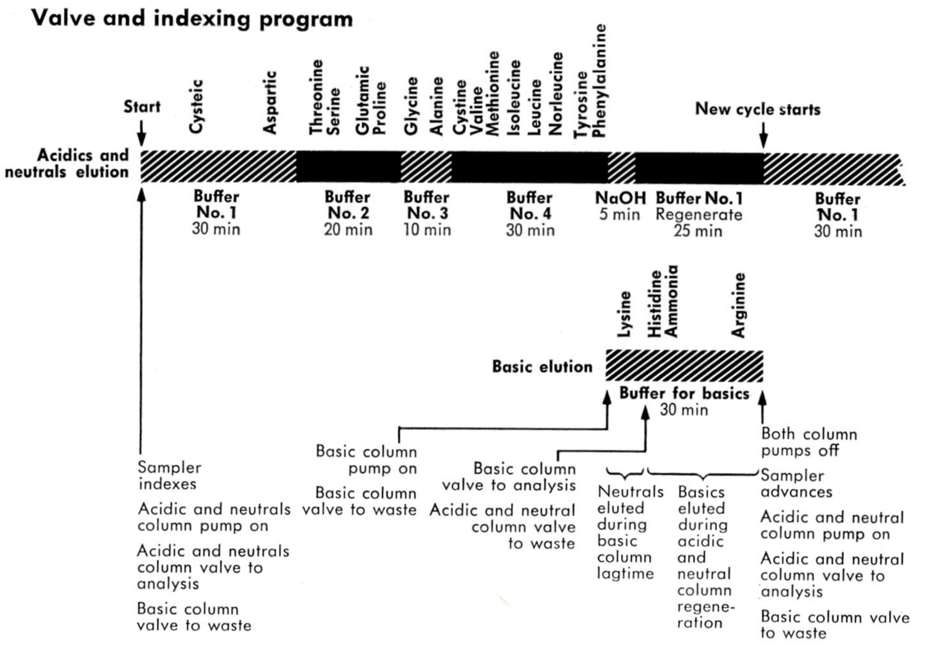

Fig. 4-12. Diagrammatic representation of buffer presentation relative to amino acid elution showing a 2 h analysis cycle

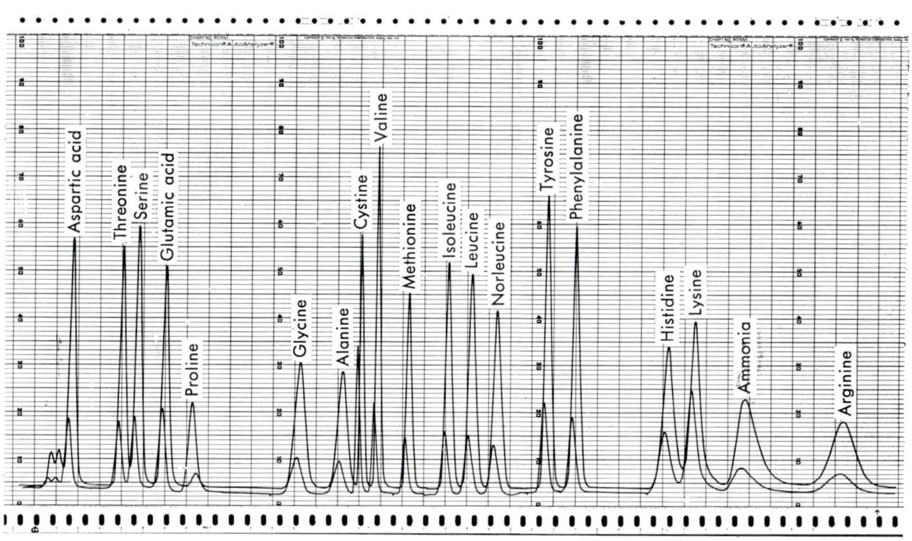

Fig. 4-13. Chromatogram showing separation achieved in a 2 h analysis of 0.05 micromoles of each of 20 standard amino acids

electrical failure or a colorimeter lamp burn-out will terminate the analysis. Similarly, if the operator fails to provide sufficient reagent and buffers, if there is a reduction in supply, or if the columns develop high back pressures due to blockage by proteins left in samples, etc., the system shuts down. There is also a synchronization check between the loop timer and the position of the peristaltic valve during every analysis.

The columns used for this type of analysis are packed with 8 μ spherical particle resin and are operated at 60 °C with buffer flow-rate set at 0.55 ml/min. A typical chromatogram is shown in Fig. 4-13.

7. Buffer Timing and Compositions

When samples are introduced to the column, as has been described, it is important that no space is left in the column above the resin. Diffusion and mixing of the sample and buffers in any such space would destroy the resolution of the system. Adjustable column fittings are used, therefore, to keep the fitting directly in contact with the resin. This adjustment feature and the fact that height of the resin bed affects the elution time of amino acids, precludes presentation of an exact buffer sequence protocol.

The elution period for each buffer is illustrative only, for this reason. Precise timing of the buffer program is performed empirically.

Elution buffer times for 2 h analysis

Valve position	Acidic-Neutral column	Basic column	Time
1.	pH 3.25 + Cellosolve		0— 30 min
2.	pH 3.10		30— 48 min
3.	pH 4.25		48— 70 min
4.	pH 5.25		70— 84 min
5.	0.2 N NaOH	pH 5.25	84— 92 min
6.	pH 3.25 + Cellosolve	pH 5.25	92—122 min

The buffers used in this two column system are similar to those employed in gradient elution except that the citrate ion concentration is increased from 0.05 M to 0.10 M. These buffers are conveniently prepared from a ten times concentrated stock solution of the following components:

210 g Citric acid

82.5 g Sodium hydroxide

Water to 1 l after cooling.

Working buffers are prepared by diluting 100 ml of stock solution plus 10 ml of BRIJ 35 to approximately 900 ml, then titrating to the required pH before making up to 1 l. In addition to the reagents shown, 6% v/v of methyl cellosolve is added to Buffer No. 1 (pH 3.25) before titration and 10 g of sodium chloride per liter is added to Buffer No. 5 (pH 5.25) to increase the sodium ion concentration from 0.2 N to 0.38 N. The pH of buffers used in very rapid analyses is far more critical than in longer analyses and the precautions regarding buffer preparation and storage, previously discussed, become particularly important. Fortunately the presence of ammonia in the solution is not usually as detrimental in a two column system as it is in gradient elution.

It should be noted that Benson and co-workers [71] have reported a variation of the conventional buffers, using lithium citrate instead of sodium citrate. They achieved separation ot the amides, asparagine and glutamine, from each other and from the early eluted amino acids with which they normally interfere when eluted with sodium buffers. This modification was primarily designed for physiological samples such as serum and urine, but it could prove extremely useful when enzymic digests of peptides and proteins are to be analyzed.

Buffer used for the dilution of samples and to wash samples on the cartridges, is prepared from the stock solution as above. After the addition of 10 ml/l of thiodiglycol, the buffer is adjusted to pH 1.5. Thiodiglycol is used in this buffer to minimize oxidation of cystine and methionine during possible long term storage on the resin in the cartridges.

8. Precision of the Method

Thomson [1054] has shown that there is excellent recovery of amino acids from the cartridge even after prolonged storage. This is also demonstrated in Table 4 by the decomposition rates expressed here as percentages of initial sample per analysis period. Table 4 also indicates the degree of precision that can be expected from a contemporary system. This order of reproducibility ($\pm$ 2 to 3%) should be obtained

Table 4. *Analysis of one sample plate — 40 sequential chromatograms*

	Coefficient of variation of retention time (% of elapsed time	Standard deviation of retention times (seconds)	Linear decomposition rate (% initial sample per 2 h)	Coefficient of variation of peak areas (%)
Aspartic acid	0.53	10.1	None	2.72
Threonine	0.60	13.6	0.009	2.12
Serine	0.65	15.2	0.006	1.85
Glutamic acid	0.48	12.2	None	1.60
Glycine	0.50	17.1	None	2.33
Alanine	0.43	16.0	None	1.55[a]
Cystine	0.14	5.4	0.2	1.79
Valine	0.15	6.2	None	0.96
Methionine	0.17	7.2	0.2	2.26
Isoleucine	0.19	8.5	None	1.64
Leucine	0.22	10.2	None	1.84
Norleucine	0.26	12.6	None	2.81[a]
Tyrosine	0.35	18.4	None	1.73
Phenylalanine	0.22	11.8	None	1.77[a]
Histidine	0.50	30.1	None	3.51
Lysine	0.50	31.2	None	2.47
Ammonia	0.22	14.7	Not calculated	
Arginine	0.25	23.0	None	2.65

[a] Buffer changes

routinely from any well designed amino acid analyzer. This statement will only apply to systems that are maintained conscientiously by an operator who is prepared to pay close attention to consistency in operation and to important, but often neglected details, such as cleanliness of resin, buffers and reagents.

C. Computation of Results

The conventional product of an amino acid analyzer is a recording of absorbance with respect to time the chromatogram. If the system is performing satisfactorily, each amino acid will be represented by a separate peak on the chromatogram and the relative position or each peak will be constant throughout a series of analyses. This fact permits qualitative analysis, i.e., identification of the amino acids, on the basis of their specific retention times, (peak position) for a given series of operating parameters. Variations in operating procedures, instrument function or reagent composition may change the position or the shape of the peaks representing the amino acids and thus make accurate interpretation difficult or impossible. Chromatographic analysis is, in principle a relative method and is completely dependent on exact repetition of given parameters throughout any series of anaylses.

1. Interpretation of the Chromatogram

Intelligent interpretation of the chromatogram will assist in the diagnosis of malfunctions or errors in operation of the system, in addition to providing an assessment of the validity of the recorded information. Many sources of error (common to all amino acid analyzers) may be diagnosed by visual inspection of the chromatogram.

Elution times of the amino acids may be valuable indicators of the stability of the system. The positions of peaks in the chromatogram will vary with changes in pH or concentration of the column buffers, or with buffer flow rate or temperature changes. In general, an increase in any of these parameters will elute the amino acids sooner and vice versa. If only a portion of the elution pattern is distorted, the problem is usually associated with a specific buffer and is most commonly the result of a buffer pH change. Long term aberrations with complete divergence from the normal pattern are usually the result of column thermostat or P.D. pump malfunctions.

The profile of the recorded peak is another indicator of system function. Asymmetry, apart from that resulting from generalized poor resolution, may result from overloading the column with sample. This condition produces very large peaks with a characteristic slowly rising frontal slope. Similarly, if the top of the resin in the column is distorted (not flat), or excessively dirt encrusted, the first several peaks on the chromatogram may be skewed, while the latter two-thirds or the record may appear essentially normal. Short duration variations in flow rate or channeling within the resin bed, are indicated by departure from the ideal gaussian distribution of absorbance values. Similarly, partial or temporary blockage of the flow cell shows characteristic distortions on the record.

Long term drift in the base-line, although esthetically undesirable, can usually be compensated for in extracting quantitative data. This phenomenon can, however, be indicative of change in reagent proportioning or of chemical deterioration. The result may be a sensitivity change or deviation from linearity in extreme cases. Short term drifting and sudden changes in baseline can be more serious, particularly if they occur during elution of an amino acid. These sudden changes, if systematic, can often be traced to inadequate buffering capacity of the ninhydrin reagent in discrete buffer elution systems. Random shifts are indicative of electrical instabilities or of dramatic changes in fluid flow rates. A wavy baseline is apparent when fluid surging occurs or the reaction bath temperature control has a large differential. Finally, particulate matter or small gas bubbles in the reaction stream produce characteristic spikes in the base-line These may be distinguished from electrical transients by their duplication, with a slight delay, in the other channels.

2. Manual Integration

The chromatogram is a differential recording of concentration with respect to time. In order to express the components in terms of relative mass, the individual peaks must be integrated. Comparison of these derived peak areas in a sample with those obtained using a calibration mixture of known content permits calculation of the absolute amounts of each component.

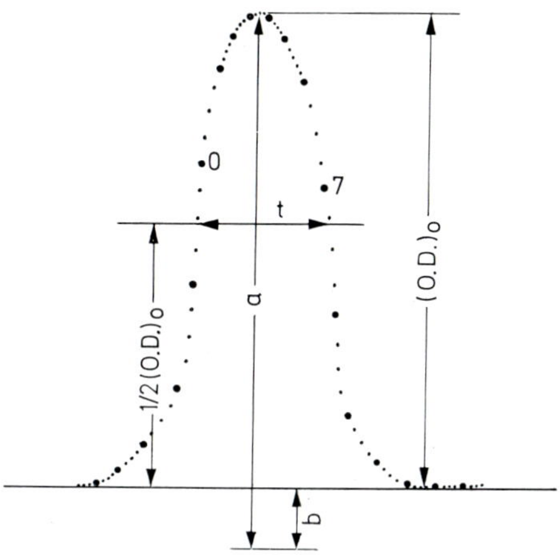

Fig. 4-14. Standard procedure for the integration of an animo acid curve

$(O.D.)_0 = a—b = 0.203—0.020 = 0.183$; $^1/_2\,(O.D.)_0 = {}^1/_2\,(a—b) = 0.091$. Therefore, height at which "t" is measured $^1/_2\,(a—b) + b = 0.111$. Every 4th dot is black, therefore count all black dots after one marked "O".

7 × 4	= 28.
All dots below "O" and "7"	= 3.
Interval in 1/10's to 0.111	= 0.4
Total	= 31.4
So $n\,(O.D.)_0 = (31.4)\,(0.183)$	= 5.72
q_s	= 0.1 μmole

and from a previous run using a standard amino acid mixture with $q_s = 0.1$ μmoles $n_s\,(O.D.)_s = 3.62$. So that

$$q = \frac{5.72}{3.62} \times 0.1 = 0.158 \text{ μmoles}$$

The simplest method of integration involves multiplication of the height ot the peak by its width. The net height of the peak in O.D. units is found by substracting the base value from the gross height, $(a - b)$ in Fig. 4-14. The width is measured at "half-height", that is one-half $(a - b)$, after marking on the chart at the value one-half $(a - b)$ plus b. This is often more conveniently calculated as one-half $(a + b)$. The width (t) can be found by counting the printed dots, using interpolation if necessary, that occur above this half-height line. The area is then given by the product of net height and width, $t\ (a - b)$. It has been found that the repetitive arithmetic used in finding net and half-heights can be readily avoided by preparing a table for a series of base-line and chart height values. Similarly, the use of a nomogram can save considerable time and effort, as demonstrated by SCHROEDER, *et al* [904]. A far more accurate, but unfortunately very tedious, method employs the trapezoidal rule as shown and explained in Fig. 4-15.

3. Analog Devices

Several devices are now available to the analyzer user to overcome the tedium of manually integrating peak areas. As an example of an "off-line" device, the Technicon Integrator-Calculator employs electrical circuitry to perform the calculations, the chromatogram being retraced manually by the operator. Sophisticated on-line electronic devices are now offered by several specialist manufacturers to supply printed area values for individual peaks as they appear on the record. These instruments are equipped with circuitry to detect the start and finish of peaks and to correct for base-line drift. In common with manual methods, unresolved peaks are calculated by essentially dropping a perpendicular from the lowest inflection value. The linearized output from the photometer of the Model TSM Analyzer will also permit the ball and disc mechanical integrator or a planimeter to be used for quantitation of the chromatograms.

4. Digital Computers

The application of digital computers to the integration of chromatograms has been reported by several workers including YONDA [1162] and KRICHEVSKY [583]. To provide a suitable output, a data acquisition system is linked to the analyzer and the transmittance values stored on punched paper or magnetic tape. These values are then assessed within the computer where a very sophisticated treatment may be applied to their integration.

A successful coupling of the integration step with the arithmetical calculation of composition tables and peptide ratios has been reported by THOMSON *et al.* [1055]. In their program, individual amino acids are manually defined by the order of appearance and an approximate peak height value. This identification plus the use of memory-stored standardization values permits a complete calculation to be performed in 30 sec.

The remarkable reproducibility of elution times observed with contemporary automated instruments, should now make it possible to relegate even this identification to the computer. Complete automation of the chromatographic method of amino acid analysis, from sample application to computerized results, is now possible. The efforts of many researchers, the majority regretfully unmentioned in this

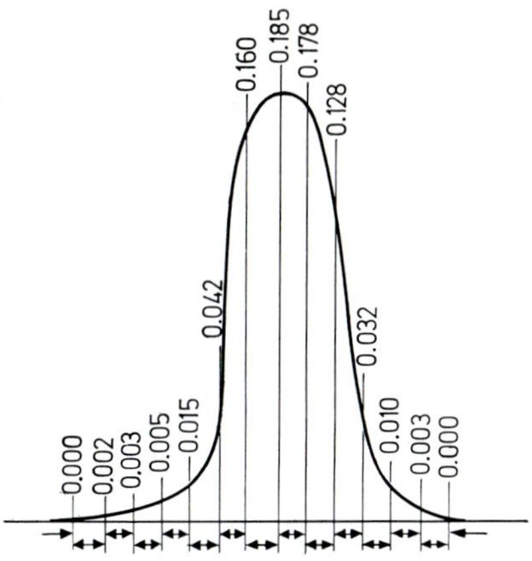

Fig. 4-15. Trapezoidal rule for integration of amino acid curve

(1) If the sample curve is integrated by the trapezoidal rule, then the standard curve must also be integrated by the trapezoidal rule. (2) Any unit of length may be used to divide the base evenly and as long as the same unit is used for integrating the standard curve, that unit may be called = 1.0 and the area will be equal to the sum of the heights. The trapezoidal rule states:

$$\text{sum of the trapezoidal areas.} = \frac{x}{2}(y_0 + 2\,y_1 + 2\,y_2 \ldots . 2y_n - 1 + y_n)$$

So that if $x = 1$ and $y_0 = y_n = O$, then sum of trapezoidal areas $= y_1 + y_2 + \ldots y_{n-1}$ which in the example below

$$= 0.002 + 0.003 + 0.005 + 0.015 + 0.042 + 0.160$$
$$0.185 + 0.178 + 0.128 + 0.032 + 0.010 + 0.003$$
$$= 0.763.$$

If the area of the standard for 0.1 μmoles is 0.482, then

$$q = \frac{0.763}{0.482} = 0.158\ \mu\text{mole}$$

contribution, have given the protein chemist a tool that is unique in analytical chemistry. Let us hope, on their behalf, that he will use it well.

Permission from the Technicon Corp., Ardsley, New York, to reproduce certain figures and data from its technical publications is gratefully acknowledged.

Chapter 5

Gas Chromatography of Amino Acid Derivatives

AMOS LINENBERG

I. Introduction

Extensive studies have been made in attempts to adapt the gas-liquid chromatography technique to amino acids analysis. Since the discovery of the gas-liquid chromatograph, the idea of using it for amino acids analysis has been enthusiastically expounded because of its great potential in providing a means by which qualitative and quantitative analysis of amino acids can be achieved quickly and accurately in comparison with other techniques currently in use. Excellent automated procedures are known at the present for analysis of amino acid mixtures. With ion exchange chromatography columns, mixtures of amino acids, in the amount of 10^{-7} mole each, can be analyzed. The disadvantages of this system, however, are the time required (up to 24 h in some cases), the cost of the instrument, and the relative complexity of the operation.

In contrast, gas-liquid chromatography technique offers the following advantages:

1. *Speed of operation*: Analysis can be done within 1 h or less.
2. *High resolution power*: A comparatively short column (6′) may have as much as several thousand theoretical plates.
3. *High sensitivity*: The gas-liquid chromatography detectors can detect at the nanogram level and, for certain derivatives, even at picogram levels.
4. *Simplicity of operation.*
5. *Comparatively low cost.*

These advantages encourage many scientists in this area to concentrate their efforts to find a procedure for routine anaylsis of amino acids using gas-liquid chromatography.

II. The Gas Chromatograph

A. General Description

The principle of the gas-liquid chromatography technique is based on the separation of volatilized samples on a liquid phase, due to differences in solubility of the various gases in a fixed liquid phase at a specific temperature and carrier gas flow rate. There are cases, when the gases are separated on a solid phase (gas solid chromatography), where the separation is due to other factors such as specific adsorption-desorption equilibrium at certain temperatures.

A large surface area of liquid is achieved by coating the required liquid phase (usually a high boiling point polymer or silicone compound) on an inactive solid support consisting of small grains in the 30 to 120 mesh range. The coated grains are then packed in glass or metal columns which can vary from approximately 2' to 20' long with an inner diameter of 2 to 15 mm. The most commonly used column sizes are from 4′ to 8′ long with an inner diameter of 2 to 4 mm.

The packed column is placed in an oven in which the temperature is very accurately controlled. Accurate temperature and a steady carrier gas flow rate are required to obtain accurate and reproducible results. The carrier gas enters the column from a flow controller at a fixed flow rate. Three separate heated zones are generally provided for the sample inlet, the chromatographic column and the detector. The temperature

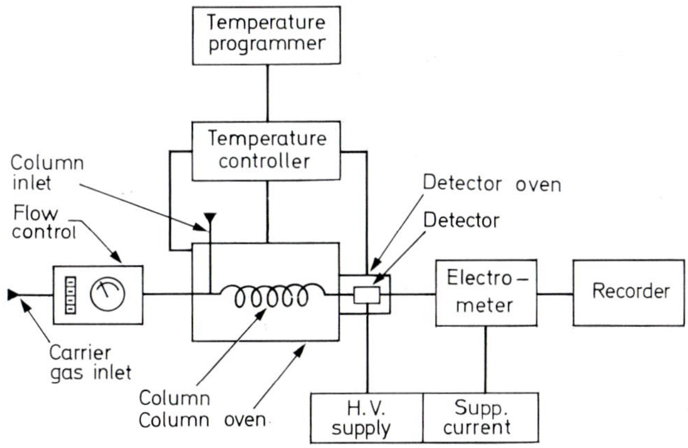

Fig. 5-1. Gas chromatograph, basic configuration diagram

of the inlet portion of the column is higher than the column temperature to assure fast vaporization of the sample. The column temperature controller can be held constant or connected to a temperature programmer which is used in cases when several compounds of widely varying boiling points are analyzed, such as for amino acids analysis. For this, an accurate and reproducible temperature programmer is required. The column is connected to a detector contained in a separate oven. Because the various detectors used in gas-liquid chromatography are sensitive to temperature changes, the detector's oven temperature must be accurately controlled.

The most widely used detectors in gas-liquid chromatography involve ionization of gases which require a stable high voltage supply. Detection consists of ionization of the sample in a high voltage electric field. The resulting ionization current is then amplified by an electrometer and passed to a recorder for analog display. This analog display is the chromatogram. The sample, usually dissolved in a volatile solvent, is injected into the column inlet and vaporization takes place immediately. The components, in the form of a gas, are then carried by the carrier gas through the column and separated by the liquid phase. On reaching the detector, each component is then ionized and generates a signal which is indicated as a peak on the chromatogram.

A schematic of a gas chromatograph is shown in Fig. 5-1.

1. Flame Ionization Detector

The most popular detector for organic compounds is the hydrogen flame ionization detector (see Fig. 5-2). Air is supplied to maintain the hydrogen flame which burns in a high voltage field. The effluent gas entering the detector is mixed with the hydrogen. When any compound in the effluent gas reaches the flame, it burns and is partly ionized. The free electrons produced are collected by a basket-shaped electrode and represented later as a peak on the recorder.

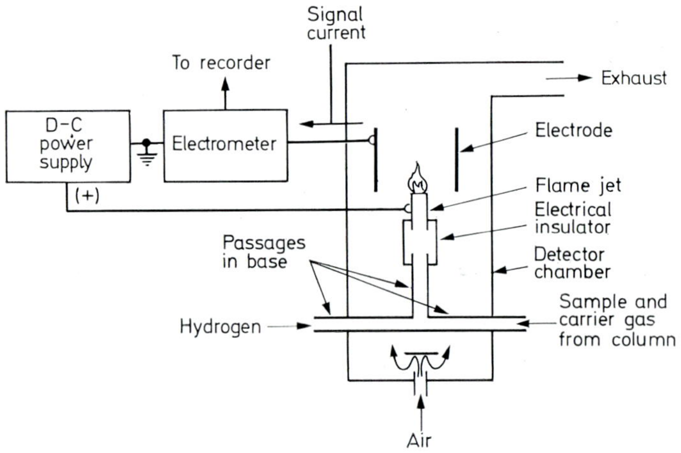

Fig. 5-2. Hydrogen flame detector, basic configuration diagram

The flame ionization detector is sensitive and has a low background noise. It will detect quantities from a few nanograms to several micrograms. Its response is linear over a wide range of concentrations and the results are highly reproducible. These features make it the most popular detector and the most commonly used in gas-liquid chromatography.

2. Electron Capture Detector

The electron capture detector is important because of its high sensitivity. This detector can be designed in many ways. A concentric design is shown in Fig. 5-3. Here, the carrier gas, nitrogen in this case, enters the detector and is ionized by a radioactive foil (tritium or nickel [63]). With voltage (D.C. or pulsed) applied, a steady standing current is produced. When any compound which has an electronegative group attached, such as chlorinated derivatives of organic compounds, enters the detector, it decreases the standing current, and causes a negative peak indication at the recorder. However, a conventional positive peak indication at the recorder may be obtained by using a negative polarity setting, *i.e.*, reversing the signal connections to the recorder. This detector is much more sensitive than the flame ionization detector and can detect quantities of a few picograms. However, since the electron capture detector will detect only electronegative compounds, many attempts have been made to convert certain compounds to chlorinated derivatives. Most of these efforts have been made in the field of steroid hormones where the determination of very small quantities is necessary.

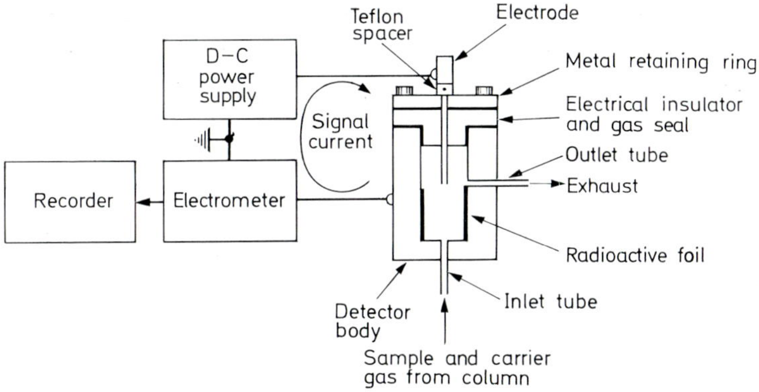

Fig. 5-3. Electron capture detector, basic configuration diagram

3. Other Gas Chromatographic Detectors

Among the other gas-liquid chromatography detectors, the thermal conductivity detector is very well known. This detector, one of the first detectors designed for gas-liquid chromatography, has the advantage of detecting every gas including atmospheric gases. However, it is much less sensitive than the other detectors.

The argon ionization detector can detect organic compounds, but is less sensitive than the flame ionization detector.

The detectors described above are the most widely used in the gas-liquid chromatography field. Other detectors are available which demonstrate specific sensitivity to phosphorus and nitrogen compounds.

B. Dual Channel Gas-Liquid Chromatography

As discussed previously, there are cases, such as in amino acids analysis, where temperature programming is used in gas-liquid chromatograph operation. This programming presents a problem that can be solved by using the dual channel gas-liquid chromatograph.

The liquid phase in the column has a vapor pressure which varies as a function of the temperature. At a constant temperature, this comparatively low vapor pressure causes a constant background indication on the chromatogram, due to the high sensitivity of the detector; this constant background can be zeroed by using a suppression current device present on the gas-liquid chromatograph. During temperature programming, the vapor pressure increases with temperature. This increases the background indication on the chromatogram or upward shift of the recorder baseline.

To avoid this, a dual channel gas-liquid chromatograph is used. Two identical columns are placed in the same oven and subjected to the same changes in the temperature and flow rate. The two columns are connected to two identical detectors. The detectors' outputs are generated in opposing polarities due to opposing high voltage supplies fed to the detectors. The detectors' outputs are combined in the electrometer and fed to the recorder. The result is a compensated, steady baseline during temperature programming. An example of baseline compensation during programming is

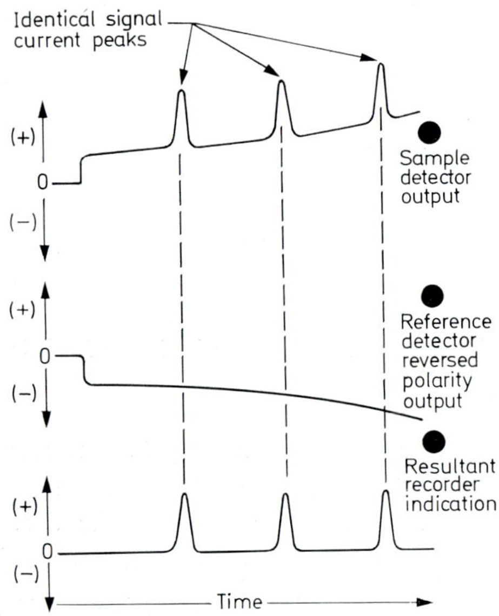

Fig. 5-4. Example of baseline shift and drift compensation technique

shown in Fig. 5-4. Baseline compensation is not required when working with large quantities because at low sensitivities, baseline shift is so slight that a single channel gas-liquid chromatograph is quite sufficient.

C. Qualitative Analysis by Gas-Liquid Chromatography

The most common method of identification of compounds by gas-liquid chromatography is to compare the retention time (the time required for the compound to go through the column following injection) of a given peak with the retention time of a known compound. Every compound has a specific retention time under specific temperature and flow rate conditions. If these conditions are repeated, the retention time will be repeated with high accuracy. Thus, to determine whether or not a specific compound is present, the standard is injected into the column and its retention time measured. The sample for analysis is then injected under the same conditions and its retention is measured and compared with that of the standard. If the retention times are not equal, the compounds cannot be the same. If the retention times are the same, the compounds can be identical. Qualifying information from the sample history will usually yield sufficient confidence.

Infrared analysis may also be used for compound identification. The most recent approach is the combination of gas-liquid chromatography and mass spectrometer for separation and identification of chemical compounds, respectively.

D. Quantitative Analysis by Gas-Liquid Chromatography

The peak area (in some cases, the peak height or other peak parameter is used) is a linear function of the sample size. Therefore, the peak area can be used for quantitative analysis. A calibration curve of peak area versus sample size can be plotted for various known standard concentrations. By evaluating the peak area of a given sample, its size can be found using the calibration curve.

Another method for determining sample size is one in which internal standards are used. First, equal amounts of a standard compound and the sample compound are injected together and the ratio of their peak areas is determined. Then a subsequent injection is made where the same amount of standard compound is added to the unknown amount of the same sample compound and the peak area ratio is again determined. A comparison of the two ratios will then enable the evaluation of the unknown sample size. Peak area evaluation may be done by direct measurement with a ruler or by mechanical, automatic integration. Recently developed complex electronic integration devices which may have an output to a computer can also be used.

III. Gas-Liquid Chromatography of Amino Acids

A. Difficulties in the Method

The analysis of amino acids by gas-liquid chromatography appears to be a simple matter, through many difficulties have been encountered in the past. Free amino acids, for example, cannot be analyzed since gas-liquid chromatography is based on the separation of gases. Underivatized amino acids are not volatile and may be decomposed at high temperatures. In addition, amino acids do not represent a homologous series making the selection of conditions for separation rather complicated.

It is implicit that amino acids must be converted into volatile derivatives for gas-liquid chromatographic analysis. This in itself presents several difficulties because the following requirements must be taken into consideration:

1. The derivatives should be volatile.
2. The derivatives should be stable.
3. Derivatives should be received with high recovery.
4. The procedure for making derivatives should be fast and simple.

While these problems have not been easy to solve, it appears that the difficulty in applying gas-liquid chromatography techniques for amino acids analysis is not in the technique itself, but in finding the most suitable derivative to use in the gas-liquid chromatograph.

B. Amino Acid Derivatives

Since the possibility of using gas-liquid chromatography for amino acids analysis was realized, many derivatives have been prepared in attempts to apply the new technique to amino acids analysis. In 1956, the conversion of leucine, isoleucine, and valine to the corresponding aldehydes by oxidation with ninhydrin, was carried out by HUNTER, DIMICK and CORSE [469]. ZLATKIS *et al.* [1181, 1182], developed this procedure into an automatic method for several amino acids. BIER and TEITELBAUM [96] converted amino acids to the corresponding volatile amines. Several amino acids, in the form of hydrochloric ester derivatives, were separated on the gas-liquid chromatograph column by SAROFF *et al.* [866]. The gas-liquid chromatography of methyl esters of amino acids was studied by BAYER *et al.* [60]. It was after YOUNGS [1166] had first introduced the N-substituted esters of amino acids by preparing N-acetyl, n-butyl esters of glycine, alanine, valine and other acids, that attention was directed toward the advantages of these derivatives. N-acetyl, n-propyl esters of amino acids were prepared by GRAFF *et al.* [364]. Many N-acetyl, n-amyl esters of amino acids were separated by gas-liquid chromatography by JOHNSON *et al.* [497, 498].

The N-trifluoroacetyl esters of amino acids appeared to be promising and attention was directed toward these derivatives. BLAU and DARBE [102] prepared the N-trifluoracetyl, n-amyl esters and SOMZELY *et al.* [1183] investigated the N-trifluoracetyl, n-butyl esters. The N-trifluoroacetyl methyl esters were investigated by CRUICKSHANK *et al.* [221], WAGNER and WINKLER [1094], HAGEN and BLACK [395], SAROFF and KARMEN [865], and others. 2,4-Dinitrophenyl methyl esters were studied by PISANO [790] and later by LANDOWNE and LIPSKY [593] who were studying those derivatives by using electron capture detectors, thus, obtaining high sensitivities. Although some of the methods were tried successfully with several amino acids, they did not completely meet all the requirements outlined previously.

1. Silyl Derivatives of Amino Acids

TMS (trimethyl silyl) derivatives of amino acids were first introduced by RUHLMANN [840]. Those derivatives were not accepted by many scientists as many difficulties were found while trying to use them.

In 1965, SMITH *et al.* [963] proposed methods to prepare the TMS derivatives by using hexamethyldisilizane and trimethyl chlorosilane, using several catalysts. They also prepared these derivations by using trimethylsilyl-diethylamine, which gave high yields in a few cases. The use of BSA (N, O-bis-(trimethylsilyl) acetamide) as a silylating reagent for amino acids as well as for several other nitrogen-containing organic compounds was suggested by KLEBE *et al.* [555], and BIRKOFER [98]. Nineteen protein amino acids were obtained by this method and the reaction looked promising as it was fast and quantitative. However, due to insufficient volatility of one of the BSA products, interference in the separation of glycine and alanine occurs.

In 1968, STALLING, GEHRKE and ZUMWALT [995] suggested a new silylating reagent which enabled this limitation to be overcome. The new reagent, BSTFA (Bis-(trimethylsilyl) trifluoroacetamide) and its products are more volatile than BSA and therefore do not interfere with the separation of any amino acids as they appeared in the solvent front. The procedure for derivative preparation as suggested by STALLING *et al.* is as follows:

1. Place 1.00 mg of dry amino acid or its hydrochloride and an exact amount (ca. 1.00 mg) of phenanthrene as an internal standard in the reaction tube.

2. Add 0.2 ml of CH_3CN and 0.2 ml of BSTFA.

3. Securely cap the tube and place it in the oil bath just to the liquid level.

4. Heat at 125 °C for 15 min, remove the tube and allow to cool.

5. Wipe the exterior of the tube with a hexane-dampened tissue to remove the oil film.

The column used for the gas chromatography separation was 1 m × 4 mm I.D. borosilicate glass packed with 3% DC-550 on 80/100 mesh H.P. Chromasorb G. The N_2 flow was 70 cm³/min and the temperature was programmed from 90 °C to 230 °C at 5 °C per min. Using this method, 18 protein derived amino acids were reproducibly obtained by silylation at 125 °C for 15 min in a closed tube. Cystine was obtained by silylation at 150 °C for 30 min. The silylation of ainine did not give reproducible results.

TMS derivatives of amino acids promise a reproducible, simple and fast analysis of protein amino acids by gas chromatography and there is much hope that problems that still exist will be solved in the very near future.

2. N-Trifluoroacetyl-n-butyl Esters

After several years of concentrated studies in the analysis of amino acids by LAMKIN and GEHRKE [592], GEHRKE *et al.* [340], STALLING *et al.* [994], STALLING

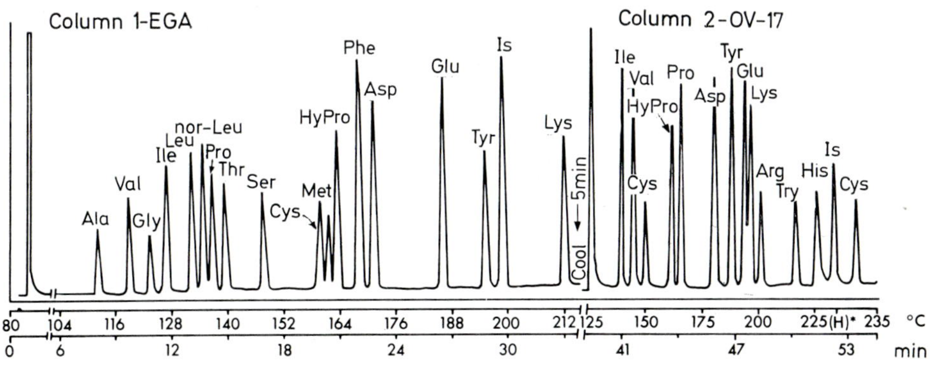

Fig. 5-5. Blood plasma amino acids (as T.F.A. derivatives) Column: 5′ × 4 mm glass. 0.325% E.G.A. on 80/100 a.w. Chromosorb G. Detector: flame ionization. Inlet temperature: 180 °C. Detector temperature: 200 °C. Programmed from 75 °C to 190 °C at 4 °C/min. N_2 flow: 30 cm³/min. Air: 400 cm³/min. H_2: 35 cm³/min. Electrometer range: 1×10^{-9} A.F.S. Packard G.L.C. Model 7301

and GEHRKE [995], and GEHRKE and SHAHROKHI [341], the N-trifluoroacetyl, n-butyl ester derivative was found to be the most promising of the required amino acids derivatives used in gas-liquid chromatography. In 1967, GEHRKE and STALLING

[342] reported a complete quantitative anaylsis of 20 natural protein amino acids with gas-liquid chromatography techniques. The sample preparation and the derivatization are described in this report in detail. The procedure is as follows:

The proteins are hydrolyzed by chemical or enzymic means and prepared for derivatization. After the water is removed to guarantee that the amino acids are dry, esterification of the amino acids to form the corresponding hydrochloride

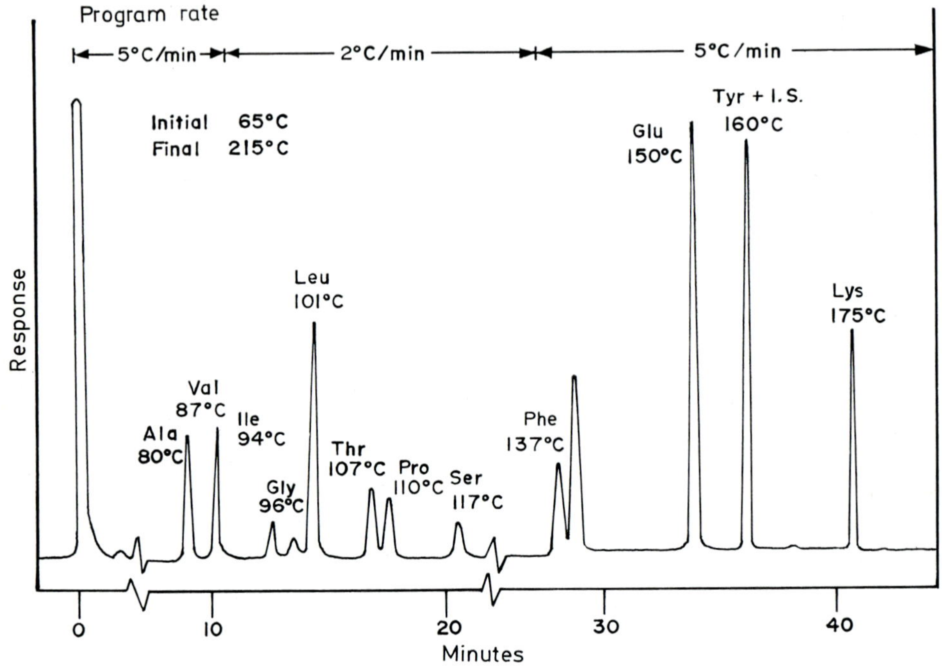

Fig. 5-6. Separation of all twenty protein amino acids with a dual column system. Column 1: 0.325 w/w% Ethylene glycol adipate on 80/100 mesh a.w. H.T. Chromosorb G, 1.5 m × 4mm I.D. glass, initial temp. 80 °C, program 4 °C/min, N_2 flow 60 m1/min Column 2: 1.5 w/w% OV–17 on 80/100 mesh H.P. Chromosorb G, 1.0 m × 4 mm I.D. glass, initial temp. 125 °C, program 6 °C/min, N_2 flow 70 m1/min. The injected mixture contained ca. 0.02 μMol of each amino acid (~ 2 μg). * Hold at 235 °C

methyl esters is carried out. This step is necessary to achieve solubility of certain amino acids in the n-butanol.

The ester is then treated with n-butanol in the presence of hydrochloric acid to form the n-butyl ester hydrochlorides. This is required to minimize volatization and provide good separation in the gas-liquid chromatograph column.

The final step is the acylation with trifluoroacetic anhydride to form N-trifluoroacetyl, n-butyl esters and a sample injected into the gas-liquid chromatography column. The amount of each amino acid injected can vary, but is usually in the range of a few micrograms.

Fig. 5-5 shows the separation of blood plasma amino acids in the form of their N-trifluoroacetyl, n-butyl ester derivatives. Fig. 5-6 presents a separation of all 20 protein amino acids, as carried out by GEHRKE *et al.* [339]. The reliability of this method is shown by the data in Table 1 which compares the gas chromatographic technique with the ion exchange method. The peak areas evaluated from the chrom-

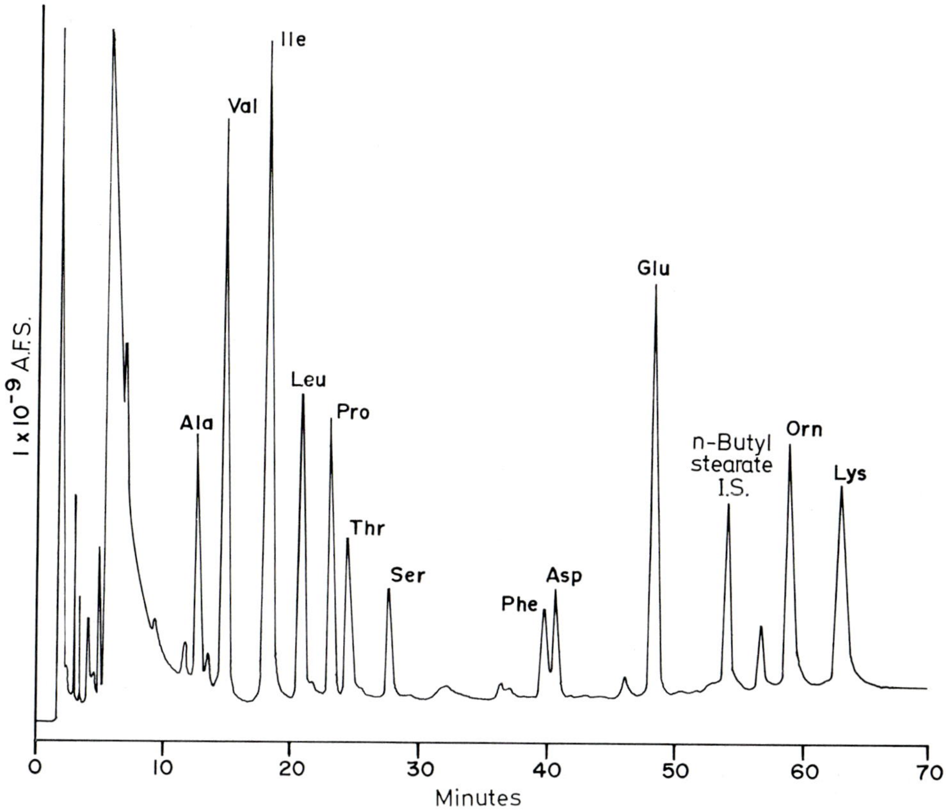

Fig. 5-7. Chromatogram of bovine serum albumin hydrolysate. Sample size: 5 μl (25 μg total amino acids). Column: 1.00 m × 4 mm I.D. borosilicate glass packed with 60/80 mesh a.w. Chromosorb W and mixed substrate phase of 0.75/0.25 w/w% of DEGS/EGSS–X

atogram in Fig. 5-7 were compared with previously prepared calibration curves for amino acids standards and the results presented in the table as compared with the results obtained with the ion exchange method.

The average yields of all derivatives exceed 97%, thereby meeting one of the primary requirements of amino acid derivatives. The results obtained from gas-liquid chromatography compared with ion exchange methods agree and prove that there is no doubt about the capability of the gas-liquid chromatography technique for the determination and evaluation of amino acids.

Table 1. *Amino acid analysis of bovine serum albumin*[a, b]

Amino acid	Gas chromatography[c]			Av.	Ion exchange[d]
Alanine	5.71	5.53	5.67	5.67	5.64
Valine	5.55	5.69	5.50	5.58	5.44
Isoleucine	2.29	2.32	2.22	2.28	2.38
Glycine	1.63	1.53	1.66	1.61	1.70
Leucine	11.61	12.26	11.85	11.90	11.68
Threonine	5.43	5.31	5.30	5.34	5.46
Proline	4.79	4.71	4.72	4.74	4.78
Serine	3.81	4.11	3.65	3.86	4.05
Methionine	0.71	0.68	0.59	0.66	0.70
Phenylalanine	6.47	6.50	6.37	6.44	6.42
Aspartic acid	10.39	10.36	10.19	10.31	10.27
Glutamic acid	17.04	16.75	16.70	16.83	16.43
Tyrosine	5.21	4.97	5.33	5.17	5.24
Lysine	12.31	12.21	12.03	12.18	12.16
Arginine	5.55	5.11	5.40	5.35	5.69
Histidine	3.38	3.44	3.30	3.38	3.64
Cystine	5.37	5.45	5.40	5.41	5.65
				106.71	107.33

[a] Adapted from Gehrke & Stalling [342].
[b] Protein hydrolyzed for 18 h at 105 °C in a sealed tube with constant-boiling HCl.
[c] n-Butyl stearate used as internal standard.
[d] Each value represents an average of two independent determinations, norleucine as internal standard.

IV. Discussion and Comments

As mentioned earlier, a complete and detailed procedure for analysis of amino acids by gas-liquid chromatography is described by Gehrke [342]. A more detailed set of instructions for the routine analysis of amino acids by gas-liquid chromatography is to be published shortly by Gehrke and coworkers.

Fig. 5-5 demonstrates the separation of several amino acids with one column programmed from 75 °C to 200 °C over a period of 70 min. Although the amino acids are shown clearly, some of them are not completely separated. For accurate peak area evaluation, improved resolution is required. Such improved resolution is indicated in Fig. 5-6, using a dual-column system. Here, the same sample is injected into each column, the second injection being made after the results are obtained from the first injection. Due to the difference in packing materials in the columns, some amino acids will be completely separated by one column and the remaining amino acids completely separated by the second column. Thus, all 20 amino acids can be completely separated and analyzed within 55 min.

The following points should be noted about the type of gas-liquid chromatograph needed for amino acids analysis. For quantitative analysis, high sensitivity is not required, since the quantities will vary in the microgram level. Flame ionization detectors may be used satisfactorily at this level. The electrometer range at this level may vary slightly from 10^{-9} A full scale and, therefore, extremely sensitive electrometers (usually defined as noise-free at 10^{-12} A full scale) are not required. However, they are required to detect smaller amounts as in qualitative analysis.

The dual channel gas-liquid chromatograph for baseline compensation during programming is not necessary, but highly recommended. Good results can be achieved with a single channel gas chromatograph. In the range of 10^{-9} A full scale, baseline drift during programming will not be critical. It may be found, however, that baseline compensation is favorable, especially for slightly higher sensitivities.

From Fig. 5-6, it appears that a dual-column, dual-detector system is definitely desirable to achieve complete resolution To utilize one electrometer and one recorder for this purpose, the output of each detector can be connected to a common electrometer input.

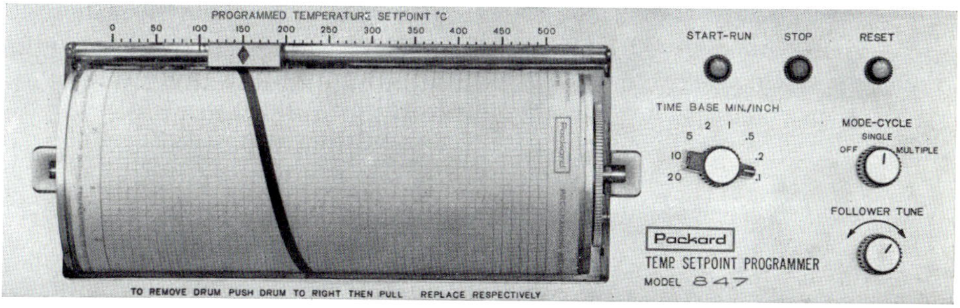

Fig. 5-8

Generally, linear temperature programmers are used for programming the temperature. However, a unique curve-follower type of temperature programmer is known that can provide variable programming rates during temperature programming, in addition to automatic cooling to a desired temperature at a desired rate. Such a programmer is shown in Fig. 5-8. Here, a standard sheet of linear graph paper, wrapped around a metal cyliner, provides a 10-in usable surface. The desired temperature program is "drawn" or plotted on the graph paper with narrow strips of black tape. A beam of light is made to follow one edge of the tape or curve at any one of 8 speeds. The light beam is associated with a mechanical linkage which adjusts the temperature in accordance with the curve. Since this programmer permits increasing and decreasing column temperature at any rate and at any time, it is very suitable for routine analysis of the type used for Fig. 5-6.

A low-mass air oven is preferred for accurate temperature programming. Adequate air circulation is required to minimize the temperature gradient in the oven. Both column and detector ovens should be controlled very accurately. Glass columns 5′ × 4 mm. should be used. Preparation of column packing and operation conditions are described in [342].

In using the gas-liquid chromatography technique, it should be noted that many factors can affect the results. Satisfactory results can be achieved by patient and careful search for the optimum operating conditions. The gas-liquid chromatograph should not be treated only as a tool, nor should it be expected to respond the same way for any analysis. To achieve the best results, complete familiarization with the instrument is essential.

V. Summary

Although amino acids analysis by gas-liquid chromatography has proven to be reliable, this technique is not yet widely accepted for routine amino acids determination. Because of the relative complexity in the preparation of derivatives, although they are reliable, many scientists still refrain from replacing the ion exchange method by gas-liquid chromatography.

However, this technique is gaining more acceptance by laboratories for amino acids research. Work is being continued to find a simpler and faster way to prepare derivatives. One might expect to see the appearance of a considerable number of reports, such as the forthcoming by GEHRKE, which will be a substantial contribution towards this effort. It is believed that the day in which a fully automatic method for derivative preparation, followed by automatic injection to the gas-liquid chromatograph, with the results automatically computerized, is not far from being achieved. This will supply scientists in all laboratories with an inexpensive, fast, and automatic device for routine analysis of amino acids.

Chapter 6

Fragmentation of Proteins for Sequence Studies and Separation of Peptide Mixtures

CHARLES B. KASPER

I. Introduction

When a protein has been judged homogeneous and the end groups and amino acid composition are known, the next step in the elucidation of the covalent structure is the degradation of the macromolecule into low molecular weight peptides which can be completely characterized with respect to composition and sequence. The entire composition of the protein should be accounted for by the summation of unique peptides. Once this is achieved, the sequenced peptides must be arranged in the same order as they occur in the intact protein. No single technique is ideally suited to accomplish this task; therefore, a combination of enzymic, or chemical and enzymic techniques, is required. The methods discussed in this chapter deal primarily with the fragmentation of proteins and the fractionation of the resulting peptide mixtures. With the exception of partial acid hydrolysis, the degradative procedures have been restricted to those which exhibit a reasonable degree of specificity and which are most often used to obtain sets of peptides necessary for the elucidation of the primary structure of a protein. Consequently, certain proteases, such as subtilisin, papain, pronase, and elastase which are extremely valuable for the hydrolysis of peptides, are discussed elsewhere in this book.

II. Non-Enzymic Cleavage of Peptide Bonds

As the structure of an increasing number of large proteins is examined, the need for methods capable of selectively cleaving the polypeptide chain at specific sites becomes a necessity. A variety of non-enzymic methods which involve peptide bonds formed by specific amino acids have been reviewed by WITKOP [1143]. N-Bromosuccinimide has been shown to cleave tryptophanyl bonds in proteins [777, 813, 846]; however, oxidation will frequently occur without peptide bond cleavage. Tyrosyl and histidyl bonds are also sensitive to this reagent, and the peptide chain is often broken at positions where these residues occur. A procedure for the selective cleavage of tyrosyl bonds in the presence of tryptophan [801] is based on the prior conversion of tryptophanyl residues to N-formylkynurenine by ozonolysis before treatment with N-bromosuccinimide. Undesirable side reactions, such as oxidation of sulfhydryl groups and disulfide bonds, are also noted with this reagent. A second chemical modification, which results in the cleavage of cysteinyl bonds, has been described [981]. This method requires S-dinitrophenylation, intermediate formation of a dehydroalanyl derivative, and mild hydrolysis with dilute HCl to cleave the peptide chain. Still another procedure has taken advantage of the ease of reduction of tertiary amides. This has resulted in the preferential cleavage of acyl-proline bonds in proteins and peptides by a variety of reducing agents [66, 67, 773, 842]. The above methods suffer from one or more of the following disadvantages: (a) low cleavage yields, (b) lack of specificity, (c) undesirable side reactions, and (d) wide variability in the reactivity of the sensitive bond. Because of these factors and since these methods have not been developed to the stage where they may be routinely applied in a predictable fashion to the chemical fragmentation of proteins, they will not be considered at this time. The most specific and generally applicable method available is the cyanogen bromide cleavage of the peptide chain at methionyl residues [378].

A. Cyanogen Bromide

Cyanogen bromide cleavage has been employed with excellent results in studies related to the elucidation of the primary structure of several proteins e.g., ribonuclease [378], myoglobin [273], azurin [24], cytochrome *c* [203, 431], subtilisin BPN′ [645], extracellular nuclease of *Staphylococcus aureus* [1039], and trypsinogen [457]. This method owes its success to its specificity for methionyl residues and the fact that methionine is usually rare in proteins. This increases the probability of obtaining large peptide fragments which provide valuable information important either for the correct positioning of peptides derived from enzymic digests or for the confirmation of previously established sequences. Other applications of this reaction include the determination of the identity or non-identity of polypeptide chains [1001], degradation of γ-globulin to obtain immunologically active fragments [174], and the location of oxidized methionyl residues [574].

1. General Properties of the Reaction

Cyanogen bromide reacts with the sulfur in the thioether side chain of methionine to yield a mixture of homoserine and homoserine lactone plus methyl thiocyanate. If this reaction occurs when methionine is in peptide linkage, the bond involving the carboxyl group of methionine is cleaved (Eq. 1). All fragments produced by the action of cyanogen bromide will contain C-terminal homoserine or its lactone except for the C-terminal peptide of the protein (unless methionine is C-terminal). Under acidic conditions, cysteinyl side chains are also attacked but peptide bond cleavage does not result. This reaction can be prevented by carboxymethylation or aminoethylation of the thiol prior to incubation with cyanogen bromide. Repetitive sequences of methionine, such as the methionyl methionine sequence in ribonuclease [378], and N-terminal methionine yield free homoserine lactone. If the methionyl side chains in a protein or peptide become oxidized to the sulfoxide or sulfone, reaction with cyanogen bromide will not occur. The presence of these oxidation products can be detected in acid hydrolysates by amino acid analysis; however, the sulfoxides are capable of reverting to methionine during the hydrolysis stage [814],

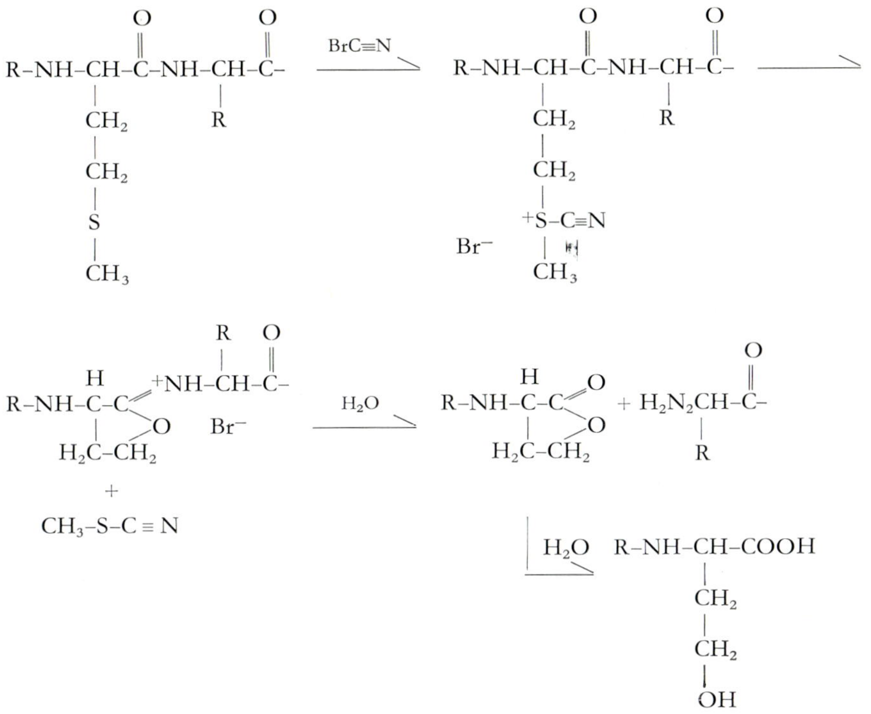

(1)

and the result may be incorrectly interpreted as incomplete reaction at certain methionyl residues.

2. Method

The cleavage reaction has been successfully performed in either 0.1 N HCl, 70% formic acid, or anhydrous trifluoroacetic acid. A difference in the extent of cleavage has been noted when using the first two solvents; in the case of β-galactosidase, 80% of the susceptible bonds were cleaved in 0.1 N HCl while 96% cleavage was achieved in 70% formic acid [1001]. This concentration of formic acid has also been found to be optimal for the fragmentation of subtilisin BPN′ (MW 27,500; 5 methionyl residues) [645] and Staphylococcus nuclease (MW 17,000; 4 methionyl residues) [1039]; greater than 90% cleavage was noted in each case. Formic acid has the decided advantage of being a better solvent than HCl for both proteins and peptides.

Cyanogen bromide[1] may be obtained from Eastman Organic Chemicals (Rochester, New York) as the analytical grade reagent and used directly. The general procedure involves dissolving or suspending the protein (approximately 10 mg/ml) in the appropriate acid solution and adding a weighed amount of the solid cyanogen bromide directly to the protein solution. Cyanogen bromide is usually present at a 30 to 100-fold molar excess with respect to methionine. A vial or tube which can be tightly stoppered makes a suitable reaction vessel. If the protein is not completely soluble, an even suspension should be maintained by using a magnetic stirrer or

[1] Caution should be exercised when using this poisonous chemical. All operations involving handling of the solid should be carried out in a well ventilated hood.

gentle shaking until solubility is achieved. After a reaction period of approximately 24 h at room temperature, excess reagent and by-products are removed by lyophilization or gel filtration. To illustrate this technique, two examples are presented with an indication as to the methods used to fractionate the resulting fragments.

a) β-Galactosidase [1001]

The lyophilized, reduced carboxymethylated protein is dissolved in 70% formic acid, in which it is readily soluble at a concentration of 1 to 5 mg/ml. A 50-fold excess of crystalline cyanogen bromide is added and the reaction allowed to proceed for 16 to 20 h at room temperature in a tightly stoppered flask. After cyanogen bromide cleavage, the reaction mixture is diluted approximately 10-fold with distilled water, shell frozen, and lyophilized to dryness. The residue is dissolved in a minimum volume of 0.01 N NaOH and applied directly to a 4 × 225 cm jacketed column of Sephadex G-75 equilibrated with 0.2 M NH_4HCO_3, pH 8.6.

b) f_2 Bacteriophage Coat Protein [569]

Protein prepared by the hot KCl treatment is freeze-dried and dissolved in anhydrous trifluoroacetic acid. A 50-fold excess of cyanogen bromide is added and the reaction allowed to stand at 25 °C for 16 h. The trifluoroacetic acid is removed and 8 M urea containing 0.05 M Tris-HCl buffer, pH 8.0, is added to the residue. The suspension is centrifuged and the supernatant solution is chromatographed on DEAE-Sephadex A-25 equilibrated with the urea containing buffer. The recovery of peptides from 8 M urea is accomplished by dialysis against 4-h acetylated 23/32 Visking membranes.

The fractionation of cyanogen bromide cleavage products by such procedures as paper electrophoresis and ion exchange chromatography is complicated by the fact that a peptide may exist in two distinct but interconvertible forms which differ only with respect to the presence or absence of a free terminal carboxyl group. This problem is circumvented by converting all C-terminal homoseryl residues to the lactone by treating the peptide residue obtained after lyophilization with anhydrous trifluoroacetic acid at 20 °C for 1 h [22]. Homoserine lactone peptides can be purified by paper electrophoresis without detectable changes during the run. In peptides, the conversion of the lactone to homoserine occurs more slowly than in the case of free homoserine lactone. For example, incomplete conversion is noted even after 12 h at 37 °C in 2 M ammonium hydroxide [22].

In order to determine the extent of cleavage, an aliquot of the digest, after removal of excess reagent, is hydrolyzed with 6 N HCl at 110 °C for 24 h in an evacuated ($< 50\ \mu$ Hg) sealed tube, and homoserine and homoserine lactone are quantitated. Homoserine lactone elutes from the 5 cm column (PA-35 Beckman resin) between the ammonia and arginine peaks while homoserine co-elutes with glutamic acid on the 60 cm column (AA-15 Beckman resin) with the pH 3.25 buffer. To separate homoserine from glutamic acid, the column is equilibrated with the usual pH 3.25 buffer, but development is started with pH 2.80 buffer (prepared by addition of acid to the pH 3.25 buffer) [908]. Under these conditions homoserine elutes approximately 15 ml before glutamic acid. Homoserine lactone may be hydrolyzed to the free acid by dissolving the residue of amino acid hydrochlorides in a small amount of pyridine acetate buffer (100 ml pyridine + 4.0 ml glacial acetic acid/1000 ml, pH 6.5) and heating at 105 °C for 1 h [22]. The volatile buffer is removed by rotary evaporation, and the sample is dissolved in the sample dilution buffer and immediately applied to the column of the amino acid analyzer. The recovery of homoserine is about 90% (uncorrected for hydrolytic losses).

B. Partial Acid Hydrolysis

The hydrolysis of proteins by acid under conditions which do not cleave all peptide bonds has proven to be a useful method in the study of protein structure. Basically two methods are employed: (a) partial hydrolysis with concentrated or constant boiling hydrochloric acid and (b) hydrolysis in dilute acid solution.

1. Specificity

The initial stages of partial hydrolysis with concentrated acid are characterized by a rapid cleavage of peptide bonds while the remainder of the breakdown becomes

Table 1. *Hydrolysis of dipeptides in acid solution* [1029]

Peptide	Relative rate of hydrolysis (Gly-Gly = 1)
Gly-Gly	1
Gly-Ala	0.62
Ala-Gly	0.62
Gly-Leu	0.40
Gly-Tyr	0.35
Gly-Val	0.31
Leu-Gly	0.23
Leu-Leu	0.048
Leu-Tyr	0.041
Val-Gly	0.015

The kinetic study was performed in a solution containing equal volumes of 10 N HCl and glacial acetic acid at 37 °C

progressively slower as free amino acids and small peptides (di- and tripeptides) accumulate in the reaction mixture [167, 1004]. Thus, the complexity of the hydrolysate in the early phase is considerably greater than at the end of the reaction; consequently, examination of the peptides near the terminal stage of hydrolysis provides the most useful information.

Various factors, such as electrostatic and steric effects, influence the rate at which a particular peptide bond is hydrolyzed. The presence of charged groups in the vicinity of the amide bond affects the approach of hydrogen ions and therefore the rate of hydrolysis. In strong acid, the carboxyl groups are unionized and the only charged species are the protonated forms of the α- and ε-amino, guanido, and imidazole groups. Positively charged groups would be expected to repel hydrogen ions and therefore decrease the rate of cleavage of peptide bonds in the immediate vicinity. Of the basic groups mentioned, only the α-amino group appears to retard hydrolysis, presumably owing to its proximity to the susceptible bond. Failure of the lysyl, arginyl, and histidyl side chains to behave in a similar fashion [362] is probably related to their increased distance from the peptide bond which would greatly diminish any electrostatic influence. The rates of hydrolysis of various dipeptides relative to glycyl-glycine are presented in Table 1. It is clear that as the bulk of the side chain increases the rate of hydrolysis decreases. The fact that glycyl-valine is less stable than valyl-glycine suggests that the bond formed by the carboxyl group of the branched chain amino acids is most resistant to hydrolysis. It is evident from a comparison of the relative rates of hydrolysis of leucyl-glycine and valyl-glycine that, when branching of the side chain occurs close to the polypeptide backbone, the reduction in rate is pronounced. Peptide bonds formed between two β-branched amino acids (e.g., Val-Ile, Ile-Ile) are extremely difficult to hydrolyze and probably form the most acid stable type of peptide bond. It is not uncommon for such bonds to require up to 60 to 70 h in 6 N HCl at 110 °C in order to obtain a quantitative cleavage. The generally accepted explanation for this phenomenon is that the approach of the hydronium ion to the peptide bond is hindered by steric factors.

Peptide bonds involving the amino groups of the aliphatic β-hydroxyamino acids, serine and threonine, are quite susceptible to hydrolysis by concentrated acid at room temperature. This is presumably due to the following sequence of reactions (Eq. 2): (a) intermediate formation of an oxazoline derivative, (b) hydrolysis of the $-\underset{|}{C}=N-$ bond to yield the ester, and (c) cleavage of the ester bond to sever the peptide chain [239].

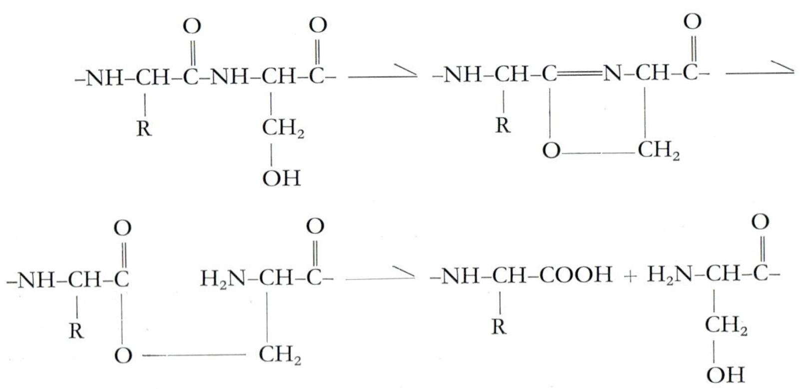

(2)

Although hydrolysis with concentrated acid has been more widely used, dilute acid hydrolysis exhibits a different type of specificity and is occasionally useful in obtaining additional information about the structure of a protein or peptide.
Partridge and Davis [767] first demonstrated that when a protein was refluxed with dilute acetic or oxalic acid, aspartic acid was preferentially released. Furthermore, the specificity is a function of the pH rather than the nature of the acid employed

[100]. This opened the possibility that a polypeptide chain could be cleaved specifically by the excision of aspartyl residues leaving other peptide bonds intact. Application of this method to proteins has resulted in the expected preferential release of aspartic acid [365, 663, 910]; however, hydrolysis also occurred at other sensitive bonds involving predominantly glycine, serine, and threonine. In fact, certain bonds formed by aspartic acid are more resistant to hydrolysis than others; this suggests that the rate of release can be markedly affected by neighboring side chains [910]. Dilute acid hydrolysis of small peptides containing aspartic acid appears to proceed in a more specific manner [468, 909, 1069], since fewer secondary points of cleavage are noted. In both peptides and proteins, asparaginyl residues are converted to aspartyl residues with the subsequent release of aspartic acid.

2. Side Reactions

Certain acid-catalyzed reactions may adversely affect the results obtained from a partial acid hydrolysis of a protein or peptide. These include (a) disulfide interchange, (b) diketopiperazine formation, (c) cyclization of amino terminal glutamine to pyrrolidonecarboxylic acid, (d) rearrangement at an aspartic acid residue, (e) destruction of tryptophan, and (f) hydrolysis of asparagine and glutamine to the corresponding carboxylic acids.

The occurrence of disulfide interchange (Eq. 3) during partial acid hydrolysis has been examined in detail by Ryle and Sanger [844].

$$R^1\text{–S–S}^1\text{–R} + R^2\text{–S–S–R}^2 \rightleftharpoons 2\,R^1\text{–S–S–R}^2 \tag{3}$$

Insulin was allowed to hydrolyze in the presence of bis DNP-cystine under a variety of experimental conditions, and disulfide interchange was detected by the incorporation of DNP-cysteine into peptides with the general structure:

$$\begin{array}{l} \text{DNP–HN–CH–CH}_2\text{–S–S–CH}_2\text{–CH–HN–} \\ \qquad\qquad\ \ |\qquad\qquad\qquad\qquad\quad\ | \\ \qquad\quad\ \text{COOH}\qquad\qquad\qquad\quad \text{C=O} \\ \qquad\qquad\qquad\qquad\qquad\qquad\qquad\ | \end{array}$$

The amount of DNP derivative remaining in the aqueous phase after ether extraction was taken as a measure of the interchange reaction. A summary of the results appears in Table 2. No appreciable interchange occurred when insulin was hydrolyzed in the presence of 2×10^{-3}M cysteine for 10 days at 37 °C with 10 N H_2SO_4; the same solution maintained at 100 °C for 2 h also showed little evidence of disulfide interchange. The notable features of this study are (a) the inhibition of the rearrangement by thiols and (b) the finding that the rate of the reaction is markedly reduced in 10 N H_2SO_4. The latter observation may be associated with the rapid increase in the activity of HCl solutions with increasing concentration; the activity coefficient of 10 N HCl is substantially greater than the corresponding value for 10 N H_2SO_4. It should be emphasized that when partial acid hydrolysis is used to obtain cystine-containing peptides, conditions must be employed which will preserve the integrity of the disulfide bridges in the native protein. The recovery of a cystine peptide in high yield is a good indication that it is not a product of an interchange reaction.

Rearrangement at aspartic acid involves the interconversion depicted by Equation 4. SWALLOW and ABRAHAM [1024] first described this reaction and demonstrated in model experiments the interconversion of ε-(α-aspartyl) lysine and ε-(β-aspartyl)

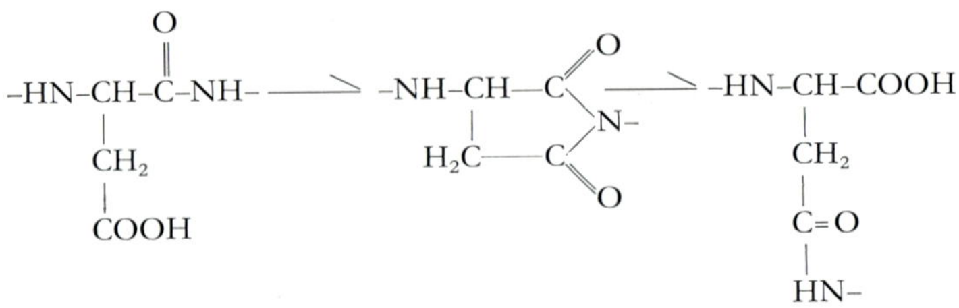

(4)

lysine *via* the α, β-aspartyl derivative. Difficulties due to this reaction were encountered by NAUGHTON *et al.* [721] when investigating the amino acid sequence around the reactive seryl residues of several proteases. Because of the aspartyl interconversion

Table 2. *Disulfide interchange during partial acid hydrolysis of insulin* [844]

Acid	Temperature	Additions	Time	Number of disulfide bonds reacting per molecule of insulin (MW = 6,000)
12 N HCl[a]	37 °C	None	1 h	0.3
			2 h	0.65
			5 h	1.2
5.7 N HCl in 50% acetic acid[a]	37 °C	10^{-3}M cysteine	1 day	0.35
			4 days	1.0
			10 days	1.35
10 N H_2SO_4 in 50% acetic acid[b]	37 °C	2 × 10^{-3}M cysteine	5 days	0
			10 days	0.01
10 N H_2SO_4 in 30% acetic acid[b]	100 °C	None	30 min	0.35
			90 min	0.7
10 N H_2SO_4 in 30% acetic acid[b]	100 °C	5 × 10^{-3}M thioglycollic acid	35 min	0.05
			2 h	0.15

[a] 2 × 10^{-3}M *bis* DNP-cystine; 1 mg insulin/ml.
[b] 2 × 10^{-2}M *bis* DNP-cystine; 10 mg insulin/ml.

reaction, each aspartic acid peptide can occur in three different forms, which greatly increases the complexity of the hydrolysate. Conditions could not be found that completely avoided the α-β-shift; however, the reaction was less marked in 5.7 N HCl than in 12 N HCl. The presence of the α, β-aspartyl or β-aspartyl structure in a peptide renders the peptide bond resistant to hydrolysis by exopeptidases such as leucine aminopeptidase and carboxypeptidase. Furthermore, the Edman degradation is blocked at the cyclization stage by the imide or β-aspartyl structure. Hydrolysis of the imide occurs readily at alkaline pH values to yield predominantly the β-peptide bond [496, 1024].

Diketopiperazine formation is another problem encountered during the course of partial acid hydrolysis. The cyclization of a dipeptide through the elimination of water between the terminal amino and carboxyl groups is favored in dilute acid solution at elevated temperatures [2] (Eq. 5).

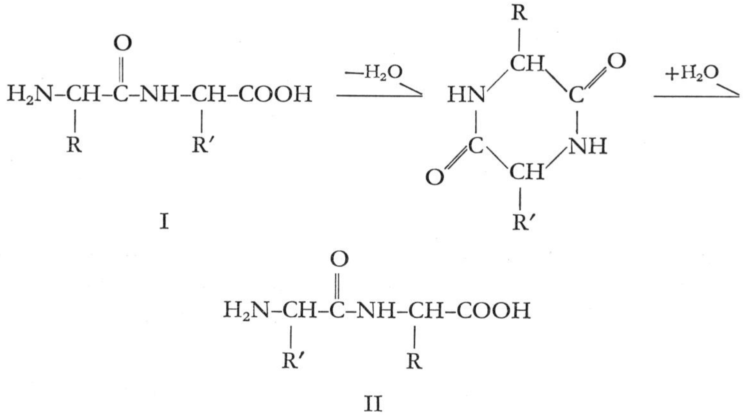

(5)

Hydrolysis of the anhydride may yield either the parent dipeptide (I) or a new dipeptide which is a structural isomer (II). In general, diketopiperazines are less stable to acid hydrolysis than the corresponding dipeptides. The possibility of obtaining an inverted dipeptide sequence under conditions of dilute acid hydrolysis markedly diminishes the reliability of a sequence deduced from a dipeptide isolated from such a hydrolysate. Sanger and Thompson [858] noted that when glycylvaline was treated with 0.1 N HCl under reflux conditions for 24 h the reaction mixture contained both glycyl-valine and valyl-glycine as well as some of the free amino acids. No inversion was detected in 12 N HCl at 37 °C.

A second type of cyclization reaction involves the conversion of amino terminal glutamine to a pyrrolidone carboxyl residue (Eq. 6).

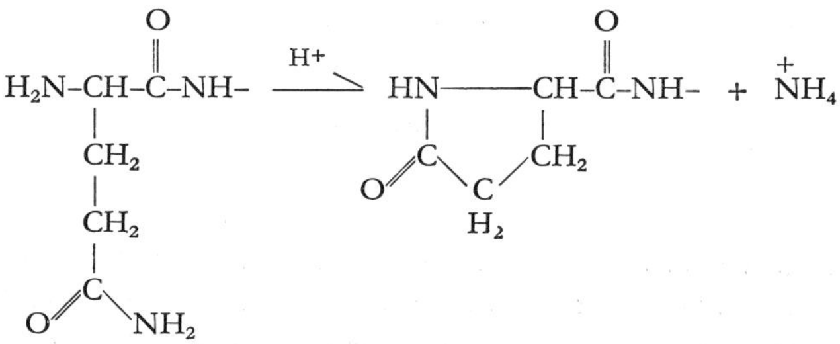

(6)

This reaction occurs in acid solutions and has been found in certain tryptic and chymotryptic peptides of ribonuclease [445b, 974]. Since peptides are frequently exposed to acid conditions, the occurrence of this cyclization reaction must be recognized in order to avoid errors. During the fractionation of the partial acid hydrolysate of papain [663] on Dowex 50–X2 the first component to be eluted was ninhydrin-negative and yielded only glutamic acid on acid hydrolysis. Such properties are typical of pyrrolidonecarboxylic acid. Glutamic acid may also form the cyclic amide, but this conversion requires vigorous conditions generally not employed in protein structure studies.

Side reactions involving the destruction of tryptophan and the conversion of asparaginyl and glutaminyl residues to the corresponding carboxylic acids require that milder degradative techniques be used where these residues are concerned.

3. Method

Various experimental conditions that have been employed for partial acid hydrolysis of proteins are presented in Table 3. Hydrolyses conducted at 100 °C are carried out conveniently in a boiling water bath while a thermostated bath suffices for

Table 3. *Conditions for the partial acid hydrolysis of some proteins*

Acid	Temperature	Protein	Time of hydrolysis	Reference
12 N HCl	37 °C	Insulin	3 days	[859]
		DIP[a]-chymotrypsin	3 days	[721, 811]
		Phosphoglucomutase	30—40 h	[694]
		IMP[b]-trypsin	3 days	[876]
		DIP-trypsin, DIP-elastase	2—3 days	[721]
		Dinitrophenylated subtilisin BPN′	6 days	[575]
		Lysozyme	4 days	[1051]
		Performic acid oxidized papain	7 days	[546]
5.7 N HCl	37 °C	DIP-trypsin, DIP-chymotrypsin, DIP-elastase	3 days	[721]
5.7 N HCl	100 °C	DIP-trypsin, DIP-chymotrypsin, DIP-elastase	20–30 min	[721]
		Phosphoglucomutase	30 min	[694]
2 N HCl	100 °C	Phosphoglucomutase	2 h	[424]
0.03 N HCl	105 °C	Ribonuclease, insulin, glucagon	8–48 h	[365, 910]
0.03 N HCl	110 °C	Reduced-carboxymethylated papain	48 h	[663]

[a] Diisopropylphosphoryl-
[b] Isopropylmethylphosphonyl-

lower reaction temperatures. In either case, a test tube equipped with a ground glass stopper makes a suitable reaction vessel. The use of concentrated HCl requires long term incubation; however, treatment with 5.7 N HCl for 10 min at 100 °C has been reported to be equivalent to a 24-h hydrolysis period with 12 N HCl at 37 °C [721]. The choice of optimal hydrolysis conditions depends to some degree on the sample, and trial experiments with small amounts of protein or peptide are recommended.

The conditions used for dilute acid hydrolysis of proteins and peptides differ from those discussed for concentrated and constant boiling HCl in that the sample is usually hydrolyzed in an evacuated, sealed, heavywalled Pyrex tube at 105 to 110 °C. When very small amounts of material are being hydrolyzed, the sample dissolved in acid may be sealed in a glass capillary tube and heated in an oven for the required period of time. Protein concentrations ranging from 1 mg [910] to 5 mg [663] per ml of acid have been used. Hydrolysates are taken to dryness either in a vacuum desiccator (< 100 μ Hg) over solid NaOH or by rotary evaporation at 40 °C. In order to remove the last traces of HCl, the residue is dissolved in distilled water and re-evaporated. The latter method is rapid and efficient, but if many samples are being processed at one time the use of a desiccator is most convenient. When submicromole amounts of sample are hydrolyzed it is advisable to carry out the hydrolytic and HCl removal steps in the same vessel to avoid loss of material on transferring.

Schroeder *et al.* [909] have employed the following conditions for the release of aspartic acid from peptides derived from the γ-chain of fetal hemoglobin. A solution of several micromoles of peptide in 10 ml of 0.25 M acetic acid was refluxed for 8 h and the solvent then removed by rotary evaporation. Under these conditions, cleavage occurred on both sides of the aspartyl residue, and little or no effect was noted on asparaginyl, glutaminyl, or glutamyl residues. Dilute acid hydrolysis of several tryptic peptides from tobacco mosaic virus protein [1069] was accomplished by dissolving the peptide in 0.03 N HCl (approximately 1 mg/ml) and hydrolyzing in an evacuated sealed tube at 105 °C for the desired time. Aspartyl and asparaginyl bonds were cleaved on both sides with the liberation of free aspartic acid. The release was considerably slower from asparaginyl peptides than from aspartyl peptides indicating that deamidation of asparagine is required prior to the release of the free acid. It has been suggested that dilute acid hydrolysis of peptides may be advantageously conducted in two stages [1069]. The first step would involve a hydrolysis period of 6 h (0.03 N HCl, 105 °C), during which time cleavage would occur at the amino and carboxyl side of each aspartyl residue, and asparaginyl residues would be deamidated. In the second step, the aspartic acid peptides (formerly asparaginyl peptides) would be isolated and subjected to another 6 h hydrolysis period. The net result of this procedure would be the stepwise degradation of a peptide with cleavage first at aspartyl residues followed by cleavage at asparaginyl residues.

Dilute acid hydrolysis (0.03 M HCl, 110 °C, 48 h) of reduced carboxymethylated papain was performed to obtain unique peptides which would permit the unequivocal assignment of overlapping sequences in the native protein [663]. No attempt was made to isolate and purify all of the peptides from the hydrolysate; instead, special attention was devoted to peptides containing arginine and tyrosine which could be readily detected by specific color tests on paper. Kinetic studies on the release of free amino acids indicated that aspartic acid was released more rapidly than any other, but extensive hydrolysis also occurred at peptide bonds involving glutamic acid, glycine, alanine, serine, and threonine. When hydrolysis was performed with 0.01 N and 0.003 N HCl, no change in selectivity was observed.

4. Assessments of the Method

Generally, the peptide mixtures obtained from a protein or large polypeptide by partial acid hydrolysis are exceedingly complex and present special problems in separation and purification. For example, the separated A and B chains of insulin (21 and 30 residues, respectively) yielded numerous di- and tripeptides which were resolved chiefly by preliminary fractionation on charcoal columns followed by paper electrophoretic and chromatographic techniques [859, 862]. Schroeder *et al.* [905] used column chromatography to fractionate the products formed when gelatin was incubated with 3.6 N HCl at 37 °C for 7 days and found predominantly dipeptides and

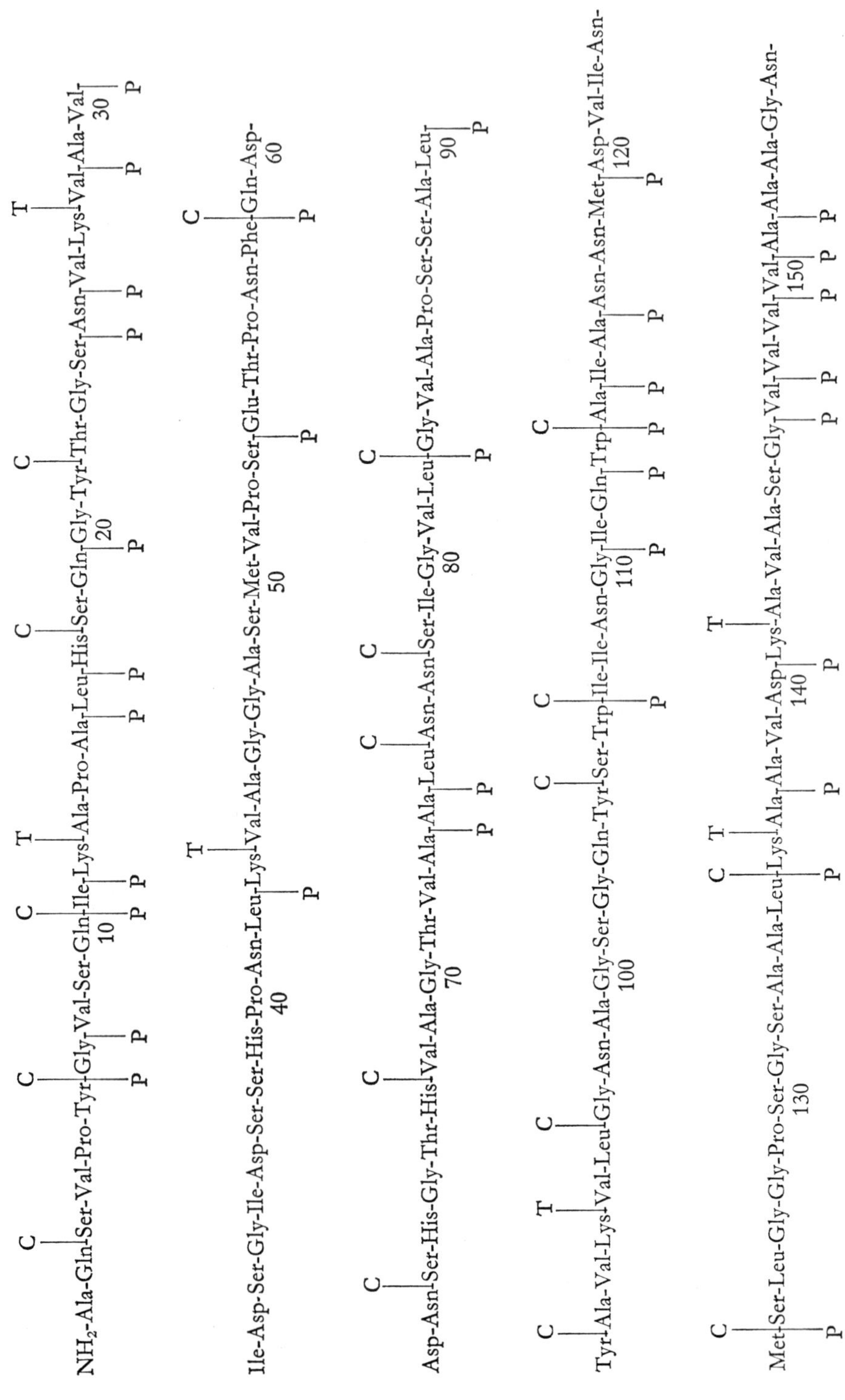
NH2-Ala-Gln-Ser-Val-Pro-Tyr-Gly-Val-Ser-Gln-Ile-Lys-Ala-Pro-Ala-Leu-His-Ser-Gln-Gly-Tyr-Thr-Gly-Ser-Asn-Val-Lys-Val-Ala-Val-
Ile-Asp-Ser-Gly-Ile-Asp-Ser-Ser-His-Pro-Asn-Leu-Lys-Val-Ala-Gly-Gly-Ala-Ser-Met-Val-Pro-Ser-Glu-Thr-Pro-Asn-Phe-Gln-Asp-
Asp-Asn-Ser-His-Gly-Thr-His-Val-Ala-Gly-Thr-Val-Ala-Ala-Leu-Asn-Asn-Ser-Ile-Gly-Val-Leu-Gly-Val-Ala-Pro-Ser-Ser-Ala-Leu-
Tyr-Ala-Val-Lys-Val-Leu-Gly-Asn-Ala-Gly-Ser-Gly-Gln-Tyr-Ser-Trp-Ile-Ile-Asn-Gly-Ile-Gln-Trp-Ala-Ile-Ala-Asn-Asn-Met-Asp-Val-Ile-Asn-
Met-Ser-Leu-Gly-Gly-Pro-Ser-Gly-Ser-Ala-Ala-Leu-Lys-Ala-Ala-Val-Asp-Lys-Ala-Val-Ala-Ser-Gly-Val-Val-Val-Val-Ala-Ala-Ala-Gly-Asn-

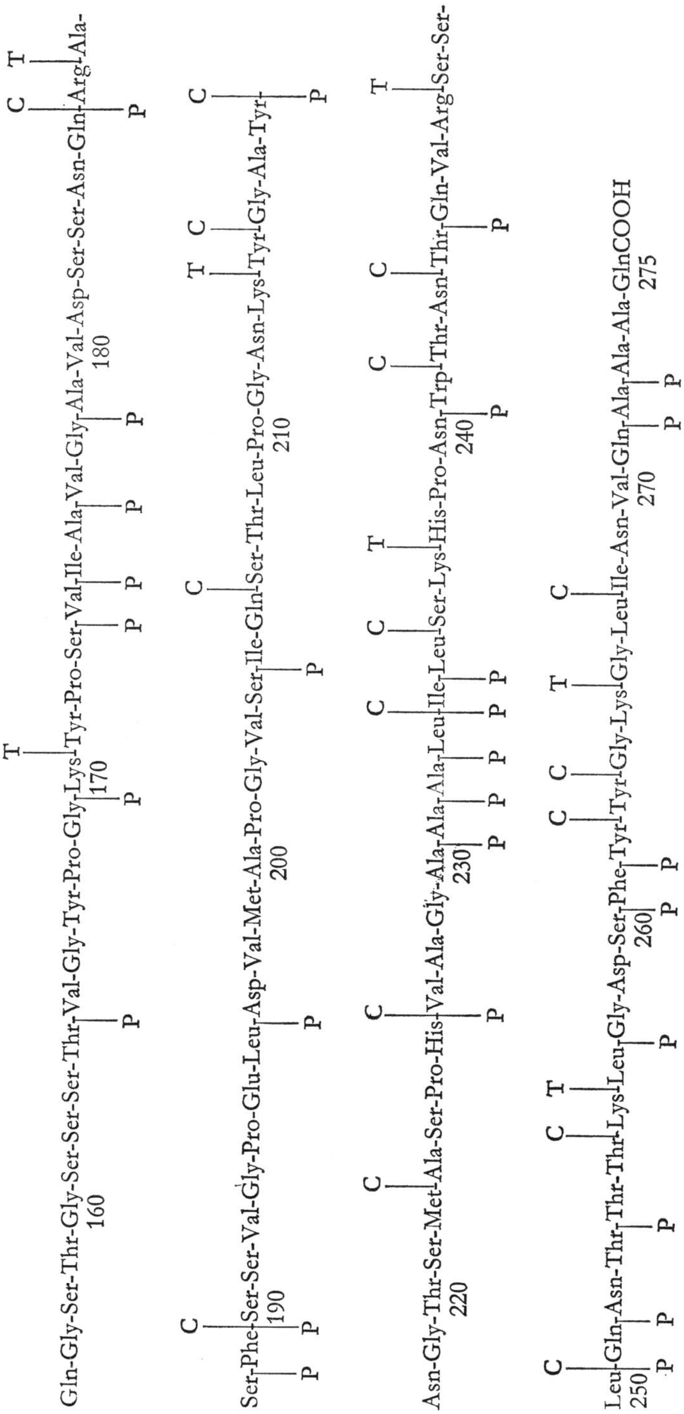

Fig. 6-1. The amino acid sequence of subtilisin BPN′ showing the bonds hydrolyzed by trypsin (T), chymotrypsin (C), and pepsin (P)

free amino acids. The partial acid hydrolysate of lysozyme (12 N HCl, 37 °C, 4 days) also yielded complex mixtures of peptides which did not permit the establishment of sequences beyond the hexapeptide stage [1051]. Although determining the sequence of small peptides is not difficult, limited information is gained regarding the unambiguous positioning of these units in the polypeptide chain.

Dilute acid hydrolysis is not a suitable method for the selective degradation of proteins. The specificity of the method is not as great as the release of free amino acids would suggest, since extensive peptide bond cleavage occurs at numerous sites along the polypeptide chain. In the case of small peptides, hydrolysis at aspartyl residues is reasonably specific, and its use is recommended.

III. Enzymic Degradation

Enzyme catalyzed cleavage of a polypeptide chain offers many distinct advantages over partial acid hydrolysis. In addition to being required in only catalytic amounts, certain enzymes exhibit a high degree of specificity, hydrolyzing only peptide bonds formed by specific amino acids; consequently, the resulting peptide mixtures are comparatively less complex and more amenable to fractionation. Peptides are usually obtained in high yields, and side reactions are negligible. Glutamine, asparagine, and tryptophan are preserved, as are other acid-labile structures such as esters of serine, threonine, and tyrosine. Carbohydrate chains and lipid moieties which are integral parts of some proteins but which may be sensitive to acid are recovered intact. In this section, the three enzymes most commonly used to obtain sets of peptides for the determination of the amino acid sequence of proteins are discussed. For the purpose of illustration, the primary structure of subtilisin BPN′ [645] is presented and the sites of attack by trypsin, chymotrypsin, and pepsin are indicated (Fig. 6-1).

A. General Considerations

The exact conditions for digestion depend upon the enzyme employed, but the following factors are important considerations irrespective of the type of digest: (a) pH, (b) temperature, (c) enzyme to substrate ratio, and (d) time of incubation. The optimal conditions for complete digestion should be determined by performing trial digests using sub-micromole amounts of protein. Examination of the digest by a two-dimensional separation technique ("fingerprint") provides useful information regarding the number of basic, neutral, and acidic peptides as well as the relative complexity of each of these electrophoretic groups. Additional information may be obtained by staining the peptide map with reagents specific for certain amino acids, e.g., the Ehrlich (tryptophan), Pauly (tyrosine and histidine), Sakaguchi (arginine), platinic acid (methionine and cysteine), and nitroprusside (cystine) stains (see Chapter 7).

Frequently the denatured protein is insoluble under the initial hydrolysis conditions, but the suspension will generally clarify during the early stages of digestion. It is desirable to obtain a suspension of finely divided particles prior to the addition of enzyme; this may be easily accomplished by either trituration or sonication. Whenever possible, it is advantageous to perform the digestion in either a volatile buffer (e.g., ammonium bicarbonate or trimethylammonium acetate), so that salts can be removed by lyophilization, or water adjusted to and maintained at the correct pH by the addition of titrant. With the latter system lyophilization of the digest is not required prior to fractionation, since salts are not present at a level which would interfere with either column chromatography or preparative paper separation methods. When digestion is carried out for an extended period of time, it is advisable to add thimerosal (1:5000) or a crystal of thymol to discourage the growth of microorganisms. Hydrolysis is terminated by the addition of a specific enzyme inhibitor, freezing, or a pH adjustment.

B. Preparation of a Protein for Enzymic Digestion

In general, native proteins are not good substrates for extensive proteolytic degradation. Peptide bonds on the surface of the three-dimensional structure of the protein in contact with the external environment are accessible and may be susceptible to enzymic attack, but peptide bonds buried in the protein matrix are not readily hydrolyzed. Since the main objective in structural investigations is to cleave as quantitatively as possible all sensitive bonds, denaturation of the substrate is a prerequisite for optimal proteolytic degradation. Denaturation disrupts the native secondary and tertiary structure of the polypeptide chain so that its conformation approaches that of a random coil.

Of the methods available, heating and trichoroacetic acid precipitation are simple and effective treatments for preparing protein substrates. When the native structure of the protein is stabilized by disulfide bridges, these methods are inadequate, and it is first necessary to cleave the –S–S– bonds by either a reductive or oxidative procedure. When reduction is used the newly formed cysteinyl side chains must be converted to a stable derivative for further manipulation. Reduction followed by S-carboxymethylation has been used in structural studies involving papain [622], lysozyme [177], f_2 bacteriophage coat protein [569], and Bence Jones proteins [439b]. The oxidative procedure, on the other hand, yields stable derivatives in a single step. Treatment of the intact protein with performic acid converts cysteine and cystine to cysteic acid, and methionine is oxidized to the stable sulfone. This method has been applied to insulin [857], bovine ribonuclease [445c], papain [544, 622], ribonuclease T-1 [1033] and azurin [24]. It should be emphasized that oxidation destroys tryptophan and also renders the protein resistant to degradation by cyanogen bromide. Procedures such as S-carboxymethylation and performic acid oxidation introduce charged groups into the protein molecule, thereby increasing the solubility properties of the peptides derived from those modified regions of the polypeptide chain. The derivatized protein is frequently insoluble in non-urea containing buffers because of the denaturing conditions employed in its preparation.

1. Heating

A Pyrex test tube containing the solution of native protein (approximately 1% w/v) is placed in a boiling water bath for usually 2 to 10 min. The tube is then transferred to an ice bath and the contents cooled to 0 °C. Frequently, the heat denatured protein will be insoluble and may be isolated by centrifugation, but it is most convenient at this stage to suspend the protein and proceed with the enzymic digestion. Care should be taken to avoid pH extremes during the heating process so as to minimize the possibility of hydrolytic cleavage of peptide bonds as well as oxidation of sulfur containing amino acids. The temperature and time of heating may be varied, depending upon the characteristics of the protein under investigation. Conditions which have proven satisfactory for the thermal denaturation of certain proteins are the following: hemerythrin, 65 °C for 5 min at pH 8 [375]; calf rennin, 100 °C for 15 min in distilled water [312]; phosphoglucomutase, 100 °C for 30 min [693]; hemoglobin F, 90 °C for 4 min at pH 8 [909].

2. Trichloroacetic Acid Precipitation

To a 1% solution of protein, an equal volume of 10% (w/v) trichloroacetic acid is added at room temperature. After 10 to 15 min, the protein suspension is cooled to ice-bath temperature, and the insoluble protein is collected by centrifugation. TCA is removed from the pellet by washing five times with 20 volumes of reagent grade acetone followed by an ether extraction (peroxide free)[2] with similar volumes. Residual amounts of organic solvent may be removed by drying the pellet in a gentle stream of nitrogen. Trichloroacetic acid precipitation is an efficient method for removing bound amino acids and peptides; this is particularly useful when a proteolytic enzyme is being examined and autolysis products are possible contaminants.

3. Reduction and S-Carboxymethylation

The preparation of reduced and alkylated proteins by the method of CRESTFIELD *et al.* [318] is described. To 5 to 100 mg of protein in a 12-ml screw cap vial maintained under a nitrogen barrier, add 3.61 g of deionized, crystalline urea, 0.30 ml EDTA solution (50 mg of disodium EDTA per ml), 3.0 ml of Tris buffer, pH 8.6, (5.23 g of Tris and 9 ml of 1.0 N HCl diluted to 30 ml with water), and finally 0.10 ml of mercaptoethanol (Eastman Organic Chemicals). The solution is made up to a 7.5 ml mark with water, and a solution 8 M in urea and 0.2% in EDTA is used to fill the vial completely. A disk of polyethylene is placed over the top to exclude air and is held in place with the screw cap. After 4 h at room temperature (22 to 25 °C), the contents of the vial are transferred to a 25-ml beaker under a nitrogen barrier. A freshly prepared solution of 0.268 g of iodoacetic acid in 1.0 ml of 1.0 N NaOH is added to the reaction mixture. The iodoacetate added is slightly less on a molar basis than the amount of mercaptoethanol. The -SH groups of cysteinyl residues react most rapidly; since the excess iodoacetate reacts faster with mercaptoethanol above pH 8 than it does with thioether sulfur, alkylation of methionine is minimized. The buffering capacity of the system is adequate to maintain the pH above 8.3.

After the addition of iodoacetic acid, the solution should be protected from light to prevent the formation of iodine which may react with tyrosine, tryptophan, and histidine residues. Fifteen minutes after the addition of iodoacetate, the sample is applied to a 4 × 40 cm column of Sephadex G-75, medium particle size, equilibrated with 50% acetic acid and wrapped in aluminum foil. Fractions of 8 ml are collected at a flow rate of 20 ml/h. Aliquots equivalent to approximately 2 mg of protein are transferred to plastic tubes (115 × 19 mm) in a rack fitting in a heated desiccator [444b]. The acetic acid is removed in 2 h with a water pump and a voltage setting

[2] Ether is shaken with moist $FeSO_4$, distilled, and stored in a ground glass stoppered bottle over dry $FeSO_4$.

such that the temperature of the desiccator wall is 70 °C. Alkaline hydrolysis is performed (Section IV.A.1.d), and the hydrolysates are analyzed by the ninhydrin method (Section III C1). The fractions containing the reduced and alkylated protein are pooled and lyophilized. The protein is stored in the cold.

4. Performic Acid Oxidation

The method as applied to ribonuclease is described [441a]. It is necessary to perform the oxidation on halide-free protein, since free halogen formed in the presence of performic acid results in the production of halotyrosines. The fact that tryptophanyl residues are converted by this oxidizing agent to kynurenine and other degradation products particularly limits the applicability of this method. Removal of halide ion is accomplished by passing a 10% solution of protein in 0.01 M acetic acid over a bed of Dowex 2-X10 in the acetate form equilibrated with the same solvent. One ml of resin bed is used per ml of protein solution. The halide-free preparation is recovered by lyophilization of the filtrate. Performic acid is prepared by adding 0.50 ml of 30% H_2O_2 (reagent Superoxol, Merck) to 9.5 ml of 99% formic acid. The resulting solution is allowed to stand at room temperature for 2 h in a closed vessel prior to use. Solutions of performic acid are not stable and must be prepared each time. In a second flask 200 mg of ribonuclease (approximately 8% moisture) are dissolved in 5.0 ml of 99% formic acid, after which 1.0 ml of anhydrous methanol is added with stirring; the methanol prevents freezing of the solution during the oxidation. The quantity of performic acid used is 12 times that required to transform all of the cystine and methionine to cysteic acid and methionine sulfone, respectively. The protein and performic acid solutions are cooled for 30 min in a bath maintained between −7 °C and −10 °C before mixing the contents of the two flasks. The reaction is allowed to proceed for 2.5 h at the indicated temperature. The contents of the tube are rinsed with 50 ml of ice water into a flask containing 350 ml of water at 0 °C. The diluted solution is shell frozen and the solvents removed by lyophilization. The residue is dissolved in 40 ml of water, transferred to a smaller flask with the aid of an additional 40 ml of water and again lyophilized.

In studies performed on papain [544] the lyophilization step was omitted and the oxidized protein was isolated by the addition of trichloroacetic acid to a final concentration of 10%. The precipitated protein was then washed with 50-ml portions of 10% trichloroacetic acid until the washing gave a negative test for peroxide with KI-starch paper. The precipitate was further washed with three 50-ml volumes of absolute ethanol followed by three 50-ml volumes of ether.

C. Methods of Measuring Hydrolysis

1. Ninhydrin Method [702]

This method is based on the colorimetric measurement of the α-amino groups released during proteolysis. The ninhydrin solution is prepared in a 1-l bottle appropriately equipped so that nitrogen may be continuously bubbled through the solution during the preparation procedure and which permits the transfer of the solution to a dark colored reagent bottle under nitrogen pressure. Dissolve 20 g of ninhydrin and 3 g of hydrindantin in 750 ml of methyl Cellosolve[3] with stirring. Add 250 ml of 4 N sodium acetate buffer (pH 5.5)[4], mix thoroughly, and transfer to a dark bottle for storage in the cold room under a nitrogen atmosphere; not more than a 2-week supply should be prepared at one time.

In order to follow the course of proteolysis, aliquots are removed from the digestion mixture at various periods of time and added to 1.0 ml of ninhydrin solution; this immediately stops the enzymic reaction. The capped tubes are shaken for a few seconds to insure adequate mixing and then heated 15 min (accurately timed) in a covered boiling water bath [699]. Transfer the rack containing the tubes to a pan of cold water and allow the reaction solution to cool to room temperature. Add 5 ml of 50% ethanol to each tube shake vigorously for 30 to 60 sec to air-oxidize the residual hydrindantin. The optical density of each solution is measured at 570 mμ[5]. The rate of hydrolysis may be expressed as the relative increase in ninhydrin color value as a function of time. Throughout the procedure, care must be taken to prevent contamination with ammonia.

2. Titrimetric Method

This procedure takes advantage of the fact that, in the pH range of 7.4 to 9.0, where the titrimetric method has the greatest application, each newly liberated carboxyl group will be fully ionized while the α-amino group will be only partially protonated; thus, a decrease in the pH of the solution will result. Titration of the unbuffered enzyme-substrate solution with standardized alkali permits a direct means of measuring the rate and extent of proteolysis. The alkali uptake is related to the average pK of the α-amino groups and the pH at which the hydrolysis is performed ($\mathrm{pH} = pK + \log[-\mathrm{NH_2}]/[-\mathrm{NH_3^+}]$). This is most easily measured with an automatic titrator such as the pH-stat [489].

The precise titrimetric monitoring of the hydrolytic reaction requires a titration vessel which allows temperature regulation and efficient stirring. When digestions are carried out in the alkaline range of the pH scale, CO_2 from the air should be excluded by continuously passing a gentle stream of nitrogen through the solution; a flat base line must be established before the enzyme is added. When a plateau region is reached, the digestion should be terminated since beyond this point a slow hydrolysis of peptide bonds occurs that usually increases the complexity of the digest and does not improve the yield of major peptides. A typical result is presented in

[3] These reagents are obtainable from Pierce Chemical Co. in a sufficiently pure state so that they may be used without further purification. The methyl Cellosolve (monomethyl ether of ethylene glycol), however, should always be tested for peroxides by adding 2 ml of the solvent to 1 ml of freshly prepared 4% aqueous KI. A colorless solution indicates the absence of peroxides. Methyl Cellosolve is quite toxic and should be handled only in a well ventilated hood. The fact that it is odorless emphasizes this precaution.

[4] Dissolve 2720 g of $NaOAc \cdot 3\,H_2O$ (reagent grade) in 2 l of water; gentle heat may be used to hasten solubility. Cool to room temperature, add 500 ml of glacial acetic acid, and make up to a final volume of 5 l. The pH of the undiluted buffer should measure 5.51 ± 0.03 If a final adjustment of pH is necessary, 5 g of NaOH correspond to about 0.04 pH unit.

[5] Although the color yield will vary with the amino terminal residue of the peptide, the molar absorption coefficient of the colored product of the ninhydrin reaction is close to 1.8×10^4 for a 1 cm light path. This value is useful in determining the size aliquot required for analysis.

Fig. 6-2. If it is assumed that the average *pK* of the liberated α-amino groups is near 7.8, then the consumption of alkali is that expected for the scission of the 13 trypsin-susceptible peptide bonds in the subtilisin molecule.

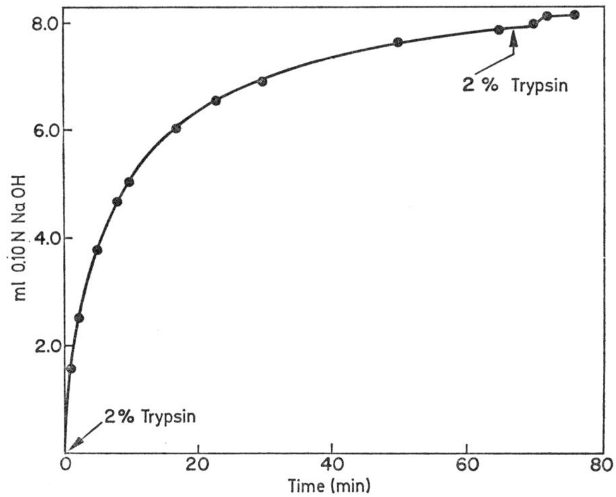

Fig. 6-2. Time course of the tryptic digestion of subtilisin BPN′ (88 μmoles) at pH 8.1. (Reprinted from Ref. [576] by courtesy of The Journal of Biological Chemistry)

D. Trypsin

1. Specificity

Trypsin exhibits the highest degree of substrate specificity known for an endopeptidase; only those bonds involving the carboxyl groups of lysyl and arginyl residues are hydrolyzed. This permits the estimation of the number of tryptic peptides obtainable from a protein when the amino acid composition is known. In theory, this number is equal to the sum of lysyl and arginyl residues present in the molecule plus a C-terminal peptide. The relative rates of cleavage of susceptible peptide bonds are influenced by the chemical nature of the side chains in the immediate vicinity. Generally, the presence of polar groups close to the susceptible bond decreases the rate of hydrolysis. Side chains with a net negative charge (e.g., aspartic acid, glutamic acid, cysteic acid, or S-carboxymethyl cysteine) which are adjacent to a lysyl or arginyl residue tend to reduce significantly the rate of hydrolysis. Thus, in oxidized ribonuclease, the sequence –Arg–$CySO_3H$– is only partially cleaved, as are the –Asp–Lys–Leu and –Asp–Lys–Ala– sequences found in the γ-chain of human hemoglobin. In the copper-containing protein, azurin, incomplete hydrolysis at lysine 92, which is positioned between an aspartic acid and glutamic acid residue, has been noted [24]. Similarly, if the α-amino group of lysine or arginine is free, as is the case for amino terminal lysine in ribonuclease [974] and lysozyme [176b], free lysine is released very slowly. Repetitive sequences of lysine and/or arginine are also attacked by trypsin at a considerably slower rate. Peptide bonds involving the imino nitrogen of proline appear to be totally resistant to tryptic hydrolysis, i.e., Lys–Pro or Arg–Pro. Examples of this type of linkage may be found in ribonuclease [974], corticotropin [615], whale myoglobin [273], lobster glyceraldehyde 3-phosphate dehydrogenase [226], Staphylococcal nuclease [1041], azurin [24], and α MSH [419].

2. Removal of Chymotryptic Activity

Commercial preparations of crystalline trypsin may contain trace quantities (less than 0.5%) of chymotrypsin, which, if present during a prolonged incubation period, could significantly decrease the yield of tryptic peptides. Early attempts to eliminate chymotryptic-like activity from trypsin preparations took advantage of the fact that trypsin is more stable than chymotrypsin when exposed to 8 M urea [414] or dilute acid [816]. The most promising technique, however, involves treating the enzyme with an inhibitor specific for chymotrypsin such as L-(1-tosylamido-2-phenyl)ethyl chloromethyl ketone (TPCK) [888]. Diphenylcarbamyl chloride [288] and phenylmethanesulfonyl fluoride [294] also inhibit chymotrypsin but do not appear to be as effective as TPCK in eliminating chymotryptic activity from trypsin preparations.

Treatment of trypsin with TPCK [576, 1099]: One gram of crystalline trypsin is dissolved in 330 ml of 0.001 M $CaCl_2$, and the pH is adjusted to 7.0 with 1 N NaOH. TPCK solution (30 mg in 6 ml of anhydrous methanol) is added dropwise to the enzyme solution with gentle stirring. The reactants are stirred for 5.5 h at room temperature during which time titrant is added to maintain the pH of the solution at 7.0. The reaction is terminated by the addition of 1 N HCl to pH 3.0. The acidified enzyme solution is dialyzed against 25 l of water (pH 3.0) for 24 h at 4 °C, and the solvent is finally removed by lyophilization. The yield of TPCK-trypsin is approximately 70 to 72% on a weight basis. The treated enzyme retains full tryptic activity as judged by its ability to hydrolyze benzoyl-L-arginine ethyl ester. TPCK-trypsin is still capable of slowly hydrolyzing N-acetyl-L-tyrosine ethyl ester but does not exhibit chymotryptic activity toward peptide substrates.

3. Chemical Modifications Altering Trypsin Specificity

Masking of the ε-amino groups of lysine restricts tryptic hydrolysis specifically to arginyl residues in the polypeptide chain. This may be accomplished by carbamylation [998], carbobenzoxylation [42], dinitrophenylation [816], succinylation [613], guanidination [1107], dithiocarbamylation [673], trifluoroacetylation [353], and maleylation [173]. Of the methods mentioned, trifluoroacetylation and maleylation have two main advantages. (a) They exhibit a high degree of specificity for the ε- and α-amino groups of the protein, and (b) the blocking groups may be removed under mild conditions to regenerate the unmodified amine functions. The advantages of such reactions are obvious. Proteins with blocked ε-amino groups, when digested with trypsin, will yield fragments with C-terminal arginine. These large fragments can be purified, the blocking groups removed, and then further digested with trypsin to cleave at the lysyl bonds. The stepwise use of trypsin in this manner provides the most specific means for the controlled proteolytic degradation of proteins.

In addition to modifications which restrict tryptic hydrolysis, it is also possible to expand the specificity of trypsin by creating new points of cleavage. The S-aminoethylation of cysteinyl residues is an example of this type of alteration.

a) Trifluoroacetylation

The preparation of trifluoroacetylated ribonuclease [353] is presented below. The basic reaction is represented by Eq. (7):

$$F_3C{-}\overset{\overset{\displaystyle O}{\|}}{C}{-}S{-}C_2H_5 + \text{Protein-}NH_2 \underset{\text{1 M piperidine, } 0^{\circ}}{\overset{\text{pH 10, 25 °C}}{\rightleftharpoons}} \text{Protein-}\underset{H}{N}{-}\overset{\overset{\displaystyle O}{\|}}{C}{-}C{-}F_3 + C_2H_5{-}SH \qquad (7)$$

In a 125-ml Erlenmeyer flask 0.5 g of ribonuclease is dissolved in 50 ml of water and the pH of the solution is adjusted to 10.0 with 1.0 N KOH. Base is added from a 25-ml syringe fitted with a No. 19 needle attached to polyethylene tubing. The flask is placed on a magnetic stirrer in a well ventilated hood, and the electrodes of a pH meter are immersed in the protein solution so that the pH can be continuously monitored during the course of the reaction. Ethyl thioltrifluoroacetate (2.5 ml), prepared by the method of HAUPTSCHEIN *et al.* [427] or obtained from the Pierce Chemical Company (Rockford, Illinois), is added to the solution with rapid stirring. The reagent, which is immiscible with water, is divided into small droplets by the revolving magnet, exposing a large surface area for reaction with protein in the aqueous phase. The pH is maintained between 9.95 and 10.00 by the addition of KOH from the syringe. The reaction is conducted at room temperature and allowed to proceed for about 1 h, after which the base uptake (approximately 18 ml) is essentially complete. A large amount of alkali is consumed because of the spontaneous hydrolysis of the thioester. To the reaction mixture is added 5.0 ml of 0.1 M sodium acetate buffer, pH 5.0. The pH is lowered to 6.0 by the addition of a few drops of concentrated HCl; this treatment causes the derivatized protein to precipitate. The reaction mixture is added to 4 volumes of cold (– 20°C) absolute ethanol. The precipitate is collected by centrifugation and washed twice with a mixture of ethanol and sodium acetate buffer (5:1 by volume) at 3 °C. The final residue is suspended in 50 ml of the ethanol-sodium acetate mixture and dialyzed against 100 volumes of distilled water at 3 °C for 24 h with repeated changes of the outer solution. The contents of the dialysis sac are centrifuged, and the supernatant, containing a few milligrams of protein, is discarded. The residue is lyophilized and stored as the dry powder. The yield of trifluoroacetylated ribonuclease is approximately 75%. An alternate procedure has been described in which excess reagents and hydrolysis products are removed from the derivatized protein by dialyzing the reaction mixture against repeated changes of 0.1 M acetic acid [1041].

The degree of trifluoroacetylation is easily determined by treating the derivatized protein with nitrous acid [41, 1041]. Only free amino groups undergo deamination; acylated amino groups are unaffected. After nitrous acid treatment, the protein is subjected to acid hydrolysis in the usual manner and the hydrolysate analyzed for lysine. The lysine level is a direct measure of the number of ε-(trifluoroacetyl)lysine residues in the modified protein.

Detrifluoroacetylation is readily accomplished by the following method [353]. Ten mg of TFA-ribonuclease are dissolved in 0.5 ml of 1 M piperidine at room temperature and the solution immediately chilled in an ice bath. After incubation at 0 °C for 2 h, the reaction mixture is added slowly, with constant stirring, to 1.5 ml of precooled 0.5 N acetic acid (0 to 3 °C). The deblocked protein is recovered by lyophilization after removal of reagents by either gel filtration or dialysis.

Caution should be exercised if trifluoroacetylation is carried out on a protein containing disulfide bonds; the formation of ethylmercaptan in alkaline solution favors both disulfide bond reduction and interchange. This problem may be circumvented by acylating the performic acid oxidized or reduced-carboxymethylated protein.

b) Maleylation

The reversible blocking of amino groups with maleic anhydride has been recently described [173]. Maleic anhydride rapidly reacts at pH 8.5 to 9.5 with the α- and ε-amino groups of proteins according to the following reaction (Eq. 8):

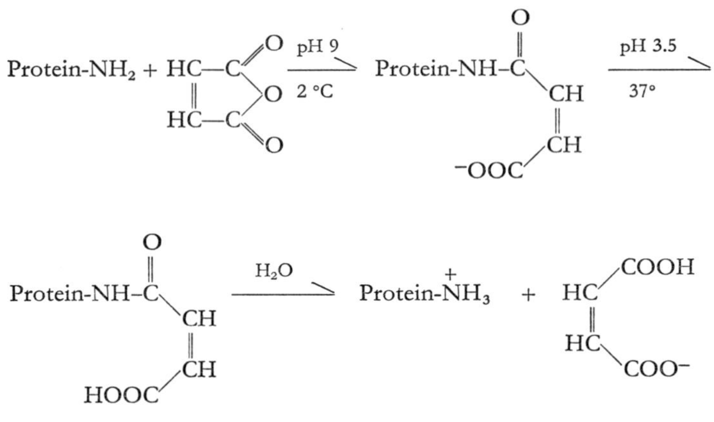

(8)

With bovine chymotrypsinogen A (0.04 mM) the reaction was 98% complete in 5 min using 30 mM reagent at pH 9 and 2 °C. Maleyl proteins are quite soluble at neutral or alkaline pH in the absence of urea or guanidine since the introduction of a negative charge minimizes aggregation through electrostatic repulsion. The maleyl amino bond is very stable above pH 6 but is readily hydrolyzed below pH 5 by intramolecular catalysis involving the unionized carboxyl group. The half-life of ε-maleyl lysine is 11 h at 37 °C and pH 3.5.

c) S-Aminoethylation

It is possible through the reaction of a reduced protein with ethylenimine to quantitatively convert cysteinyl side chains to the S-(β-aminoethyl) cysteinyl derivative [806] (Eq. 9):

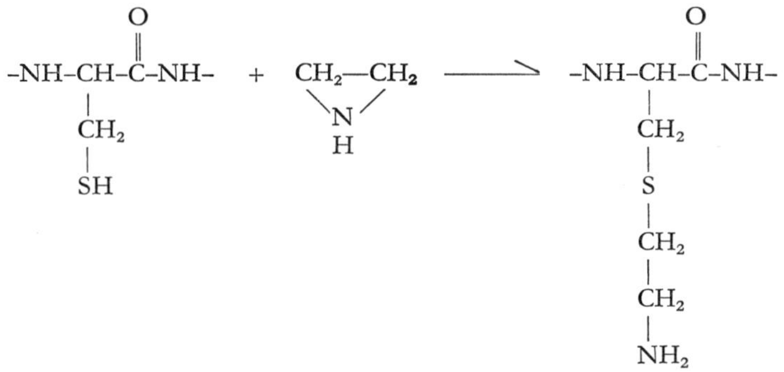

(9)

Reactions employing β-bromoethylamine [624] yield the same product; however, the intermediate formation of the imine is required before the reaction with the mercaptide anion occurs. Since the cyclization step has been shown to be rate limiting [806], the use of ethylenimine is most efficient and direct. The introduction of this basic side chain into the protein creates a new point for tryptic hydrolysis [624, 806]; the rate at which this bond is cleaved is markedly reduced and in order to insure maximal hydrolysis the level of trypsin and time of incubation must be increased. It should be remembered that prolonging the digestion time increases the possibility that non-tryptic cleavages will occur. A quantitative study on the tryptic digestion of S-aminoethylated ribonuclease revealed that the cleavage of S-aminoethylcysteinyl bonds averaged approximately 56% while the corresponding hydrolysis of lysyl and arginyl bonds was close to 83% [791]. The derivatized cysteinyl residue at position 40 in the ribonuclease molecule apparently was not cleaved; this is not too surprising since the sequence is

–Lys–Asp–Arg–Cys–Lys–Pro– .
|
AE

Aminoethylation has been successfully employed in structural investigations on trypsinogen [457], tryptophan synthetase [380], ferridoxin [652, 1035], and tobacco mosaic virus protein [1070]. SCHROEDER *et al.* [908] have recently reported that aminoethylation of apocatalase resulted in the modification of methionyl residues when excess reagent was removed under acidic conditions. This alteration interfered with the subsequent cyanogen bromide degradation of the protein.

Procedure: The preparation of aminoethyl trypsinogen is described [805]. Disulfide bonds are reduced by the addition of 200 µl (2.88 mmoles) of β-mercaptoethanol to 200 mg of the protein dissolved in 20 ml of 8 M urea containing 1.0 ml of EDTA solution (2 mg/ml) and 4 ml of 2-amino-2-methyl-1,3-propanediol buffer (4 M, pH 8.6). Alternatively, Tris buffer may be employed at a final concentration of 1.0 M and at pH 8.6. The mixture is allowed to stand at room temperature for 4 h. Alkylation is accomplished by the addition of ethylenimine in 3 portions of 0.250 ml each (4.8 mmoles) at 10 min intervals; the progress of the reaction may be measured by examining aliquots of the reaction mixture for the presence of free thiol with the nitroprusside reagent. Thirty minutes is usually ample time for the completion of the reaction. After a negative nitroprusside test is obtained, the pH is adjusted to 5.0, and the entire mixture is dialyzed for 20 h against changes of distilled water and finally lyophilized.

The degree of substitution is determined from the recovery of S-(β-aminoethyl) cysteine from an acid hydrolysate of the modified protein. Hydrolysis is performed in 6 N HCl at 110 °C for 20 h in a sealed evacuated glass tube ($< 50\ \mu$ of Hg). S-(β-aminoethyl) cysteine is eluted between lysine and histidine under the standard conditions of SPACKMAN, STEIN, and MOORE [987]. The length of the column must be increased to 15 cm, however, in order to sufficiently resolve the alkylated cysteine derivative from lysine and histidine. The color value of S-(β-aminoethyl) cysteine is close to 91% that of lysine [791].

4. Method of Hydrolysis

A variety of conditions have been employed for the tryptic hydrolysis of proteins with respect to enzyme-to-substrate ratios, temperature, pH, and time of digestion. Trypsin preparations of a high degree of purity, which have been treated with TPCK, should be used in structural investigations in order to assure the highest possible yields of tryptic peptides. Trypsin is usually employed at the level of 1 to 4% by weight of substrate with the temperature and pH ranging from 23 °C to 40 °C and 7 to 9 respectively. A fingerprint of the digest quickly indicates whether the predicted number of peptides is being produced and also reveals any undigested material. Staining the peptide map with the Sakaguchi reagent will identify the position and number of arginine-containing peptides, which should correspond to the value expected from the arginine content of the intact protein.

The following examples are presented to illustrate conditions employed for tryptic degradation.

a) Hydrolysis of Egg White Lysozyme [176a]

Reduced-carboxymethylated lysozyme is suspended in distilled water at a 1% concentration. A few drops of phenol red are added together with sufficient NH_4OH to adjust the suspension to a color corresponding to approximately pH 8. Trypsin, in an amount equal to 2% of lysozyme by weight, is added and the solution stirred for varying periods of time at either 23 °C or 37 °C. Aliquots of dilute NH_4OH are added to keep the phenol red color constant. The suspension usually clears in 10 to 15 min; however, the enzymic digestion is generally allowed to proceed for 120 min. A small amount of insoluble material is occasionally present at the end of the digestion which has an amino acid composition similar to that of reduced-carboxymethylated lysozyme. More complete digestion is obtained if the 1% suspension of the derivatized protein is first adjusted to pH 10 with NH_4OH, heated to 60 °C, and then digested with trypsin at 37 °C. Proteolysis is terminated by adjusting to pH 2.8 with formic acid.

b) Hydrolysis of Diisopropylphosphoryl-subtilisin BPN' [576]

The TCA denatured protein is triturated with a small volume of water until it is evenly wetted and all lumps are dispersed. Additional water is added so that the suspension is equivalent to a 0.8% protein solution. The denatured protein is uniformly suspended by use of a magnetic stirrer, and 0.1 N NaOH is added to pH 10 to 10.5 in order to solubilize the substrate. The pH is immediately adjusted to 8.1 with 30% acetic acid, and the solution of denatured protein (323 ml), in an appropriate flask, is placed in a constant temperature bath at 30 °C. A Radiometer automatic titrator equipped with a magnetic valve to regulate delivery of titrant (0.10 N NaOH) is used to maintain the pH at 8.1. Stirring is accomplished with a teflon-coated magnetic bar and an external magnetic stirrer. The protein solution is thoroughly flushed with nitrogen gas prior to the addition of trypsin and during the digestion period. Digestion is initiated by adding 4.3 ml of trypsin solution (12 mg/ml Worthington twice crystallized trypsin in 0.001 N HCl). Hydrolysis is followed by measuring alkali consumption (Fig. 6-2). After 67 min, an additional 51 mg of trypsin is added to give a total enzyme concentration of 4%. The digestion is terminated after 76 min by adding 17 ml of glacial acetic acid to pH 2.5. The digest is not lyophilized but applied directly to a column of AG 50W-X2.

c) Hydrolysis of f_2 Bacteriophage Coat Protein [569]

The alkylated protein is suspended in an aqueous solution of ammonium hydroxide, pH 10.5, and sonically treated for 30 sec to give a finely divided colloidal suspension. A 1% solution of trypsin, pretreated with TPCK, is added to give a final trypsin concentration of 0.01% and a total protein concentration of 1%. Solid CO_2 is used to adjust the pH to 8.5. The digestion is allowed to proceed for 8 h at 37 °C, with the addition of another equal portion of trypsin after 4 h. The digest is lyophilized twice prior to column fractionation of the tryptic peptides.

E. Chymotrypsin

1. Specificity

Although chymotrypsin exhibits a rather broad specificity, peptide bonds involving the carboxyl groups of tyrosine, phenylalanine, tryptophan, and leucine are hydrolyzed preferentially. Protein structure studies involving proteolytic degradation with chymotrypsin have revealed that cleavage may also occur at the carboxyl group of methionine, glutamine, asparagine, histidine, and threonine; however, many of these bonds are split slowly and do not represent major points of hydrolysis. Cleavage has also been reported to occur at bonds formed by serine [622], glycine [603], valine [27, 603], and isoleucine [603], but the extent of hydrolysis was low, and cleavage at these bonds is unexpected. The use of high ratios of enzyme to substrate and long digestion periods may increase the likelihood of cleavage at sites normally not susceptible to chymotrypsin. Frequently, within the same substrate certain peptide bonds formed by the same amino acid are susceptible while others are not attacked. In the case of subtilisin BPN′, hydrolysis occurred at a His–Ser (17 to 18) and two His–Val (67 to 68 and 226 to 227) sequences while a His–Gly (64 to 65) bond was not noticably attacked (Fig. 6-1). In the same protein chymotryptic cleavage was detected at Met–Ser (124 to 125) and Met–Ala (222 to 223) sequences but three other methionyl bonds were apparently unhydrolyzed (Met–Val, 50 to 51; Met–Asp, 119 to 120; Met–Ala, 199 to 200). Thus, it appears that factors other than the immediate environment of the susceptible bond may influence chymotryptic action. Neighboring groups are known to have a marked effect; the presence of a basic residue (Lys, Arg, or His) adjacent to a weakly susceptible bond seems to enhance cleavage at that site. The chymotryptic hydrolysis of the Thr–Lys bond (255 to 256) in subtilisin BPN′ [643] as well as hydrolysis of the threonyl bond in the sequence Arg–Lys–Thr–Gly which occurs in human cytochrome *c* [653] illustrates this point. Furthermore, cleavage at the carboxyl group of lysine has been noted when it occurs in a sequence of basic amino acids [46, 603, 653]. Peptide bonds involving the imino nitrogen of proline appear to be resistant to chymotryptic hydrolysis; only the Phe–Pro bond in ovine corticotropin [603] has been reported to be hydrolyzed to a small extent by chymotrypsin. Even though the sites of attack of chymotrypsin on proteins of known primary structure have been examined, it is not possible to clearly predict the course of chymotryptic hydrolysis except in the case of the aromatic amino acids.

2. Method of Hydrolysis

The experimental conditions used for the chymotryptic digestion of a protein are very similar to those employed with trypsin. Chymotrypsin has been used in structural investigations with a number of proteins some of which are the following: insulin [859], ribonuclease [445c], lysozyme [176b], papain [622], hemoglobin [436a, 909], cytochrome *c* [46, 641, 653], subtilisin BPN′ [643], Staphylococcal nuclease [1040], azurin [24], and myoglobin [274]. Three examples are presented.

a) Hydrolysis of the α-Chain of Human Hemoglobin [436a]

The α-chain (1 g) is dissolved in 90 ml of water and 10 ml of 10% (w/v) NH_4HCO_3, and 10 mg of chymotrypsin, dissolved in 1 ml of 0.001 N HCl, are added. After the pH is adjusted to 8.0, digestion of the milky suspension is allowed to proceed at 25 °C for 6 h with constant stirring. The precipitate is removed by centrifugation, and 10 ml of glacial acetic acid are added to stop the reaction. The solution is lyophilized twice to remove ammonium acetate.

b) Hydrolysis of Diisopropylphosphoryl-subtilisin BPN′ [643]

About 60 μmoles of denatured DIP-subtilisin are suspended in 250 ml of water and the pH is raised to 10 by dropwise addition of 40% NaOH. The solution is stirred overnight to completely wet and evenly disperse the protein that remained insoluble. Acetic acid (1 N) is slowly added to lower the pH to 8.15. Digestion is performed in a jacketed vessel at 40 °C under a steady stream of nitrogen. The pH is maintained at 8.15 by the addition of 0.10 N NaOH with a Radiometer automatic titrator. Hydrolysis is initiated by the addition of 35 mg of α-chymotrypsin (Worthington, three times crystallized), and the reaction is followed by measuring alkali consumption. After 70 min, an additional 35 mg of α-chymotrypsin are added. The reaction is terminated after 130 min by placing the reaction vessel in a boiling water bath for 15 min. The pH of the digestion mixture is then raised to 9.8 with 1 N NaOH in preparation for chromatographic fractionation on AG 1-X2.

c) Hydrolysis of Calf Rennin [312]

Crystalline rennin (140 mg) is dissolved by dialysis against distilled water and diluted to 8 ml in the reaction vessel of a pH-stat. The solution is adjusted rapidly to pH 10.6 with 0.1 N NaOH and incubated for 30 min at room temperature to denature the rennin. The pH is then adjusted to 8.5 with HCl and 2 ml of a solution containing α-chymotrypsin (1 mg/ml) and soya-bean trypsin inhibitor (1 mg/ml) are added. The solution becomes turbid after 2 h digestion and after 5 h the alkali uptake indicates that about 20 bonds per molecule of rennin have been split. The pH is then adjusted to 6.5 with HCl and the insoluble precipitate is removed by centrifugation. The soluble chymotryptic peptides containing cystine were separated by the appropriate diagonal paper electrophoretic procedure [160].

F. Pepsin

Commercial preparations of crystalline pepsin are inhomogeneous on the basis of a number of physicochemical criteria [807]; this is primarily the result of autolysis products produced throughout various stages of preparation. Several of the main components possess proteolytic activity, and in practice this mixture of native and modified pepsin may be used without the need of further purification being apparent. However, to achieve the highest degree of specificity and the most reproducible hydrolysis possible, the use of the purified enzyme seems desirable.

1. Preparation of Pepsin from Pepsinogen [807]

A 75 mg sample of pepsinogen is dissolved in 8 ml of glass-distilled water to yield a slightly turbid solution, pH 5 to 6. The solution is transferred to a jacketed vessel equipped with a magnetic stirrer and fitted with electrodes from a Radiometer TTT-1 autotitrator. Chilled water from a thermostated bath is circulated through the jacket of the vessel to maintain the solution at 14 °C ± 1 °C. Temperature equilibrium is attained within 2 to 3 min. The solution is brought to pH 2.0 by the addition of 0.75 ml of a chloroacetic acid – HCl mixture (9 ml of 2 N chloroacetic acid + 1 ml of N HCl). The acid must be added rapidly, in 20 to 25 sec. The amount of acid needed is determined in a trial experiment with a small sample. With preparations which contain a large amount of inorganic salt, more acid is needed; when this is the case, the concentration of HCl is increased, rather than that of monochloroacetic acid. The solution is maintained at pH 2.0 and 14 °C for exactly 20 min, at which time 0.75 ml of 4 N acetate buffer, pH 5.0, is added to bring the mixture to pH 4.40. The activation mixture is applied promptly (within 5 min) to a column, 2.5 × 30 cm, of sulfoethyl Sephadex C-25 equilibrated with a 0.4 N sodium acetate buffer, pH 4.38 ± 0.03 (40.29 g of sodium acetate·3 H_2O plus 29 ml of glacial acetic acid, made to a volume of 2 l). The column is operated at 4 °C under gravity at a flow rate of about 60 ml/h. Fractions of 4 ml are collected. The effluent fractions which contain the material emerging at the void volume, as shown by the recorder, are pooled within a few minutes after elution. The solution is brought to pH 5 by the addition of a few drops of 2 N sodium acetate. The pepsin solution is promptly frozen in 10 ml lots and stored at – 20 °C. The frozen samples are thawed immediately before use and converted to a solution of the desired pH and ionic composition by passage at 4 °C through a column, 2 × 30 cm, of Sephadex G-25.

2. Specificity

Pepsin has proved to be an extremely useful enzyme for the degradation of proteins. Peptic digests are usually used to supplement information derived from a characterization of the tryptic and chymotryptic peptides. Studies with synthetic substrates [51, 333, 411] have demonstrated that pepsin can hydrolyze peptide bonds formed by either the amino or carboxyl groups of phenylalanine, tyrosine, glutamic acid, cystine, and cysteine. In addition to protease activity, pepsin is also capable of hydrolyzing ester bonds [476]; in both proteolysis and esterolysis, activity is favored by hydrophobic side chains on both sides of the sensitive bond. An assessment of the susceptible bonds in proteins reveals that hydrolysis has occurred at bonds formed by the carboxyl groups of all L-amino acids except proline. It should be emphasized, however, that pepsin acts mainly at the amino and carboxyl side of aromatic residues and leucine. The presence of an α-carboxylate group or a protonated α-amino group adjacent to the sensitive bond is inhibitory to peptic cleavage [478]. Thus, residues in the amino-terminal or carboxyl-terminal position of the polypeptide chain would be released slowly by pepsin. Recent studies employing p-bromophenacyl bromide [377], diazoacetyl norleucine methyl ester [808], and L-1-diazo-4-phenyl-3-tosylamidobutanone [236] have indicated that the esterification of a single carboxyl group in the pepsin molecule results in loss of proteolytic activity.

With low concentrations of enzyme and short digestion periods (0.002% pepsin, 4 min, 25 °C), it was found that certain tryptic peptides from the β-chain of hemoglobin were hydrolyzed in a very specific manner by pepsin to give high yields of the resulting peptic peptides [354]; when higher enzyme concentrations and longer hydrolysis times were employed, more extensive cleavage was obtained. Our knowledge of pepsin action on large protein substrates has been extended appreciably by the detailed characterization of the peptic digest of subtilisin BPN′ [644]. Of particular interest is the hydrolysis of an Ile–Leu bond (234 to 235) and of several Ala–Ala bonds. Cleavage at peptide bonds involving basic amino acids also occurred in several instances, e.g., between Leu–His (16 to 17), Leu–Lys (42 to 43; 135 to 136) Asp–Lys (140 to 141), and Gly–Lys (169 to 170).

3. Method of Hydrolysis

For synthetic substrates pepsin has a pH optimum of 4.0, whereas for protein substrates the maximum hydrolysis rate occurs near pH 2. The general procedure is to dissolve or suspend the protein in an acidic solvent which will provide a pH of approximately 2 (5% v/v aqueous formic acid, 30% acetic acid, dilute hydrochloric acid, or formic acid-acetic acid-water, 1:4:45 by volume) and then to start the reaction by the addition of an aliquot of a freshly prepared enzyme solution. Proteolysis is terminated by raising the pH above 5. Conditions for the peptic digestion of several proteins appear in Table 4, and the use of pepsin is illustrated below.

Hydrolysis of diisopropylphosphoryl-subtilisin BPN′ [644]: Subtilisin (75 μmoles in a volume 280 ml) is digested in a thermostated vessel at 25 °C and pH 2.25. The pH is maintained at 2.25 by the automatic addition of 0.1 N HCl by a Radiometer titrator (type TTT 1C) equipped with a magnetic valve. Hydrolysis is initiated by the addition of 20 mg of pepsin (Worthington, twice crystallized) in 1.4 ml of 10^{-3}M HCl, and the reaction is followed by measuring acid consumption. After 58 min the digestion is stopped by raising the pH to 6.0 with pyridine and the solution is allowed to stand at room temperature for 2 h. Acetic acid is added to pH 4.0

Table 4. *Conditions for peptic hydrolysis of some proteins*

Substrate	Enzyme: substrate Molar ratio[a]	Conditions of hydrolysis				
		pH	Solvent	Temperature	Time	Reference
DIP-subtilisin BPN′	1:123	2.25	Dilute HCl	25 °C	58 min	[644]
Reduced-carboxy-methylated lysozyme	1:195	—	5% aqueous formic acid	23 °C	1 h	[177]
Ribonuclease	1:230	2.2	0.2 M Na citrate buffer	25 °C	4 h	[50]
	1:50	1.9			24 h	
Human hemoglobin, α-chain	1:188	2.8	0.01 N HCl	Room temp.	16 h	[568]
Human hemoglobin, β-chain	1:188	2	0.01 N HCl	25 °C	1 h	[566]
Human hemoglobin, γ-chain	1:100	2	—	Room temp.	2 h	[909]
Papain	1:71	2	—	40 °C	10 h	[620]
Insulin, A chain	1:330	—	0.1 N acetic acid	37 °C	48 h	[860]
Reduced-carboxy-methylated γ-globulin	1:100	2	30% acetic acid	38 °C	18 h	[439b]

[a] The molecular weight of pepsin is taken as 32,700 [1139]

followed by hydrochloric acid to a final pH of 2.2. A slight precipitate is removed by centrifugation, and the digest (350 ml) is applied to a Dowex 50-X2 column.

IV. Fractionation of Peptides

The successful separation of peptide mixtures resulting from either the chemical or enzymic degradation of a protein frequently requires the application of more than one technique. Several different approaches are possible. Initial fractionation may be performed on a cation exchange resin such as Dowex 50, followed by rechromatography of impure fractions on the anion exchanger Dowex 1 as recommended by SCHROEDER *et al.* [906]. Further subfractionation of unresolved peptide mixtures may be accomplished by gel filtration, chromatography on derivatized cellulose or dextran gels, or paper electrophoresis and chromatography. Alternatively, the digest may be "sized" by a molecular sieving process to yield peptide fractions of varying chain lengths which in turn can be further resolved by ion-exchange chromatography.

In fractionating peptide mixtures, it is useful to know the number of peptides present, their relative electrophoretic and chromatographic behavior, and the distribution among the various peptides of those amino acids detectable by spot tests (e.g., tyrosine, histidine, tryptophan, methionine, cysteine, arginine). This information is obtained by fingerprinting the mixture and using the appropriate method of staining. The results are often instructive in selecting that system capable of best resolving the constituent peptides, since properties other than net charge are important considerations in the separation process. This also provides a means of bookkeeping, for all peptides noted on the peptide map should be recovered from the column.

There are certain problems encountered in handling peptides that can produce difficulties in sequence studies. Several modifications which may result during the purification of a peptide are cyclization of N-terminal glutamine to pyrrolidone-carboxylic acid, the α, β-transformation at an aspartyl residue, and deamidation; each of these reactions has been discussed in Section II.B.2. Recently the conversion of N-terminal S-carboxymethylcysteine to a thiazane carboxyl residue has been described (Eq. 10) [976].

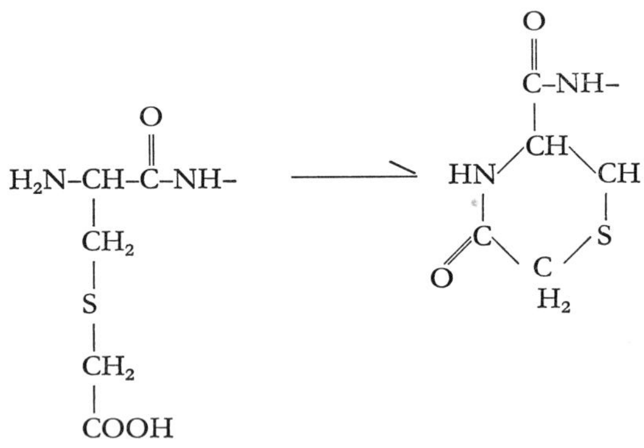

(10)

Ring closure occurred under rather mild conditions, e.g., chromatography at pH 3.2.

Difficulties are frequently encountered during the fractionation of peptides containing sulfur in thioether linkage. Exposure to mild oxidizing conditions can result in the formation of the corresponding sulfoxide, which markedly influences the chromatographic properties of the parent peptide. Thus, when multiple forms of the same peptide are encountered, low yields of the pure methionyl, S-carboxy-methyl cysteinyl, or S-aminoethyl cysteinyl peptides result. Two corrective measures are available. Either the native protein can be oxidized with performic acid to convert methionine to the stable sulfone and cysteine and cystine to cysteic acid, or an antioxidant such as thiodiglycol may be incorporated into the buffer systems used for fractionation of the peptides. As mentioned previously, performic acid destroys tryptophan as well as rendering methionyl linkages insensitive to cyanogen bromide cleavage. Thiodiglycol, at a level of 0.1 to 0.3%, is quite effective in maintaining the thioether sulfur in the reduced state [344, 383, 625] and its use is recommended when fractionating peptides with oxidizable sulfur.

A. Ion-Exchange Chromatography

1. Dowex 50

This resin is a strongly acidic cation exchanger which is prepared by the sulfonation of a styrene-divinylbenzene copolymer. The per cent crosslinkage is indicated by an "X" number following the name of the particular Dowex resin. For example, Dowex 50W-X2 is made from a copolymer containing 2% divinylbenzene. Because of the highly ionized character of the sulfonic acid resin, ion-exchange behavior is exhibited over a wide pH range.

a) Buffers

Only volatile developers will be considered. The composition of aqueous buffers used with Dowex 50 resins appears in Table 5. Pyridine (reagent grade) must be distilled prior to use owing to the presence of ninhydrin-positive contaminants. Distillation from phthalic anhydride reduces the level of these contaminants, but the pyridine so obtained is frequently not of adequate purity for use in chromatographic solvents. The usual procedure involves distillation of reagent grade pyridine after refluxing with ninhydrin [72]. Not more than a 2 month supply of pyridine should be purified at one time, since some deterioration occurs with time. Amber bottles that can be tightly sealed should be used for storage and exposure to direct sunlight avoided. Glacial acetic acid is reagent grade and does not require further purification.

Table 5. *Volatile buffers for Dowex 50 chromatography*

Pyridine concentration M	pH	Pyridine	Glacial acetic acid	Reference
		ml/1000 ml solution		
0.20	3.1	16.1	260	[544]
0.5	4.1	40.2	138	[544]
2.0	5.1	161	145	[544]
4.0	5.6	318	90	[653]
8.5	5.6	684	180	[567]

b) Preparation of Resin and Chromatographic Column

Dowex 50-X2 may be obtained as the analytical grade resin (AG 50W-X2) from the Bio-Rad Laboratories (Richmond, Calif.). The most commonly used particle size range has the dry mesh designation of 200 to 400 mesh. This highly purified resin is ready for use after it has been cycled through the following regeneration scheme. The resin contained in a Büchner funnel is washed successively with 5 to 6 volumes of 2 N NaOH, distilled H_2O, 4 N HCl, and distilled H_2O. Before use, the resin is converted to the pyridinium form by washing with several volumes of 2 M pyridine followed by a final equilibration with 0.2 M pyridine-acetic acid, pH 3.1.

Heavy-walled, jacketed, chromatographic columns of varying diameters and lengths as well as the appropriate connectors are available from Beckman Instruments (Palo Alto, Calif.). Manufacturers of other suitable columns are Technicon Corp. (Chauncey, New York) and Phoenix Precision Instrument Co. (Philadelphia, Pa.). The column is mounted vertically and the bottom outlet closed. Water from a constant temperature bath maintained at 40 °C is circulated through the jacket, and a small amount of the pH 3.1 buffer is poured into the column to establish a column height of approximately 10 cm. A slurry of resin (3 volumes of buffer to 1 volume of settled resin) is introduced down the inside of the column and about a 1- to 2 cm layer allowed to settle by gravity before the column outlet is opened. The remainder of the packing operation is carried out under a pressure of 10 to 15 cm Hg as directed by

MOORE and STEIN [701]. Additional portions of slurry are added to obtain the desired bed height. The resin bed is washed with 1 to 2 column volumes of starting buffer before the sample is applied. It is advisable not to prepare a column more than a few days in advance, since discoloration of the resin may occur and a nonvolatile colored material is eluted with the gradient.

c) Chromatographic Procedure

If the sample has been lyophilized, the residue is dissolved in the minimum volume of either 0.2 M, pH 3.1 equilibrating buffer or 2% acetic acid and the pH adjusted to 2.2 to 2.5 with HCl before application to the column. When lyophiliza-

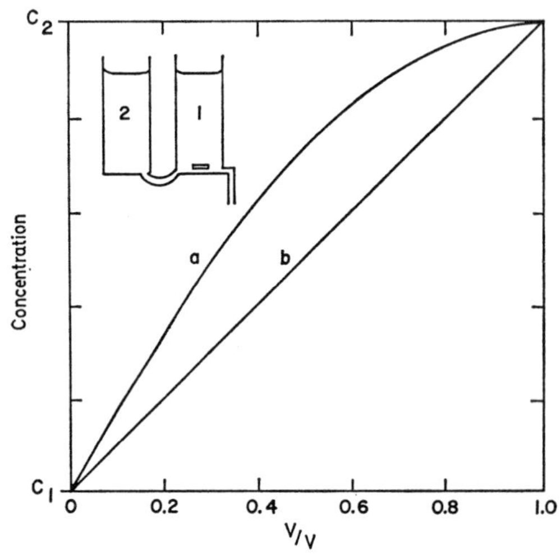

Fig. 6-3. Concentration gradient curves. Curve *a*: cross-sectional area of vessel *2* is twice that of mixing vessel *1*. Curve *b*: vessels *1* and *2* have the same cross sectional area. The abscissa indicates the fraction of the total volume of the system that has been delivered. (Reprinted in part from Ref. [111] by courtesy of Analytical Chemistry)

tion, is not necessary, the digest may be acidified directly by adding glacial acetic acid. The buffer above the surface of the resin is carefully removed and the sample layered on top of the column. Air pressure may be used to facilitate sample application; however, care should be exercised to prevent air from entering the resin. Whenever pressure is applied to the column, it should be released slowly to prevent cracking of the resin bed or the introduction of air at the bottom of the column on decompression. The sample is rinsed in with several small volumes of equilibrating buffer before the column is filled.

Generally, a convex gradient or a linear gradient will be used to develop the chromatogram (Fig. 6-3). A system that generates a linear gradient of increasing pH and pyridine concentration will be described. Cylinders of equal cross sectional area, similar to those illustrated in Fig. 6-3, are joined by two short pieces of Tygon tubing with a 5-cm section of heavy-walled glass capillary tubing (2 mm internal diameter) in the center. The connecting tubing is filled with the pH 3.1 buffer and clamped shut on the side distal to the mixing cylinder, which is then partially filled with the starting buffer. Next, the tubing leading from the mixing cylinder to the pump[6] and to the column is filled with the same buffer. Connection is made to the top of the column, and elution is begun by pumping approximately 1 to 2 column volumes of the pH 3.1 buffer through the column at the appropriate flow rate. To establish the gradient, equal volumes of 0.2 M pyridine-acetic acid, pH 3.1 and 2 M pyridine-acetic acid, pH 5.1 are placed in the mixing and adjoining cylinders respectively. The tops of the two buffer columns are adjusted to the same height by raising or lowering one of the cylinders. With the pump shut off, the clamp between the two cylinders is carefully opened and a final adjustment is made at this time to attain hydrostatic equilibrium. Magnetic stirring is started and pumping is resumed. At the end of the gradient, limiting buffer (2 M pyridine-acetic acid, pH 5.1) is added to the mixing cylinder, and an additional three column volumes of effluent are collected. Finally, the column is eluted with either 4 M or 8.5 M pyridine-

Table 6. *Conditions for operation of Dowex 50–X2 columns*

Column size	0.9 × 160 cm	1.9 × 160 cm	3.8 × 160 cm
Volume of buffer contained in each cylinder at start of gradient (liters)	0.3	3.5	10
Flow rate (ml/hr)	15	60	220
Fraction volume (ml)	2	10	20

acetic acid, pH 5.6 in order to remove tightly adsorbed peptides. During this phase, the resin will contract appreciably and the column back pressure will tend to increase. It may be necessary to compensate for this by decreasing the flow rate. The general operating conditions for columns of varying dimensions are presented in Table 6. These are intended as guidelines and may be modified within limits. An example illustrating this chromatographic procedure is presented in Fig. 6-4.

d) Examination of Column Fractions

The progress of the fractionation procedure is followed by analyzing the column effluent for ninhydrin-reactive material after alkaline hydrolysis [445a]. Aliquots from alternate fractions are pipetted into clean polypropylene tubes. One ml of 2.5 N NaOH (freshly prepared) is added to each tube, and the rack containing the unstoppered tubes is placed into a boiling water bath for 2 h. After the hydrolysis period the tubes are removed and cooled to room temperature. Acetic acid (1.0 ml of 30% by volume) is added to each tube and the contents thoroughly mixed. Next, 2.0 ml of ninhydrin reagent are added to the neutralized hydrolysate and the remainder

[6] The pump must be capable of maintaining a constant volume output over a wide range of flow rates. Two pumps which are satisfactory for this purpose are the Accu-Flo Pump (Beckman Instruments, Palo Alto, Calif.) and the Mini pump (Milton Roy Co., Philadelphia, Pa.). It is convenient to equip the pump with a gauge so that operating pressures can be monitored.

of the analysis is carried out as described in Section III. C. 1. With the peptide elution pattern as a guide, the appropriate fractions are combined and concentrated to dryness on a rotary evaporator at 40 °C. The residue is taken up in a small volume of water and re-evaporated to remove residual pyridine acetate. The peptide(s) is dissolved in an appropriate solvent (e.g., aqueous solutions of pyridine, trimethylamine, acetic acid, or formic acid) and transferred to a labeled vial for further study. The homogeneity of each fraction is assessed by the two-dimensional separation technique described by INGRAM [475] and further purification carried out as indicated.

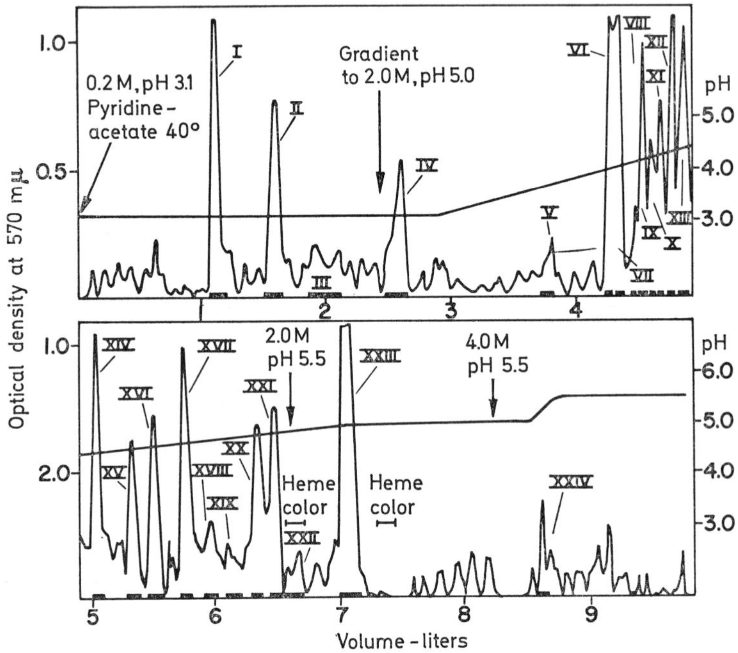

Fig. 6-4. Elution pattern of peptides derived from the tryptic digest of human cytochrome c. Thirty-one μmoles were chromatographed on a 2 × 165 cm column of Dowex 50-X2. The solid bars on the abscissa indicate the fractions pooled from which peptides were isolated. (Reprinted from Ref. [653] by courtesy of The Journal of Biological Chemistry)

e) Comments

The amount of sample (total μmoles) which can be chromatographed on any given size column will depend primarily on the complexity of the peptide mixture. For example, 100 micromole quantities of various chymotryptic digests of cytochrome C [143 to 145] have been separated on 3.8 × 150 cm columns of Dowex 50–X2. Digests of lysozyme (25 μmoles) [176a] and subtilisin BPN′ (60 μmoles) [643] were chromatographed on 1.9 × 150 cm columns. The 0.9 × 150 cm column will usually handle between 8 to 10 μmoles of an enzymic digest. As a first approximation, peptides are expected to elute from an ion-exchange column on the basis of their net charge. This is not always observed, since non-ionic interactions with the resin matrix are also influencing factors. Peptides containing hydrophobic amino acids, particularly the aromatic residues, exhibit a greater degree of retention as a result of the hydrophobic interaction of these side chains with the polystyrene backbone. Behavior of this type can be illustrated by considering the following example. Tryptic peptide T–2 from subtilisin BPN′ [576] was digested with chymotrypsin and the following points of cleavage were noted:

C–2 ←———→ C–1 ←———→ C–3 ←———→

Ala–Pro–Ala–Leu–His–Ser–Gln–Gly–Tyr–Thr–Gly–Ser–Asn–Val–Lys

←———→ C–4

The chymotryptic peptides are numbered in order of their elution from AG 50W-X2 (Fig. 6-5). The neutral tetrapeptide C–1 did not emerge until the gradient had taken

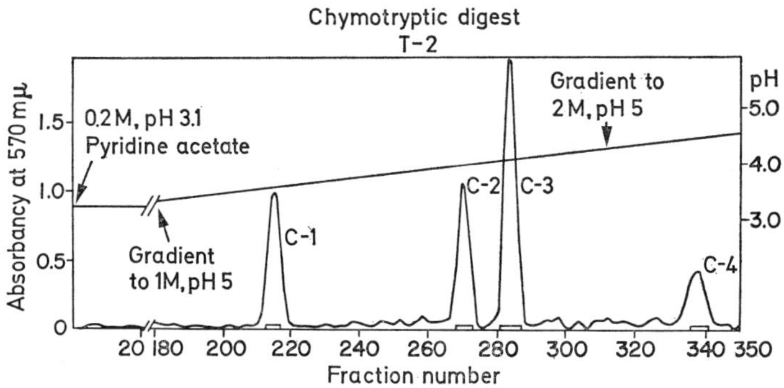

Fig. 6-5. Elution pattern of chymotryptic peptides derived from 11 μmoles of Peptide T-2 of subtilisin BPN′. Chromatography was performed at 40 °C on a 0.9 × 160 cm column of Dowex 50-X2. (Reprinted from Ref. [576] by courtesy of The Journal of Biological Chemistry)

effect; this retardation is attributable to the presence of the tyrosyl residue since neutral non-hydrophobic peptides usually elute before the gradient is established. Furthermore, the basic aromatic peptide C–4 was retained to a much greater extent than the basic hexapeptide C–3, which lacks an aromatic residue. The occurrence of leucine in both C–2 and C–4 also contributes to their hydrophobic character. Pyridine in the eluting buffers tends to disrupt the non-ionic peptide-resin interactions. This point was nicely illustrated by Kimmel *et al.* [544] who compared the chromatographic behavior of tryptic peptides derived from oxidized papain on Dowex 50-X2 using pyridine-acetic acid and sodium acetate systems. Marked differences were noted between the two elution patterns even though the experiments were conducted under essentially identical conditions with regard to pH, cation concentration, and temperature. In general, peptides were less retarded when columns were eluted with the pyridine buffers, but this effect was most notable with peptides containing tyrosine, phenylalanine, and tryptophan.

2. Dowex 1

This resin is a strongly basic anion exchanger with a cross-linked polystyrene network similar to that found in Dowex 50. The basic properties derive from the quaternary ammonium function attached to the polymer backbone; specifically, Dowex 1 is the trimethylbenzyl ammonium derivative.

a) Developers

The improved gradient system of Schroeder and Robberson [907] for ion-exchange chromatography of peptides on Dowex 1 will be described. This elution schedule has a decided advantage over the previously used developers [906] in that it produces a gradual, slightly sigmoid pH gradient. Thus, the abrupt pH change which results from the addition of acetic acid to the starting buffer is avoided. Five developers are required. The following quantities are for the preparation of 4 l of the pH 9.4 starting buffer: N-ethyl morpholine, 60 ml; α-picoline, 80 ml; pyridine, 40 ml; glacial acetic acid, 0.5 ml. Buffers of pH 8.4 and 6.5 are prepared in a similar manner except for the addition of increasing amounts of acetic acid to achieve the desired pH; approximately 3 and 37 ml of acid, respectively, are needed. Aqueous solutions of 0.5 N and 2 N acetic acid are also required. All reagents except the acetic acid must be distilled before use. Redistilled N-ethylmorpholine is best stored in the cold. It is advisable to use freshly prepared buffers, since they are somewhat unstable and decompose to give non-volatile products on standing. Also, it is important to prepare the alkaline buffers with CO_2-free water so that bubbles will not form in the column as the pH drops during the gradient. This is best accomplished by thoroughly flushing the water to be used for the preparation of buffers with nitrogen before adding the reagents.

b) Preparation of the Resin and Chromatographic Column

The analytical grade of the anion-exchange resin Dowex 1-X2 is obtained from Bio-Rad Laboratories (Richmond, California) as AG 1-X2 (200 to 400 mesh) in the chloride form. Fines are removed by suspending the resin in approximately 10 volumes of distilled water, allowing the resin to settle for about 30 min, and siphoning off the supernatant. This step is repeated until the supernatant is essentially clear. The following method of regeneration and equilibration of the resin is patterned after Schroeder *et al.* [906] and applies to the processing of 1 lb of AG 1-X2. The resin is washed on a Büchner funnel with 2 l of water at 60 °C, and at room temperature with 1 l of carbonate-free 0.5 N NaOH, 3 l of H_2O, 0.5 l of 1 N HCl, and 3 l of H_2O. The resin may be stored wet as the chloride in a dark bottle. Before pouring the column, the required amount of resin is placed in a Büchner funnel, and the following solvents are allowed to flow through under gravity: 2 volumes of H_2O, 2 volumes of carbonate-free 0.5 N NaOH, 5 volumes of H_2O, 2 volumes of 1 N acetic acid, 2 volumes of H_2O, and 5 volumes of the pH 9.4 equilibrating buffer.

The chromatographic columns are of the same design as employed for Dowex 50 separations. The column is clamped in a vertical position and water at 40 °C is pumped through the jacket from a constant temperature bath. A small amount of the pH 9.4 buffer is allowed to pass through the column, and the outlet is closed in time to retain a column of buffer of approximately 10 cm. The resin is suspended in the pH 9.4 buffer in the proportion of 1 part of resin to 3 parts of buffer, warmed to

40 °C, and degassed. The slurry is poured into the column and a gentle vacuum is applied to the top of the column while the resin settles. When settling is complete, excess buffer is removed except for about a 10-cm layer above the resin, and the procedure just described is repeated until a resin bed of the desired height is obtained. The column is equilibrated by washing with the pH 9.4 starting buffer (Table 7).

c) Chromatographic Procedure

The buffer is removed from above the resin bed, and the sample, which has been dissolved in a small volume of pH 9.4 buffer and adjusted to pH 10.4 with 0.2 N NaOH, is carefully applied to the column. After the sample is rinsed into the resin, pH 9.4 buffer is added to fill the column. The gradient device employs a constant volume

Table 7. *Conditions for chromatography on Dowex* 1 [907]

Column size	0.6 × 60 cm	1 × 100 cm
Equilibrating volume (ml)	60	100
Volume of buffer (pH 9.4) in mixing flask (ml)	40	135
Fraction size (ml)	1	3[a]
Volume of developer (ml)		
pH 9.4 buffer	10	40
pH 8.4 buffer	30	120
pH 6.5 buffer	40	160
0.5 N acetic acid	60	240
2 N acetic acid	100	400
Flow rate (ml/h)	60[a]	60[a]
Re-equilibrating volume (ml)	75	150

[a] These values differ from those given in the original table and have been kindly provided by Dr. Walter A. Schroeder (personal communication). The flow rates are attained with a constant volume pump

mixing flask connected to a buffer reservoir by a short section of polyethylene tubing (Fig. 6-6). It is convenient to use the same type of pump that was recommended for use with the Dowex 50 columns, but the application of a hydrostatic head has been suitable for obtaining the desired flow rate in certain instances. The volume of buffer placed in the mixing flask, as well as the volume of other developers used during the run, will depend upon the size of the column (Table 7). A CO_2 trap (Ascarite, A. H. Thomas Co.) is installed at the air intake of the buffer reservoir, and the elution is begun with pH 9.4 buffer in both the mixing chamber and reservoir. After the required volume of starting buffer has flowed through the column, the pH 9.4 buffer in the reservoir is replaced with the appropriate volume of pH 8.4 buffer which in turn is replaced by the pH 6.5 buffer and so on until the elution is completed with the 2 N acetic acid developer. The column effluent fractions are examined as described for Dowex 50 (Section IV. A. 1d). A peptide elution pattern illustrating this method of separation is shown in Fig. 6-7. It is possible to re-equilibrate the resin in the column by washing with the pH 9.4 buffer until the pH of the effluent matches that of the influent. Under certain conditions, however, this may not be desirable. When an enzymic digest of a whole protein is fractionated on Dowex 1-X2, certain peptides may not be eluted by the developers, and a regeneration cycle would be required to adequately clean the resin.

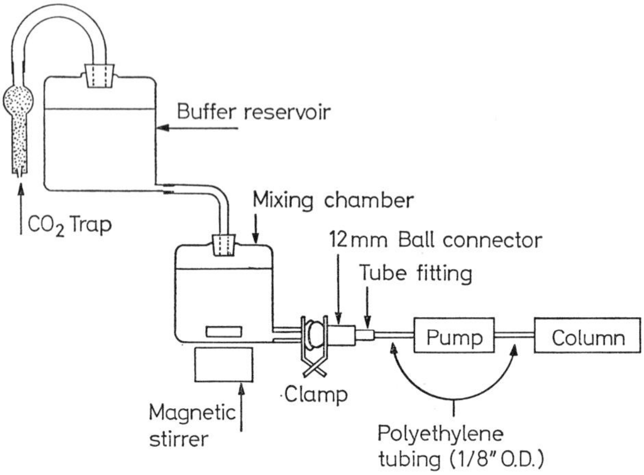

Fig. 6-6. Constant volume gradient apparatus for Dowex 1 chromatography. The 12 mm ball connector (Part No. 312937) and the tube fitting (Part No. 313285) are obtained from Beckman Instruments, Palo Alto, California

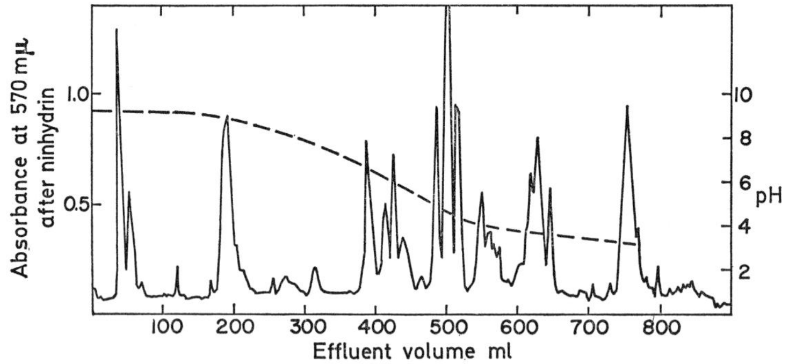

Fig. 6-7. Separation of tryptic peptides obtained from approximately 7 μmoles of amino-ethylated-γ-chain of bovine fetal hemoglobin (1 × 100 cm column). The dashed line indicates the pH of the effluent fractions. (Reprinted from Ref. [907] by courtesy of Analytical Chemistry)

d) *Comments*

Peptides containing histidine, tyrosine, phenylalanine, and tryptophan are eluted from Dowex 1 more slowly than would be predicted on the basis of charge. Generally the order of emergence of peptides from this anion exchanger is approximately the reverse of that noted with Dowex 50. Whereas the method of elution from Dowex 50 employs a gradient of increasing ionic strength and pH, peptides are eluted from Dowex 1 by continuously increasing the acetic acid concentration. With certain peptide mixtures, it may be desirable to modify the pH gradient by changing the volumes of the different developers. For example, if several basic peptides are being separated, resolution may be improved by increasing the volume of pH 9.4 buffer in the mixing flask to provide a more gradual change in pH in the alkaline range.

3. Phosphocellulose

a) Chromatographic Procedure

Phosphorylated cellulose (Whatman Chromedia, P-11) with a capacity of 7 meq. per gram is obtained from H. REEVE ANGEL (Bridewell, New Jersey). The cellulose is successively washed on a Büchner funnel with 4 volumes of 1 N NaOH, distilled

Table 8. *Elution buffers for phosphocellulose columns* [177]

Buffer system	Chamber no.	Volume ml	Acetate, concentration, M	pH
Ammonium acetate	1	1100	0.02	3.95
	2	1100	0.07	4.03
	3	1100	0.13	4.30
	4	1100	0.20	5.06
Pyridine-acetic acid	1	500	0.05	3.91
	2	500	0.10	4.02
	3	500	0.15	4.21
	4	500	0.20	4.42
	5	500	0.25	4.63
	6	500	0.25	4.80
	7	500	0.30	5.01
	8	500	0.45	5.40

H_2O until the filtrate is below pH 9, 4 volumes of 1 N HCl, and distilled H_2O until the pH of the filtrate is above 4. The resin may be stored as an aqueous suspension in the cold room until ready for use. The conditions to be described are basically those employed by CANFIELD and ANFINSEN for the separation of the chymotryptic peptides of egg white lysozyme [177]. All chromatographic steps are carried out in the cold room, and the salt and pH gradient is generated by a multichambered VARIGRAD [783]. Prior to pouring the column the cellulose is equilibrated in a Büchner funnel with the starting buffer (Table 8). The column outlet is closed and buffer is added to fill approximately 1/5 of the tube. A degassed slurry of phosphocellulose (1 volume cellulose to 3 volumes of buffer) is poured down the side of the column (2.4 × 25 cm) and a 1- to 2-cm bed of exchanger is allowed to form under gravity. The bottom outlet is then opened and additional portions of slurry are added to obtain the desired bed height. Final equilibration of the column is achieved by washing with several bed volumes of starting buffer. The sample is dissolved in 10 ml of distilled water, adjusted to pH 3.4 with formic acid, and applied to the column. A four-chambered Varigrad with a capacity of 1100 ml per chamber is used with ammonium acetate buffers (Table 8). A constant volume pump maintains the flow rate at 60 ml/h and 10-ml fractions are collected. When the Varigrad is nearly empty, 500 ml of 0.2 M ammonium acetate are added to Chamber 1 and titrated to the same pH as the solution in the top of the column (pH 4.91). To

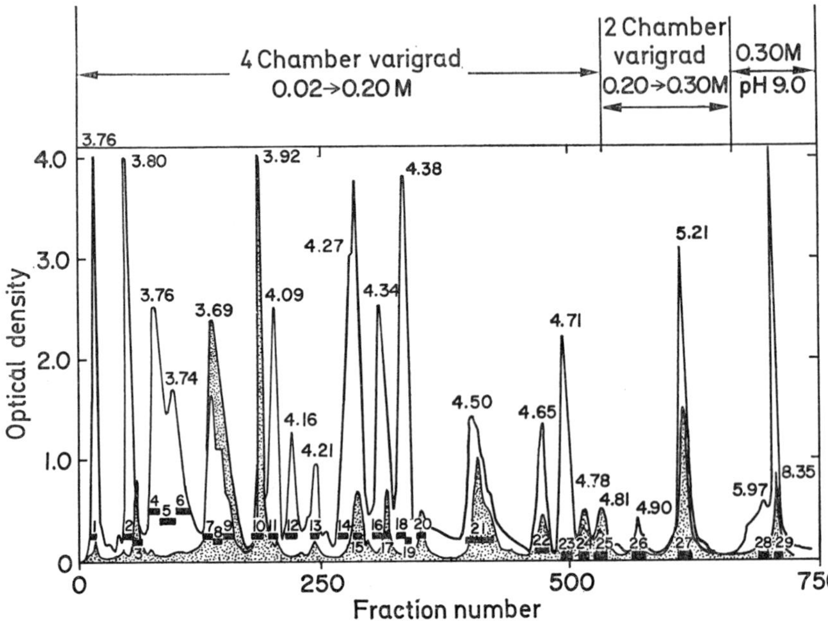

Fig. 6-8. The chromatographic separation of peptides produced by chymotrypsin digestion of carboxymethylated lysozyme. The solid lines represent the optical density at 570 mμ of aliquots subjected to ninhydrin analysis following alkaline hydrolysis. The stippled areas represent the optical density at 280 mμ of the column effluent. The regions taken for analysis are numbered, and the fractions pooled for a given region are symbolized by black bars. The pH of the peak tubes for each pooled fraction is noted. (Reprinted from Ref. [177] by courtesy of The Journal of Biological Chemistry)

Chamber 2, 500 ml of 0.3 M, pH 6.3 ammonium acetate is added and the instrument is now operated as a two-chambered Varigrad. When these buffers are exhausted, the column is washed with 0.3 M ammonium acetate, pH 9.0. The effluent fractions are examined with respect to pH, 280 mμ absorption, and ninhydrin-reactive material after alkaline hydrolysis. Samples of 200 μl are removed from alternate fractions, transferred to polypropylene tubes, and dried in a vacuum oven at 80 °C in the presence of P_2O_5 for 24 h prior to hydrolysis. The results of the chromatographic separation are presented in Fig. 6-8. The appropriate fractions are pooled and lyophilized. To remove excess ammonium acetate, the lyophilized powder is heated to

50 °C for 24 h over P_2O_5 in a vacuum oven, dissolved in a small volume of water, and lyophilized a second time.

Pyridine-acetic acid buffers have also been used for fractionating peptides on phosphocellulose (Table 8). The effluent cannot be monitored at 280 mμ owing to the strong absorption of pyridine, but the pyridine-acetic acid system has two main advantages over ammonium acetate buffers: (a) the pyridine salts are more easily removed by rotary evaporation or lyophilization and (b) aliquots of effluent fractions may be analyzed directly, since the buffer yields a low ninhydrin value in the colorimetric analysis.

b) Comments

Of the derivatized cellulose exchangers, phosphocellulose is best suited for the fractionation of peptide mixtures. It has a high capacity and functions well in the adsorption and elution of peptides in the range of pH 3 to 5; the two pK values for the phosphate group are near pH 2.2 and 6.5, respectively [784]. Carboxymethyl (CM-)cellulose has a pK value of approximately 4 and is therefore only partially ionized in the pH region that favors peptide binding. Consequently, the capacity of this exchanger is markedly reduced, and the titration of the carboxyl groups during an increasing pH gradient can exceed the buffering capacity of the low ionic strength eluting buffer and thus result in an undesirable pH change within the column. Similar limitations also apply to diethylaminoethyl (DEAE-) cellulose. The most promising application of CM- and DEAE-cellulose is for the purification of larger peptides, using a salt gradient at a constant pH. From the available results, it appears that peptides are usually eluted in order of their net charge from cellulose exchangers. The interaction of aromatic side chains with the cellulose matrix is not so pronounced as is the case with the Dowex resins.

4. Derivatized Dextran Gels

Ion exchangers prepared from cross-linked dextran gels are commercially available as the DEAE-, CM-, and sulfoethyl (SE-) derivatives of Sephadex (Pharmacia Fine Chemicals, Piscataway, New Jersey). Each exchanger may be obtained in two porosity grades which are assigned the numbers of 25 and 50; the prefix A- or C- is used to indicate either an anionic or cationic exchanger. The type 25 has a porosity which corresponds approximately to G-50 Sephadex, whereas the type 50 has a lower degree of cross linkage and is comparable to G-200 Sephadex in porosity. Thus, for peptide fractionation the type 25 exchanger is preferable since exclusion from the interstices of the gel does not occur until the molecular weight of the solute exceeds 10,000.

a) Chromatographic Procedure

The cation and anion exchangers are supplied in the sodium and chloride forms respectively and are prepared for use in the following manner. The dry beads are allowed to swell in a large volume of starting buffer at room temperature for 24 to 30 h with gentle stirring. Several times during this period, the exchanger should be permitted to settle, the supernatant removed, and fresh buffer added. In this way, equilibration is accomplished while the gel is hydrating. The column is packed in much the same way as described for phosphocellulose and is washed with several bed volumes of starting buffer before the sample is applied. The bed volume of the gel is markedly influenced by pH, ionic strength, and the specific buffer employed. In general, the gel will shrink during the conversion of the exchanger from the charged to the uncharged form. This effect is also noted upon increasing the salt concentration in the eluant. It is desirable to use a starting buffer with an ionic strength of near 0.05 since below this value the volume change associated with increasing salt concentration is quite pronounced. The most common type of gradient is one producing a linear increase in salt concentration; however, both pH and ionic strength may be varied simultaneously, but it is desirable to use a pH gradient only outside of the titration range of the exchanger. Because contraction of the gel bed will occur during the development of the chromatogram, it is advantageous to use a constant volume pump to provide an even flow rate of approximately 6 to 8 ml/cm²/h. Alternate effluent fractions are examined for ninhydrin-reactive material after alkaline hydrolysis as described in Section IV.B.1. Cation exchangers are regenerated by successive washings on a Büchner funnel with 0.5 N HCl, H_2O, 0.5 N NaOH, and H_2O to remove excess alkali. For DEAE-Sephadex, the reverse order is recommended.

b) Comments

Derivatized dextran gels can be profitably employed in situations where Dowex columns do not prove satisfactory. For example, extremely basic peptides, particularly those containing aromatic residues, can be so tightly adsorbed to Dowex 50-X2 that sodium hydroxide is required to elute them, whereas these same peptides are likely to pass through Dowex 1–X2 almost unretarded, with the result that there is poor resolution and additional fractionation is necessary. Accordingly, sulfoethyl-Sephadex was used by KONIGSBERG *et al.* [566] to further purify peptide fractions obtained by countercurrent distribution of a peptic digest of the β-chain of human hemoglobin. This was of particular value in the case of two peptides (Pβ1, Leu–Ala–His–Lys–Tyr–His; and Pβ18, Tyr–Pro–Trp–Thr–Gln–Arg–Phe) which were completely retained on Dowex 50-X2 and which also failed to separate on Dowex 1-X2. When mixtures containing these peptides were chromatographed on SE-Sephadex C-25 with the usual pyridine acetate gradient employed for Dowex 50–X2 columns, both peptides Pβ 1 and Pβ 18 were obtained in high yield and in a reasonably homogeneous state. DEAE-Sephadex A-25 proved useful for the fractionation of a mixture of insoluble tryptic peptides derived from the f_2 bacteriophage coat protein. The insoluble fraction was dissolved in 8 M urea and chromatographed on the anion exchanger equilibrated with 0.05 M Tris chloride, pH 8.0, containing 8 M urea. A linear gradient of increasing chloride concentration (0.05 M to 0.5 M Cl^-) was used to develop the chromatogram.

B. Gel Filtration

Two different types of gel filtration media are currently used to fractionate peptide mixtures. These are cross-linked dextrans (Sephadex, Pharmacia Fine Chemicals, Inc., Piscataway, New Jersey) and the polyacrylamide gels (Biol-Gel P, Bio-Rad Laboratories, Richmond, California). Both are prepared in bead form, and for a given type the beads have pores of carefully controlled dimensions which permit resolution based on molecular size. A list of the various types of gels applicable to the separation of peptides is given in Table 9 along with the fractionation range for each gel. Gels of a lower degree of cross-linkage capable of accommodating solutes of higher molecular weight have been omitted. Sephadex G-10 is a tightly cross-linked gel of relatively small pore size, whereas Sephadex G-50 contains fewer cross-links and is considerably more porous. The same trend in physical properties holds for the Bio-Gel series.

A gel filtration column contains two types of water volumes: V_0, the outer or void volume and V_x, the interstitial volume of the gel bead. The total volume of the gel bed, V_t, is described by Eq. (11), where V_g is the volume occupied by the gel matrix.

$$V_t = V_0 + V_x + V_g \tag{11}$$

V_i increases as the porosity of the gel increases. Thus, upon swelling in water, 1 g of G–50 will occupy a volume 3 to 4 times greater than an equivalent amount of G–10. The distribution of the solute between V_0 and V_x may be expressed in terms of

Table 9. *Gel filtration media for peptide fractionation*

Type	Fractionation range (molecular weight of solute)
Sephadex G–10	Up to 700
Sephadex G–15	Up to 1,500
Sephadex G–25	1,000— 5,000
Sephadex G–50	1,500—30,000
Bio-Gel P–2	200— 2,000
Bio-Gel P–4	500— 4,000
Bio-Gel P–6	1,000— 5,000
Bio-Gel P–10	5,000—17,000
Bio-Gel P–20	10,000—30,000

a coefficient, K_d. If total exclusion from the interstices of the gel occurs, K_d is equal to zero. If the diffusion of the solute is not restricted by the pore size of the gel, K_d is equal to 1. Therefore, the elution volume, V_e, of a solute with a K_d in the range of 0 to 1 will be (Eq. 12).

$$V_e = V_0 + K_d V_i \tag{12}$$

Certain molecules may have K_d values greater than 1, indicating that a process other than sieving occurs. For example, aromatic compounds are retarded on Sephadex to a greater extent than are non-aromatic compounds of the same molecular size. Thus, peptides containing tyrosyl, phenylalanyl, or tryptophanyl side chains do not exhibit normal elution behavior [794]; this adsorption effect is most pronounced with tryptophan. A second property of Sephadex that produces anomalous behavior during the filtration process is charge. The presence of a small number of carboxyl groups in the dextran matrix retards positively charged solutes and is responsible for the partial exclusion of acidic substances from the interstices of the gel. Both of these charge effects are neutralized by increasing the salt concentration in the eluting solvent. Gel-solute interactions frequently prove useful in separating peptides of similar size but different composition. Adsorption effects appear to be minimal with polyacrylamide gels, although hydrogen bonding is possible with the amide bonds of the matrix.

1. Procedure

The choice of the appropriate gel for a particular separation depends on the nature and complexity of the mixture. For the best resolution, a fine particle size is recommended. The gel is prepared by adding the dry beads to a large excess of solvent with constant stirring at room temperature. For the materials listed in Table 9, approximately 3 to 5 h should be allowed for complete hydration. The amount of gel required for a given column may be estimated from the bed volume (ml/g dry gel) specified for that gel. The gel suspension (3 parts of solvent to 1 part of settled gel) is degassed to remove trapped air and poured down the side of the column which is filled about one-fifth full with equilibrating solvent. A 6- to 8-cm layer of gel is allowed to form before the outlet of the column is opened. The gel is not allowed to settle completely before the buffer above the settling zone is removed and additional slurry added with stirring. This step is repeated until a continuous gel bed of the required height is obtained. Note that an unsectioned column of gel results from this method of pouring, i.e., separate layers of gel are not observed. Several bed volumes of solvent are passed through the column prior to use. The surface of the gel is protected by the insertion of a filter paper disk. Before starting an experiment, the column packing should be examined for imperfections by passing a solution of colored material (e.g., hemoglobin or Blue Dextran-2000, Pharmacia Fine Chemicals) through the bed and visually inspecting for air bubbles, zone distortion, or channeling. If any of these faults are noted, repacking is indicated. This also permits the determination of the void volume, provided the substance used is totally excluded from the inner portion of the gel particle.

Prior to the application of the sample, the column outlet is closed and the solvent above the gel is removed. The sample is carefully distributed over the surface of the gel and the bottom outlet of the column is opened. If the flow is too rapid, a piece of flexible tubing and a clamp attached to the outlet will control the flow rate. Several small portions of eluant are used to rinse the wall of the column and wash the sample into the gel bed before filling and connecting the column to the eluant reservoir. For optimal separation, the width of the starting zone should be kept at a minimum. The flow rate should be maintained in the range of 6 to 8 $ml/cm^2/h$ by use of a constant volume pump or a Mariotte flask. Gel columns may be reused without repouring or regenerating.

2. Comments

Gel filtration is an exceedingly mild fractionation procedure which is a useful adjunct to the common ion-exchange methods. Since the sieving phenomenon is relatively insensitive to pH, salt concentration, and eluant, the use of a wide variety of solvents is permitted. This feature is particularly valuable when dealing with peptides that are insoluble in the usual chromatographic developers. For example, the B- and C-chains of aminoethylated α-chymotrypsin, which are exceedingly difficult to separate owing to a strong tendency to aggregate, were easily separated on Sephadex G-50 when hexafluoroacetone was used as the solvent [170]. Aqueous solutions of formic acid containing either 8 M urea or 4 M guanidine hydrochloride have also been useful in fractionating insoluble peptides [645]. The purification of the polypeptide hormone oxytocin has been effectively accomplished by partition chromatography on Sephadex G–25 [1160]. Similar partition columns have proved useful in the fractionation of DNP-peptides [636]. Excellent recoveries are obtained from gel columns, and problems with the irreversible adsorption of solutes have not been reported.

C. Preparative Paper Electrophoresis and Chromatography

The fractionation of peptide mixtures by preparative paper techniques is widely employed in protein sequence studies. The methods are versatile, permitting the processing of small as well as relatively large quantities of material. The separation can involve chromatography or electrophoresis or a combination of both. It is wise to keep the number of steps to a minimum, since yields decrease as the number of manipulations increases. Thus, complex mixtures of peptides are best resolved into simpler mixtures by ion-exchange chromatography, which may be then subjected to a final one-step purification on paper.

1. Electrophoresis

Separations can be conducted with either a flat plate or tank type electrophoresis apparatus. The latter piece of equipment is described in detail by Katz *et al.* [527], and is essentially a modification of the original Michl design [683]. Commercial

Table 10. *Buffer systems for paper electrophoresis*

pH	Components	Ratio (by volume)	Reference
7.9	0.068 M N-ethyl-morpholine, 0.025 M acetic acid		[382]
6.5	Pyridine-glacial acetic acid-H_2O[a]	100:4:900	[845]
4.7	Pyridine-glacial acetic acid-H_2O[b]	1:1:48	[684]
3.6	Pyridine-glacial acetic acid-H_2O	1:10:89	[845]
1.9	Glacial acetic acid-formic acid (98%)-H_2O	87:25:888	[44]

[a] Dilute with an equal volume of water before use.
[b] The paper is wetted with a 1:10 dilution of this buffer.

versions are available from Savant Instruments (Hicksville, N.Y.), Gilson Medical Electronics (Middleton, Wisconsin), and O. Hiller Co. (Madison, Wisconsin). Whatman 3MM is most commonly used and is obtainable as sheets ($18^1/_4 \times 22^1/_2$ in) or rolls ($18'' \times 300'$). The appropriate buffer system (Table 10) and the time of electrophoresis necessary to separate the peptides in a mixture are determined by trial runs with about 0.05 micromoles of material. It is important that the peptide solution be free of salts in order to achieve satisfactory separations. Small amounts of pyridine acetate or ammonium bicarbonate will not interfere, since these salts volatilize during the drying process. If the sample is dilute, it is convenient to remove all solvent by rotary evaporation at 40 °C and redissolve at a concentration of about 10 μmoles/ml so that the application of a large volume is unnecessary. The material to be fractionated is applied as a narrow uniform zone across the width of the paper to within 6 cm of each edge; overloading should be avoided since streaking and poor resolution can result. Approximately 0.4 μmole of peptide is applied per cm and usually not more than 10 μmoles are applied to a single sheet. The area of application should be supported so that it is not in direct contact with a solid surface. Drying is facilitated by directing a stream of unheated air from a hand hair dryer across the sample zone. The position of the origin with reference to the anode and cathode can be adjusted depending on the electrophoretic properties of the respective peptides so as to provide optimal distance for migration. The paper is then transferred to a flat glass surface and a glass rod is inserted under the sample zone. The paper is wetted with buffer to within about 1 cm of each side of the applied sample. Buffer is allowed to flow evenly and slowly into the sample zone from both sides, thus concentrating and sharpening the peptide band. This is a most important step

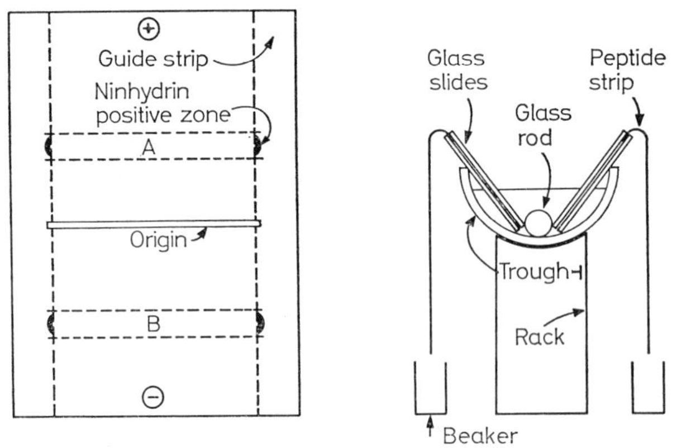

a b

Fig. 6-9 a and b. (a) Reconstructed (actual size, $22^1/_2 \times 18^1/_4''$) electrophoretogram. Guide strips have been stained with ninhydrin. Zones A and B enclosed by the horizontal dashed lines indicate the areas containing peptide material. (b) End view of peptide elution apparatus; the cover is not shown

and must be performed with great care; repeated small applications of buffer are usually necessary. The paper on both sides of the sample is carefully blotted to remove excess buffer and to provide a paper of reasonably uniform wetness. Electrophoresis is carried out at a potential gradient of approximately 50 V/cm for the required period of time. After electrophoresis, papers are clipped to a glass rod and dried in a forced air oven at 40 °C. To locate the position of the peptides, guide strips which include approximately 6 mm of each end of the sample zone are cut from the paper, sprayed with ninhydrin solution, and heated at 80 °C for 20 min to develop the stain. If necessary, a narrow strip may also be cut from the center of the paper and stained, thus providing a third point of reference for establishing the peptide zone. Tryptophan peptides are detectable under ultraviolet light as a fluorescent band and may be located directly without staining. The electrophoretogram is reconstructed by attaching the stained strips to the unstained center section with transparent tape at the extreme ends of the paper (Fig. 6-9a). The regions of the paper containing the peptide bands are outlined with a ruler and pencil, labeled, and cut with scissors. Rubber gloves or forceps should be used to handle the paper strips containing the peptide. If the peptide bands are well separated, an additional 6 to 8 mm on each side of the band is included in order to compensate for any irregularities in the shape of the zone. Peptides are eluted from the paper with 20% acetic acid or some other suitable solvent (Fig. 6-9b). The homogeneity of the purified peptide should be ascertained by electrophoresis or chromatography in at least two different systems prior to amino acid analysis.

2. Chromatography

Descending chromatography is performed with full-sized sheets of Whatman 3MM. The origin, guide strips, and lines to indicate folding are drawn on the paper as illustrated in Fig. 6-10. The bottom edge of the paper is serrated with pinking

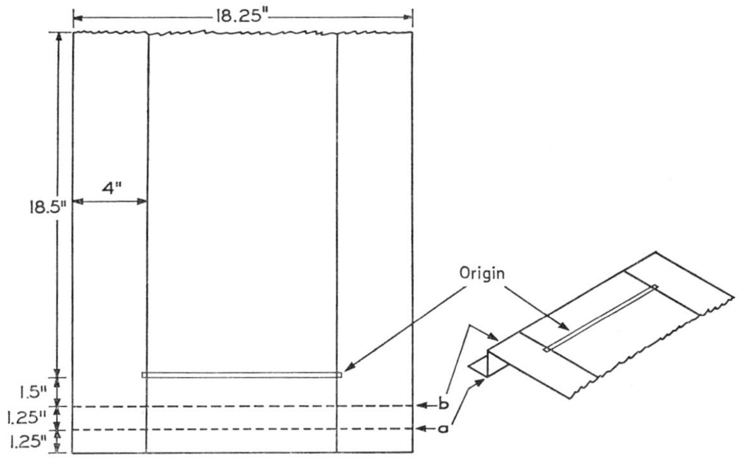

Fig. 6-10. Diagram illustrating preparative paper chromatogram. See text for details

shears so that the solvent will run off evenly should the solvent front progress that far. The paper is folded along lines *a* and *b* as indicated (Fig. 6-10), and the salt-free sample is applied evenly to the origin, which should not exceed 6 mm in width. The chromatogram is developed for a period of 16 to 20 h with the appropriate solvent (Table 11). After chromatography the paper is dried, and the guide strips are removed and stained. The peptide zones are located, cut out, and eluted as described for preparative electrophoresis. Generally, 6 to 10 μmoles of a peptide mixture may be fractionated on a single sheet.

3. Assessment of Paper Techniques

Electrophoretic separations require considerably less time than do the chromatographic procedures. Generally, the complete fractionation of a peptide mixture by electrophoresis can be accomplished easily in 1 day, whereas the chromatographic method consumes about twice this amount of time. Recoveries from paper will depend upon the properties of the peptide but usually range between 70 to 90%. Some peptides exhibit very strong adsorption to paper and are obtained in rather low yields. If small amounts (less than 1 μmole) of material are being fractionated, Whatman No. 1 paper may be used, in which case approximately 0.1 μmole of peptide per cm is applied. The main advantages of paper techniques are speed and simplicity. Recently, the development of the diagonal paper electrophoretic technique

Table 11. *Solvent systems for paper chromatography of peptides*

Components	Ratio (by volume)	Reference
1-Butanol-glacial acetic acid-H_2O	200:30:75	[622]
1-Butanol-glacial acetic acid-H_2O[a]	4:1:5	[527]
1-Butanol-pyridine-glacial-acetic acid-H_2O	75:50:15:60	[622]
Pyridine-*iso*-amyl alcohol-H_2O	7:7:6	[909]
sec-Butyl alcohol-formic acid-H_2O	7.0:0.9:2.1	[380]

[a] The upper phase is used for chromatography and the lower aqueous phase is placed in the bottom of the chromatographic chamber.

has permitted the location of disulfide bridges in proteins [160, 161, 312, 962], the selective purification of methionine peptides [1038], and the preferential isolation of lysine peptides from tryptic digests of trifluoroacetylated proteins [781].

Acknowledgement

The preparation of this chapter was aided by a grant (CA 07175) from the National Cancer Institute and by the Alexander and Margaret Stewart Trust Fund.

Chapter 7

Identification of Specific Amino Acid Residues

ERNESTO SCOFFONE and ANGELO FONTANA

I. Introduction

The determination of amino acid residues in proteins has been emphasized in recent years because of the significance of sequence studies, the importance of homologies and the evaluation of genetic changes in evolution.

Concomitant with the discovery of the amino acids, methods were explored for the determination of each amino acid. During the early period, gravimetric and then colorimetric methods were extensively employed. Only a limited number of amino acids could possibly be estimated by these procedures and, furthermore, the precision of analysis was far from satisfactory. In the same period, microbiological techniques of measurement were introduced which permitted the determination of most amino acids.

The complete separation of all common amino acids by a simple procedure and their determination awaited the development of ion exchange column procedures. With the present techniques a complete amino acid analysis can be performed in a short time on micro-quantities with a minimum of effort and provides values of high precision. The use of the amino acid analyzer of SPACKMAN, STEIN and MOORE [987] plays a fundamental role in protein chemistry.

Determination of the content of a particular amino acid in the intact protein is feasible only in the case when a reactive side chain is implicated. In such instances, with application of appropriate procedures, a quantitative measure can be obtained without the requirement of prior manipulation of the protein sample. Such reactive functional groups include the indole nucleus of tryptophan, the thiol group of cysteine, the disulfide linkage of cystine, the phenolic group of tyrosine, the imidazole ring of histidine, the guanidyl group of arginine.

The limitation inherent in such analytical methods lies in the nature of the protein, since its secondary and tertiary structure allows different degrees of reactivity of the functional groups; reaction will occur only with those groups which are openly available or unhindered. Therefore for analytical purposes the reactions are carried out in denaturing conditions for the protein, in which loss of organization is accompanied by changes in the reactivity of groups.

Besides the analytical interest, the reactivity of functional groups in a protein is a powerful tool in correlating structure and biological function. For example selective modification of a particular side-chain group allows conclusions about the participation of an amino acid residue in the active site or about the availability or the degree of burial of such a residue in the protein molecule. For comprehensive reviews about this topic see PUTNAM [804], WITKOP [1143] and TIMASHEFF and GORBUNOFF [1058].

During acidic hydrolysis a substantial loss of tryptophan and, to a lesser extent, of sulfur amino acids occurs [619]. On the other hand alkaline hydrolysis accompanies complete destruction of cysteine and cystine and is restricted to the determination of tryptophan, which is stable under these conditions. Accordingly, analytical methods for estimation of these residues in the intact protein are of great interest and of needful use.

In this chapter we shall be concerned primarily with analytical methods for sulfur amino acids and tryptophan and only procedures giving reliable results will be discussed in detail; other amino acids methods will be reported more briefly.

II. Presentation of the Results

The results of the quantitative determination of single amino acid residues in proteins are best expressed as numbers of residues per molecule of protein. However, in general, an exact determination of the molecular weight of the protein is not always available and, therefore, the results usually are expressed as "residues/10^5 g of protein" or "grams of amino acid per 100 g of protein". In this way, any ambiguities due to uncertain molecular weights are removed and a direct comparison of results of various workers is possible [145, 1068].

If the weight of solid protein is taken as the basis for the calculation, consideration must be given to the salt, ash and moisture content of the sample. Before analysis, salts are removed by exhaustive dialysis against 0.1 M KCl in order to displace other polyions by K^+ ion and then by dialysis against deionized water to remove the KCl; the protein is then recovered by lyophilization. Analytical determinations are best carried out on the air-equilibrated samples rather than with anhydrous proteins, which may be very hygroscopic. The moisture content of the sample is determined by drying *in vacuo* at 105 °C to constant weight [147, 200]. The ash content is determined by combustion in a platinum crucible over an open flame until constant weight is attained.

The total protein nitrogen is indicative of the state of the protein and its accurate estimation is a good indication of the purity of the protein sample. The percentage of nitrogen should be 13.5 or above. If the nitrogen content falls below this limit, it is essential to remove non-protein substances. The standard micro-analytical procedures of KJELDAHL is the most widely used for the determination of nitrogen. For the description of the standard procedure adopted the reader is referred to the original literature [131, 200, 782] or to the review by BLOCK [103] on this matter.

The protein concentration in solution can be referred either to the nitrogen content of the solution [105] or by using methods which estimate particular groups. Several sensitive and rapid colorimetric analysis are usefully employed, which allow high precision, when appropriate controls and careful calibration of the methods are made. For this purpose the phenol reagent of Folin [308] following the Lowry method [628] and the biuret procedure of GORNALL [363] are the most extensively used.

A. Lowry's Method [628]

Reagents. Reagent A, 2% Na_2CO_3 in 0.1 N NaOH. Reagent B, 0.5% $CuSO_4 \cdot 5H_2O$ in 1% sodium or potassium tartrate. Reagent C, alkaline copper solution. Mix 50 ml of Reagent A with 1 ml of Reagent B and discard after 1 day. Reagent D, diluted Folin reagent. Titrate Folin-Ciocalteu phenol reagent [308] (commercially available, *e.g.*, Fluka AG, Basle, Switzerland) with NaOH to a phenolphthalein end-point. On the basis of this titration dilute the Folin reagent (about 2-fold) to make it 1 N in acid.

Procedure. To a sample (25 to 500γ of protein) in 1 ml solution, 5 ml of Reagent C is added. Mix well and allow to stand for 10 min or longer at room temperature. Reagent D (0.5 ml) is added very rapidly and mixed within a second or two. After 30 min or longer, the sample is read spectrophotometrically at 750 mμ. If the protein is present in a dilute solution any convenient fraction of the total volume may be employed for analysis. Calculate from a standard curve obtained using a test protein (crystalline bovine albumin).

B. Biuret Reaction [363]

Reagent. Weigh 1.5 g cupric sulfate ($CuSO_4 \cdot 5\,H_2O$) and 6 g of sodium potassium tartrate ($NaKC_4H_4O_6 \cdot H_2O$); transfer to 1 l volumetric flask and dissolve in about 500 ml of water. Add with constant swirling 300 ml of 10% NaOH. Make to volume with water, mix, and store in a paraffin-lined bottle. This reagent should keep indefinitely but must be discarded if, as a result of contamination or faulty preparation, it shows signs of depositing any black or reddish precipitate.

Procedure. To the sample dissolved in 1 ml solution add 4 ml of the reagent. Allow the solution to stand for 30 min at room temperature and the absorbance at 540 mμ is then determined. Calculate the concentration from a calibration curve obtained using a test protein (albumin, casein).

It must be emphasized that the choice of proteins for testing analytical methods of amino acids requires caution, since it is clear that comparison of analytical data becomes ambigous when test proteins of poorly established purity and homogeneity are used. For this reason, the discrepancies in values reported in the literature often may be regarded as arising from the use of improperly defined test proteins, rather than in differing accuracies. Many workers have used casein as a test protein, but this material is by no means of sufficiently constant composition to warrant its use for this purpose [195, 1068].

Other methods for determining the protein concentration in solution involve the Sakaguchi reaction for arginine [402, 539] and the coupling of histidine and tyrosine with diazotized sulfanilic acid [402, 448c, 1108].

At the present time, however, the best and most precise procedure is the convenient and rapid estimation of the amino acid content by automatic amino acid analysis of a protein hydrolysate (6 N HCl, 110 °C, 24 h) obtained from an aliquot of the solution. By such means the analytical data are referred to the recovery of one or more amino acids in the protein hydrolysate. When the amino acid composition of the protein is known, its concentration in solution is calculated and the data are directly reported as residues per molecule.

III. Estimation of Tryptophan

Although the amino acid composition of many proteins has been reported, evaluation of the number of tryptophan residues often has been lacking. This situation reflects the difficulty associated with quantitative determination of tryptophan residues in a protein since acid hydrolysis results in the destruction of tryptophan [448c]. On the other hand, alkaline hydrolysis, though less destructive [990], allows determination of tryptophan by chromatographic separation of the protein hydrolysate [1111b], but generally does not give quantitative recovery. Indeed, in the presence of other protein constituents (sulfur amino acids), the procedure gives unsatisfactory [758, 1068] results; *e.g.*, lysozyme was believed to have 3 to 4 residues until sequence studies revealed 6 [176b, 177]. Enzymatic hydrolysis of proteins can give quantitative yields of tryptophan but this method may not be generally valid [436b].

The hydrolytic problem can be circumvented by measuring tryptophan in the intact protein.

The indole groups of proteins have been determined by a variety of methods, some of which have proved more useful than others. The discrepancies noted for tryptophan values in the literature should probably be considered the result of approximation.

Currently, the spectrophotometric method of Goodwin and Morton [360], the color reaction with the Ehrlich reagent (*p*-dimethylaminobenzaldehyde) [990] and the oxidation with N-bromosuccinimide [772, 813, 1068] are widely used.

More recently Barman and Koshland [55] reported the use of 2-hydroxy-5-nitrobenzyl bromide as a reagent for a quantitative estimation of tryptophan. In the author's laboratory has been developed a method involving the use of sulfenyl halides for the estimation of tryptophan as well as of the cysteine content in proteins (see Section IV.C.2) [316, 317, 935].

Herewith these methods are discussed in more detail.

A. Spectrophotometry

In the absence of prosthetic groups or nucleic acids, the absorption of proteins in the ultraviolet region between 250 to 310 mμ is due to the contribution to absorption of the three aromatic residues, namely, tryptophan, tyrosine and phenylalanine (cf. also Beaven and Holiday [61]). Since phenylalanine has a low molar extinction ($\mathscr{E} = 195$ at 257.5 mμ) and the maxima of absorption of the other two aromatic residues are shifted to higher wavelengths, the absorption of proteins at 280 to 290 mμ is due only to tryptophan and tyrosine. Therefore, if the small contribution of phenylalanine is neglected, the absorption of the solution of a protein can be considered as arising from a mixture of the two amino acids.

This fact was used by Goodwin and Morton [360] as the basis in their spectrophotometric determination of both these amino acids in proteins.

The procedure they followed was to dissolve the protein in 0.1 N NaOH and make up to a convenient volume in order to obtain a absorption of 0.2 to 0.5 near 290 mμ. The total molar concentration (*a*) of tryptophan and tyrosine is obtained [Eq. (1)] from the absorption at 294.4 mμ, one of the wavelengths at which the absorption curves of the two amino acids intersect [61] and therefore have the same extinction ($\mathscr{E}$).

$$A_{294.4} = a\mathscr{E} . \tag{1}$$

From the extinction at another wavelength, 280 mμ, the concentrations of the two components can be calculated. If at 280 mμ the molar extinction of tryptophan ($\mathscr{E}_1$) and of tyrosine ($\mathscr{E}_2$) is known, the observed absorption A_{280} is given by Eq. (2)

$$A_{280} = b\mathscr{E}_1 + (a-b)\mathscr{E}_2 \tag{2}$$

(where b is the concentration of tryptophan and $a-b$ the concentration of tyrosine) and b by Eq. (3).

$$b = \frac{A_{280} - a\mathscr{E}_2}{\mathscr{E}_1 - \mathscr{E}_2} . \tag{3}$$

After determination of the protein concentration by standard methods (see Section II), the tryptophan residues/protein (mole/mole) is obtained from the ratio of the two concentrations.

However, the method has not gained general acceptance because of the possibilities of anomalies.

Tyrosine residues in proteins sometimes behave abnormally [1058], and incubation at pH 13 is necessary for maximal ionization of the phenolic group [216]. Therefore the method must be tested after exposure of the protein to pH 13 for varying lengths of time. On the other hand it was found by Spies and Chambers [990] that a slight destruction of tryptophan occurs in 0.1 N alkaline solution.

The original application of the method to proteins was based on the assumption that the absorption of amino acids was unaffected when linked in a polypeptide chain. Later work has suggested that the use of the amino acids in peptide combination is more satisfactory, *e.g.*, glycyl-tryptophan [276], since a shift (1 to 3 mμ) in the maxima of absorption toward longer wavelengths of the free amino acids when incorporated in proteins occurs. Bencze and Schmid [64] modified the procedure of Goodwin and Morton [360] in order to overcome this problem.

More recently Edelhoch [261] further improved the method performing the absorbance measurements at 288 and 180 mμ on the protein, dissolved in 6 M guanidine hydrochloride in order to destroy its secondary and tertiary structure. Blocked tryptophanyl and tyrosyl compounds (N-acetyl-tryptophan amide, glycyl-tyrosyl-glycine) were selected as model substances to represent the absorbance of the two chromophores in proteins. However, although the method was shown to give values of tryptophan content in very good agreement with those known from sequence studies (trypsinogen, chymotrypsinogen, lysozyme), a imperfect fit of the spectra of the solutions containing the model compounds to those of the proteins was observed. The latter are shifted to the red in all three cases examined.

Great difficulties were encountered in the determination of tryptophan by the spectrophotometric method, when this amino acid was present in the protein at a low level or when the tyrosine content was high. As an example, the A protein component of tryptophan synthetase of *Escherichia coli* was thought to contain tryptophan [434] until the absence of this amino acid was proved by other methods (N-bromosuccinimide [771], alkaline hydrolysis, microbiological assay).

B. Colorimetry

1. p-Dimethylaminobenzaldehyde

The long known reaction of the indole nucleus of tryptophan in strong acid conditions with aldehydes to give colored products [1091, 1140] was employed by SPIES and CHAMBERS [990] in a colorimetric procedure for the determination of tryptophan. In this method, the tryptophan-containing compound (peptides or proteins) is dissolved in concentrated sulfuric acid containing *p*-dimethylamino-benzaldehyde (DAB) and a blue color is obtained after a prescribed time by adding sodium nitrite solution.

Procedure [990]. The sample (for lysozyme, 1 mg) to be analyzed is dissolved in 0.5 ml water and added to a mixture of 4 ml of 23.8 N H_2SO_4 and 0.5 ml of DAB solution in 2 N H_2SO_4 (30 mg/ml). After about 15 h at 22 to 24 °C in the dark, to the reaction mixture is then added 0.1 ml of a freshly prepared 0.045% aqueous solution of $NaNO_2$. After standing further for 30 to 60 min in the dark, the optical density of the solution at 590 mμ is measured.

The tryptophan content is calculated from a standard curve obtained with tryptophan or with a tryptophan-containing peptide, after correction for a blank value.

This procedure has been extensively used and gives reliable results. However, the reaction was found to be sensitive to SH groups, and gives for different proteins variations in the maximum of absorption (from a normal value of 590 to 600mμ to 545 to 560 mμ). In order to avoid such differences the procedure was applied to a partial digest of the protein obtained by the action of trypsin and chymotrypsin [423].

BOYD [139] showed that color formation was influenced by oxidizing and reducing agents, exposure to light and by the presence of compounds such as toluene.

Reactions of the indole nucleus of tryptophan with xanthydrol [243, 244] and with glyoxylic acid [940] also have been employed for analytical purpose, but the procedures have thus far received only limited application.

2. 2-Hydroxy-5-nitrobenzyl Bromide

KOSHLAND *et al.* [467, 573] introduced 2-hydroxy-5-nitrobenzyl bromide (HNBB) (I) as a tryptophan-modifying reagent. It is specific for tryptophan in proteins which

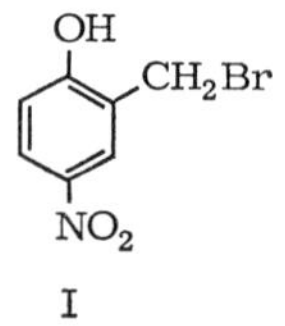

I

do not contain free SH groups and has been used by BARMAN and KOSHLAND [55] for the quantitative determinations of tryptophan residues in lysozyme, α-chymotrypsin, trypsin and glucagon.

The reagent reacts with tryptophan-containing proteins so that a chromophore is generated in the protein which absorbs radiation in the visible part of the spectrum.

Assuming that the chromophore covalently bound at the tryptophan residue has the same molar extinction as the model compound 2-hydroxy-5-nitrobenzyl alcohol ($\mathscr{E} = 18{,}000$ at 410 mμ in alkali solution), the reaction is easily quantitated by spectrophotometry. After reaction with HNBB, the absorption of the HNB-protein, free from excess reagent, in alkaline solution allows calculation of the number of tryptophan residues modified by HNBB, provided that the protein does not contain free SH groups.

Procedure [55]. The protein (about 1 μmole) to be analyzed is incubated for 16 to 20 h at 37 °C in 1 ml of 10 M urea which had been adjusted to pH 2.7 with concentrated HCl. The urea solution in then cooled at room temperature and approximately 5 mg of HNBB in 0.1 ml of dry acetone are added under vigorous stirring to the protein solution. The mixture usually remains clear but occasionally a slight precipitate forms (usually 2-hydroxy-5-nitrobenzyl alcohol) which is removed by centrifugation. The labeled protein is separated from excess reagent on a column (23 × 1.1 cm) of Sephadex G-25 which had previously been equilibrated with 0.18 M acetic acid (pH 2.7) or with 10 M urea (pH 2.7). The 10 M urea is used when necessary to keep the modified protein in solution.

The protein in the pooled protein fractions is precipitated by addition of 50% trichloroacetic acid to a final concentration of 5% trichloroacetic acid. If urea is used instead of acetic acid, 5-fold dilution with water is necessary before trichloroacetic acid addition. The precipitate is collected by centrifugation, washed twice with 5 ml of ethanol-HCl (2 ml of concentrated HCl to 98 ml of 95% ethanol), and finally dissolved in 1 ml of 11.2 M HCl. A portion of the HCl solution (usually 0.1 ml) is adjusted to pH 12 by addition of 2.5 M NaOH and diluted if necessary to a final volume of 2.5 ml. The concentration of the reacted tryptophan in this solution is determined by spectrophotometric assay at 410 mμ by using an extinction coefficient of 18,000 M^{-1} cm^{-1}. The remaining aliquot can be analyzed for protein content (see Section II).

The Koshland's analytical method has been shown to be fairly accurate [55], but appears to have some disadvantages.

On the basis of the findings of Spande, Wilchek and Witkop [988] the reaction of HNBB with the indole nucleus of tryptophan is a complicated one.

When the model compound N-acetyl-tryptophan methyl ester was reacted with two equivalents of HNBB a mixture of three products was obtained. The major component (37%), isolated after acetylation with acetic anhydride-pyridine, was shown to be II.

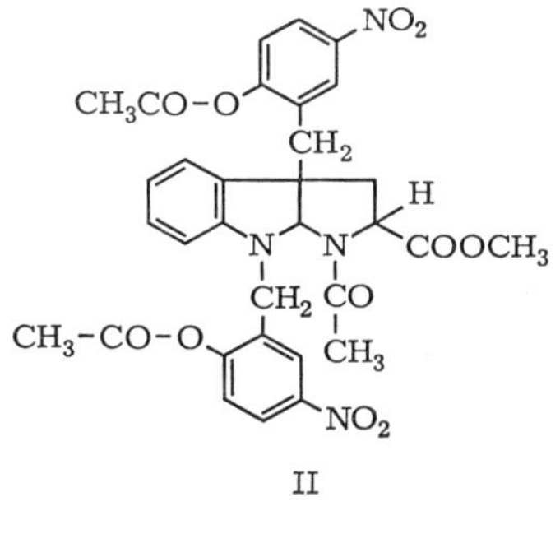

II

This fact agrees with the known polyalkylation of the indole ring by alkylating agents [506] and explains the excessive incorporation of chromophore into the protein when using a high concentration of the reagent. Accordingly, positive errors in the determination of the tryptophan content of proteins can arise [55].

The reaction of HNBB takes place also with cysteine and a procedure of blocking this residue by carboxymethylation is needed prior to treatment of the SH-containing protein with the reagent.

Finally, a quantitative reaction is achieved only under conditions of unfolding of the protein by dissolving it in 10 M urea or at best by cleavage of the S–S bridges. The use of 5 M guanidine hydrochloride as an unfolding reagent causes incorporation of the yellow label not only in tryptophan containing proteins, but also in ribonuclease, a tryptophan- and cysteine-free protein [55].

3. Sulfenyl Halides

Recently the use of sulfenyl halides as a selective modifying reagent for tryptophan and cysteine residues in proteins was reported [316, 935]. In acidic solution, tryptophan is attacked at the 2-position of the indole nucleus to give a thioether (Eq. 4), whereas cysteine is converted to a mixed disulfide (Eq. 5).

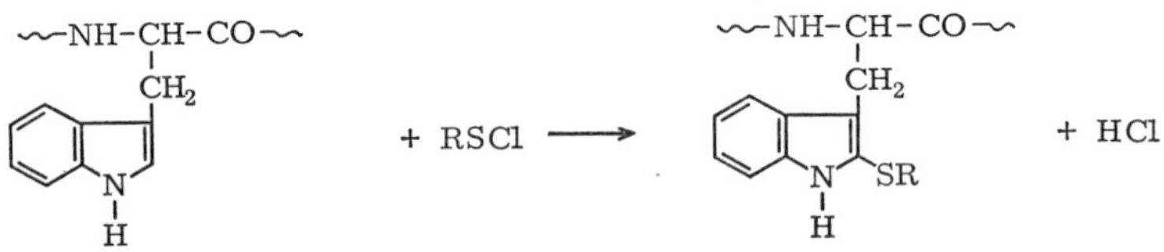

(4)

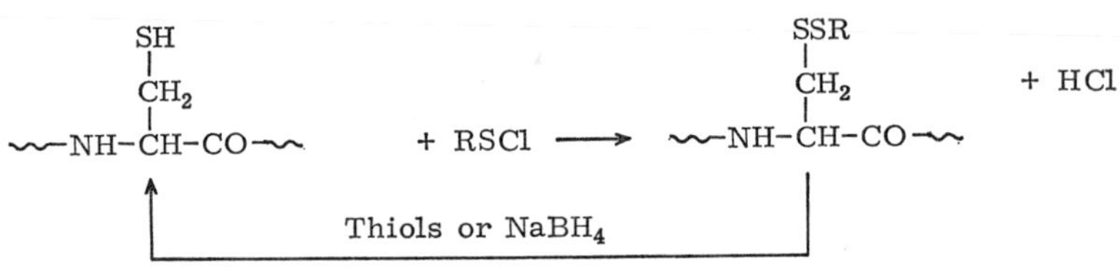

(5)

The unsymmetrical disulfide at the cysteine residue can be easily cleaved by means of reducing agents, such as β-mercaptoethanol, thioglycolic acid or sodium borohydride, or by simply exposing it to alkaline treatment (0.1 N NaOH) [316], thereby allowing specific labeling only of the tryptophan residue. For the analytical application of the reaction to cysteine the reader is referred below (Section IV.C.2).

By using nitrophenylsulfenyl halides a chromophore is generated in the protein which absorbs in the visible part of the spectrum. This fact offers an easy quantitation of the reaction by spectrophotometry. At the present time the most widely used sulfenyl halide is 2-nitrophenylsulfenyl chloride (NPS-Cl) (III).

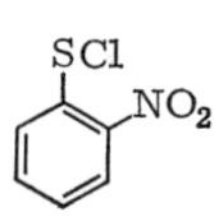

III

It was shown [1090] that the sulfenyl halide reacts with the indole nucleus of tryptophan much faster when this residue is linked in the protein than in free tryptophan and in small peptides. This enhanced reactivity enables one to perform the sulfenylation reaction in aqueous acetic acid (50%), in spite of the fact that a rapid (2 to 3 min) decomposition of the reagent occurs, provided that the halide is first dissolved in glacial acetic acid. Quantitative modification of tryptophan residues in proteins is achieved readily, by using a low molar ratio of reagent to protein [316, 1090].

The 2-thio-(2-nitrophenyl)-tryptophan residue, obtained by reaction of NPS-Cl with tryptophan-containing proteins absorbs at 365 mμ with a molar absorptivity $\mathscr{E} = 4{,}000\ M^{-1}cm^{-1}$ in aqueous acetic acid. The accuracy of the procedure was tested on a number of proteins whose tryptophan contents were known from sequence studies, *e.g.*, lysozyme, α-chymotrypsin and trypsin [935].

Procedure [935]. To a solution of about 1 μmole of the protein in 1 to 2 ml of 50% aqueous acetic acid, 20 to 30 equivalents of NPS-Cl (Fluka AG, Basle, Switzerland) in 0.5 ml of glacial acetic acid are added at room temperature under vigorous stirring. When the protein to be analyzed is insoluble in aqueous acetic acid, formic acid can be used [313]. The modified protein is precipitated at − 5 °C with acetone-1 N HCl (39:1), separated by centrifugation, and carefully washed several times with acetone and ethyl ether. In some instances, the precipitation and wash procedure is repeated in order to remove any excess reagent. The product is dried *in vacuo* over P_2O_5. By the procedure reported above, it was shown that practically no products arising from the excess of reagent are present in the samples of the modified proteins. In any case the sample can be further purified by gel filtration on a column of Sephadex G-25 using aqueous acetic acid as the eluent, following known procedure.

If the protein to be analyzed for tryptophan content is supposed to contain cysteine residues, reactive toward sulfenyl halides, a rough spectrophotometric test is made. By dissolving the sample of the modified protein in 0.1 N NaOH, the 2-nitrothiophenol released absorbs at 420 mμ and cause a evident variation in the spectrum of the modified proteins [313].

In order to remove the cysteinyl chromophoric group in a cysteine- and tryptophan-containing protein, an aliquot of the modified protein (20 mg) is dissolved in 2 ml of 8 M urea solution, adjusted to pH 8.6 with 5% methyl/amine after flushing the solution with a nitrogen stream, treated with 20 μl of β-mercaptoethanol, following the standard procedure for cleaving disulfide bonds in proteins (see Section V.B). After 4 h at room temperature, the entire solution is applied to a 1 × 40 cm column of Sephadex G-25 which had previously been equilibrated with aqueous acetic acid (0.2 M). After elution, the protein is located in the effluent (spectrophotometry) and recovered by lyophilization. Similar results can be obtained by cleaving the mixed-disulfide bond by dissolving the sample of the modified protein in 0.1 N NaOH and recovering the protein by acid precipitation or by gel filtration [313].

Calculations. In order to estimate the number of 2-thio-(2-nitrophenyl)-tryptophan residues per molecule of protein, the sample of the modified protein (about 0.1 μmole) is dissolved in 80% acetic acid (5 ml) or, for less soluble samples, in 99% formic acid and the concentration of the covalently bound chromophore determined by spectrophotometric assay at 365 mμ with the use of an extinction coefficient of 4000.

The protein content in the solution can be analyzed in several ways, the most accurate being the use of the amino acid analyzer (see Section II).

C. Other Methods

The use of N-bromosuccinimide (NBS) as a specific reagent for the modification of tryptophan in proteins has led to a rapid and convenient spectrophotometric method for determining the tryptophan content of a protein [772, 813, 1143].

The procedure is based on the large decrease in absorbance at 280 mμ which accompanies the oxidation of tryptophan to an oxindole chromophore.

The action of NBS with tryptophan-containing proteins in dilute aqueous solutions (2×10^{-4}M) is followed *in situ* with a recording ultraviolet spectrophotometer, following the disappearance of the indole absorption at 280 mμ and the concomitant appearance of a new band at 240 to 250 mμ and a low-intensity band at 307 mμ. The effect of added NBS on the indole spectrum is instantaneous and linear up to the consumption of ca. 1.5 moles/mole of tryptophan with optimal condition at pH 4 in aqueous acetate buffer. Multiplication of the decrease in optical density at 280 my by an empirical factor (1.31) gives the extinction due to tryptophan in the protein. In these estimations the amount of tryptophan in the sample solution is calculated using an extinction coefficient for tryptophan at 280 mμ of 5,500 [813].

It was noted that the amount of the oxidizing agent taken up to obtain minimum absorbance at 280 mμ varies with the nature of the protein and the medium used [813].

A tentative procedure for determining the tryptophan content in a protein was recently proposed by PREVIERO *et al.* [802]. The method involves formylation of the 1-position of the indole nucleus of tryptophan by dissolving the protein in formic acid-HCl solution (Eq. (6)].

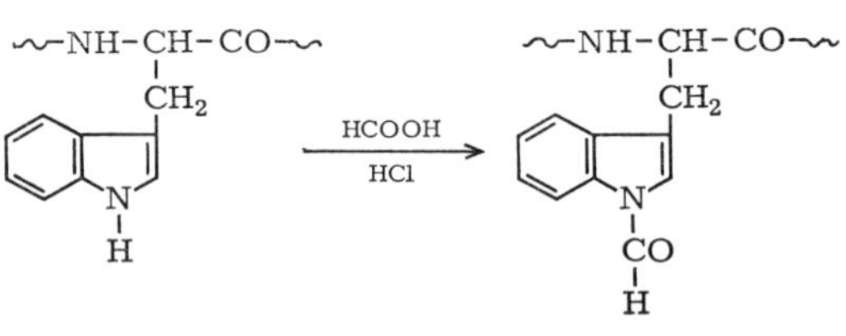

(6)

The high yield (over 90%) of specific formylation of tryptophan, concomitant to the shift of absorption to longer wavelengths (298 mμ), allows a spectrophotometric determination of tryptophan in proteins.

Another procedure involves the selective photochemical conversion of tryptophan to kynurenine [63, 338]. Since kynurenine is stable to acid hydrolysis, the tryptophan content of a protein may be related to the recovery of kynurenine on the amino acid analyzer, after proflavine-sensitized photooxidation of the protein sample in 99% formic acid followed by acid hydrolysis. The procedure was tested with lysozyme in which case all the six tryptophan residues were converted to kynurenine [338].

IV. Estimation of Sulfhydryl Groups

The SH group has undoubtely received more attention than any other reactive group in the protein molecule for analytical purposes as well for structure-activity relationship studies. This is due primarily to the relatively large number of chemical methods available for modifying this highly reactive group. Further, since SH groups play, in many cases, a prominent role in enzymic catalysis, concomitant to their wide chemistry, a large number of investigations in this area have been developed.

In this section the meaningful results of the SH chemistry will be reviewed only with respect to their analytical interest.

A large number of methods for the estimation of thiols are available to biochemist, but no single method has emerged which is superior to all others. Since values are influenced by such factors as the reagent, pH, buffer, availability of the SH group to the reagent and, therefore, on the extent of denaturation, as a rule one compares the sulfhydryl content of a protein by different methods.

In addition to the references cited the reader may consult the excellent work of Boyer [141], which reviewed this topic particularly with respect to the functions of SH and S–S groups in enzymes. Also see Cecil and McPhee [187], Cecil [184] and Hamilton [402].

A. Mercaptide Formation

1. Silver and Mercury Titration

Reactions of heavy metals with SH groups carried to mercaptide formation [Eq. (7)] are among the most specific and the most useful ones.

$$\text{Protein-SH} + \text{Me}^{+} \longrightarrow \text{Protein-SMe} + \text{H}^{+} \qquad (7)$$

The marked affinity of certain metal cations for RS^- results from their tendency to accept a pair of electrons with the resultant formation of undissociated sulfides. Silver, mercury, copper, cadmium, zinc, lead, arsenic and antimony form mercaptides. The reagents which have perhaps received greatest attention are the silver and mercuric salts and organic metallic derivatives of the type R-Hg-X.

For an excellent discussion of the interaction of metallic ions with proteins, peptides and amino acids see the review of Gurd and Wilcox [384].

The equivalence point of the reaction of SH groups with metal ions is conveniently measured either polarographically or amperometrically. A detailed account of these techniques can be found in the review article of Cecil and McPhee [187] and in the standard books on the subject by Kolthoff and Lingane [560] and by Meites [681].

Polarography, as analytical tool for SH groups, in addition to the suitable application with heavy metal reagents, was employed also for following the reaction of SH groups with alkylating reagents (iodoacetic acid as well as α-bromoacetophenone) [284, 343, 768].

Since in the polarographic technique the magnitude of the current is controlled by the rate of diffusion of the reagent and it is directly proportional to its concentration, the measurement of the current allows the estimation of the reagent concentration and therefore of the extent of the reaction with SH groups. However, a more precise and widely used procedure consists of amperometric titration in which the amount of reagent added is plotted against the current.

Two types of electrodes are in general use, namely, the dropping mercury electrode, and the rotating platinum electrode. This last can be used at very positive potentials and therefore enables estimation to be made in the presence of oxygen. However, the method lacks reproducibility since the surface of the platinum can become contaminated [187].

The polarographic technique has found the most extensive application with salts of silver and mercury. The procedure is occasionally complicated by the possibility of errors.

It has been shown that the silver mercaptides of simple thiols have a strong tendency to bind additional silver ion [172, 186, 562, 959]. If this occurs to any extent under the conditions used for estimation of the SH groups in proteins, a positive error would result. This can be minimized by varying the pH of the reaction or by using complexing agents for the heavy metal, which forms an electro-reducible complex with the ion, so that all the heavy metal reagent is present either in the complex form or as the mercaptide. Examples of this procedure are the use of ammonia [562] or tris(hydroxymethyl)aminomethane buffer [70] with silver ion. Thiols were titrated with $Ag(NH_3)_2^+$ and the reaction was followed with a rotating platinum electrode, using ammonia and ammonium nitrate in 95% alcohol [563] as the supporting electrolyte. The protein is titrated until there is a slight excess of the titrating ion, shown by the increase in the current passing through the solution. The method was improved replacing the ammonia with tris(hydroxymethyl)aminomethane buffer as the complexing agent [70]. This method has been widely used and it appears to be reliable and to give good results.

The titration of thiols with Hg^{2+} have the disadvantage that ambiguities can arise from the valence of mercury. While simple thiol compounds form mercaptides of the type $(RS)_2Hg$, proteins give half-mercaptides (Protein-SHgX). Simple thiols only were estimated accurately using Hg^{2+}. One outstanding property of Hg^{2+}-complexes with certain proteins is the easy crystallizability, as in the case of lactic dehydrogenase [584] and papain [966].

2. p-Hydroxymercuribenzoate

At the present time the most widely used mercaptide forming reagent is *p*-hydroxymercuribenzoate (PMB),[1] first introduced by Hellerman *et al.* [433] as a specific reagent for thiol groups [Eq. (8)]. Mercury, linked through one covalent

$$\text{Protein-SH} + \text{R–Mg–X} \longrightarrow \text{Protein-SHgR} + \text{HX} \quad (8)$$

$$\text{Protein-SHgR} + \text{R}_1\text{SH} \longrightarrow \text{Protein-SH} + \text{R}_1\text{S–Hg–R} \quad (9)$$

bond to an organic moiety, reacts essentially with one SH group according to Eq. (8). For certain enzymes the reaction leads to inactivation and it can be reversed by addition of an excess of a mercaptan [Eq. (9)] [952].

[1] *p*-hydroxymercuribenzoate ($HO–Hg–C_6H_4COO^-$) is frequently sold as *p*-chloromercuribenzoate. When this name appears on a commercial label, care must be exercised, since the chlorine atom has, in many marketed lots of this substance, been replaced with the OH group during purification.

Like inorganic Hg^{2+}, organic mercurials have a strong affinity to bind, not only to SH, but, also to groups with an available pair of electrons (*e.g.*, amino and carboxyl groups [393]), although at low concentrations of reagent the reaction with SH dominates. In the use of mercurials for enzymic studies, attention must be given to their possible lack of specificity [141].

Boyer [140] described a spectrophotometric procedure to follow the reaction of PMB with SH and the method has gained wide acceptance because its relative simplicity and rapidity. The method is based on the fact that mercaptide formation by PMB at pH 4.6 or 7.0 is accompanied by a considerable increase in the ultraviolet absorption, and measure of the increase in the 250 mμ region offers a useful analytical tool.

Since measurements must be made in a region where proteins absorb, precautions need be taken to ensure corrections for the absorption of proteins. But, since practically no spectral changes may be observed in the presence of proteins lacking -SH groups, the specificity of the reaction is evident.

The use of detergents [647, 745, 764] and urea [618, 1027] was reported to allow measurements of less reactive SH groups.

Neuman *et al.* [731] has modified slightly the analytical method first described by Boyer [140] and this experimental procedure is herewith reported.

Reagent. It is difficult to obtain the reagent in pure form, so that solutions of PMB must be standardized spectrophotometrically either against standard solutions of cysteine hydrochloride or mercaptosuccinic acid [140]. The molar extinction of PMB at 232 mμ is 16,900 in water solution, pH 7. A suitable concentration of the reagent is $5\text{–}7 \times 10^{-4}$M [140].

Procedure [731]. Aliquots of a solution of PMB, adjusted to pH 7.0, is added to the protein solution in 0.1 M phosphate buffer, pH 7.0. The change in absorbance at 250 mμ is measured after each addition. The observed absorbance is corrected for dilution and plotted against the volume of reagent added. A blank, in which there is no protein, is treated in an identical manner The SH content of the protein solution is obtained from the equivalents of the added reagent at the end point of the titration (no further increase in absorption at 250 mμ).

B. Alkylating Agents

Alkylating agents (iodoacetate, bromoacetate) [323, 352, 408, 432] have been extensively used in protein chemistry as inhibitors of SH enzymes [141, 545, 633, 756]. Further important use of the alkylation reaction of SH groups lies in the stabilization of these residues in proteins prior to enzymic or chemical degradation (see Chapter 6).

The reaction also has been applied to the estimation of the thiol content. There is a wide choice of SH-alkylating reagents and they may be divided into two classes: i) those carrying a halogen which reacts with the elimination of halogen acid, ii) those with an active double bond to which the thiol groups add.

1. Haloacetates

The more extensively studied and the most useful reagents of type i) are iodoacetamide or the more soluble iodoacetic acid [323, 352, 408, 432] [Eq. (10)].

$$\text{Protein-SH} + \text{I-CH}_2\text{COOH} \longrightarrow \text{Protein-S-CH}_2\text{COOH} + \text{HI} \tag{10}$$

The reactions of thiols with active halogens compounds are bimolecular, nucleophilic substitution reactions, in which the nucleophilic reagent is the highly reactive mercaptide ion [57, 141]. However attack on the amino, phenolic and imidazole groups can occur, but alkylation of SH proceeds more rapidly and side reactions can be minimized by avoiding high concentrations of reagent and strongly alkaline conditions.

The reaction of iodoacetate with proteins is measured by ion exchange chromatographic analysis of S-carboxymethyl-cysteine after acid hydrolysis of the protein (recoveries over 90%) [207, 698) (see also Section V.B). When iodoacetamide is employed for the alkylation step, the S-carboxamido methyl-cysteine is converted by acid hydrolysis to the S-carboxymethyl-derivative.

Anfinsen and Haber [40] used iodoacetate-1-C^{14} for the coupling reaction and the S-carboxymethyl-cysteine content in the modified protein was determined directly by radioactivity measurements, after separation of excess reagent through Sephadex G–25.

Alternatively, the direct titration of the halogen acid formed by the reaction allows the determination of the thiol content [69]. This procedure is very sensitive, but is feasible only with simple model thiols.

2. N-Ethyl Maleimide

Thiol add to double bonds and the reaction has many features in common with the addition of halogen acids [659]. N-Ethyl maleimide (NEM) reacts rapidly and quantitatively with glutathione and thiolactic acid at pH 7.4 [Eq. (11)] [327, 328,

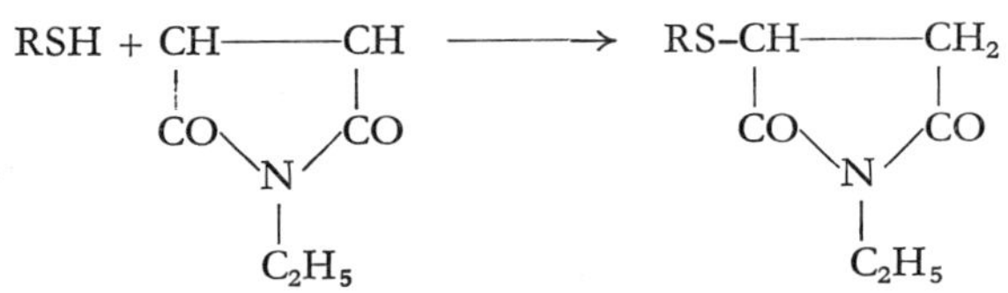

(11)

648]. This reaction has found useful applications in the inhibition of a number of enzymes [141] and for a quantitative estimation of SH groups in proteins [19, 237, 373, 605, 829].

The extent of reaction was measured by titration with iodine [328], or, in a better way, following the decrease in absorption at 300 mμ that occurs when NEM combines with thiols [19, 237, 373, 605, 829]. Alternatively in certain instance the amount of the acid hydrolysis product, S-succinyl-cysteine, has been quantitatively determined [383, 972, 989, 1067]. Excess NEM is required to obtain a stoichiometric reaction. The sensitivity is high; 1 μmole of cysteine/ml yields an absorbancy change of 0.60.

A disadvantage of the NEM reaction for analytical purpose in the intact protein is that a quantitative reaction is not always easily achieved, since proteins react more slowly with NEM than do simple thiols, as cysteine. Indeed it was found that, at pH 7, cysteine reacts rapidly but the enzyme myokinase [373] required 30 min for complete reaction and bovine serum albumin 55 min [829]. Therefore a pretreatment of the protein sample with urea solution is suggested in order to expose the less reactive SH groups [1161].

The reaction of NEM toward SH groups in proteins is not specific, but reaction occurs also with imidazole and α-amino groups [156, 973]. These side reactions however are not of great significance under the experimental conditions for thiol group determinations [156]. On the other hand Guidotti and Konigsberg [383], after treating carbon monoxyhemoglobin with NEM, noted a discrepancy between the decrease in absorbance at 300 mμ and the appearance of S-succinyl-cysteine in the hydrolysate. Their data suggested that the amino terminal valine of the α-chain was alkylated.

The reaction also has found useful application for the chromatographic determination of SH-containing compounds. The product of the reaction of NEM with thiols has been found to give an intense red color in alkaline solution [68].

C. Colorimetry

1. Ellman Reagent

The use of 5,5′-dithiobis-(2-nitrobenzoic acid) (DTNB) for the estimation of sulfhydryl groups was introduced by Ellman [279] and the procedure is now extensively used because of its ease and accuracy. The method involves reaction of thiol groups with the reagent at pH 7 to 8 to produce one mole of 4-nitro-3-carboxy-thiophenol ($\mathscr{E} = 13{,}600$ at 412 mμ) [Eq. (12)].

$$\text{Protein-S}^- + \text{R–S–S–R} \longrightarrow \text{Protein-S–S–R} + \text{RS}^- \qquad (12)$$

R = 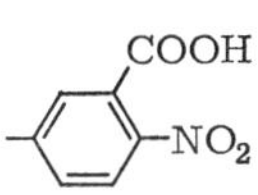

The color obtained in the Ellman assay requires metal-free reagents; it is modified by metal ions directly bound to the proteins (ferric, calcium) [637, 891]. Therefore 10^{-4}M EDTA is generally included in the reagent solution. In addition, in some instances, the reaction is performed under denaturing conditions (urea, sodium dodecylsulfate) to expose buried SH groups [220, 735, 1006].

Reagents. A choice of buffers is available, depending on the specific analytical requirements. Some of the most frequently used are listed below. The DTNB is dissolved (10^{-2}M) in the buffer and the solution flushed with nitrogen.

0.1 M phosphate, pH 7.9, 10^{-4} EDTA, 6 M urea
0.1 M Tris buffer, pH 8.0, 4 to 6 M urea
0.1 M Tris-maleate buffer, pH 7.0
0.1 M phosphate, pH 7.9, 5 M guanidine hydrochloride
0.1 M phosphate, pH 7.9, 1% sodium dodecylsulfate.

Procedure. The protein sample (0.1 ml) is added to 2.9 ml of the reagent solution and the absorbancy at 412 mμ is followed until no further increase is observed (about 30 to 60 min). The total SH content in the sample is calculated assuming

$\mathscr{E} = 13{,}600$ for the reagent. For more reliable results standardization of the method is recommended using a cysteine solution, standardized by iodometric titration.

After determination of the protein concentration by standard methods (see Section II), the SH-groups/protein (mole/mole) is obtained from the ratio of the two concentrations.

2. Sulfenyl Halides

The selective reaction of sulfenyl halides [316, 317, 935] with the thiol group of cysteine to form a mixed disulfide can usefully be employed for the estimation of the SH groups of a protein (for the estimation of tryptophan using this reaction see Section III.B.3).

Taking advantage of the ready cleavage in alkaline solution of alkyl-aryl disulfides with formation of arylthiophenol [766], a method was suggested [316] involving labeling of the SH-containing protein with *p*-nitrophenylsulfenyl chloride (p-NPS–Cl) (IV) in acetic or formic acid the p-nitrothiophenol moiety, covalently bound to the

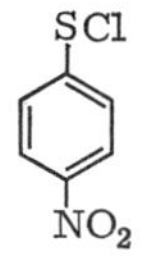

IV

protein by a disulfide linkage, is then released by exposing the sample of p-NPS-protein to alkaline media (0.1 N NaOH). The highly colored chromophore ($\mathscr{E} = 13{,}600$ at 412 mμ in 0.1 N NaOH) was shown to be quantitatively released from the S-p-NPS-derivatives, if deaerated solutions are employed.

The method was tested with model compounds (*e.g.*, N-acetyl-S-p-NPS-cysteine, S-p-NPS-glutathione) as well as with reduced ribonuclease. In this case the SH groups per molecule of enzyme were found to be 7.8 (theory 8 residues) [316].

When a protein containing both tryptophan and cysteine has to be analyzed for SH content, the procedure involving the reaction of pNPS-Cl with SH groups can still be employed, since the 2-thio-(4-nitrophenyl)-tryptophan residue (absorption peak near 323 mμ) [935] shows almost no absorption at the λ max (412 mμ) of *p*-nitrothiophenol [313]. In any case higher wavelengths can be selected for absorption measurements, *i.e.*, λ 450 mμ, at which wavelength the thiol still strongly absorbs ($\mathscr{E} = 8500$) [313].

Alternatively, the selective reaction of azobenzene-2-sulfenyl bromide (ABS–Br) with cysteine residues in proteins can be used [Eq. (13)] [317].

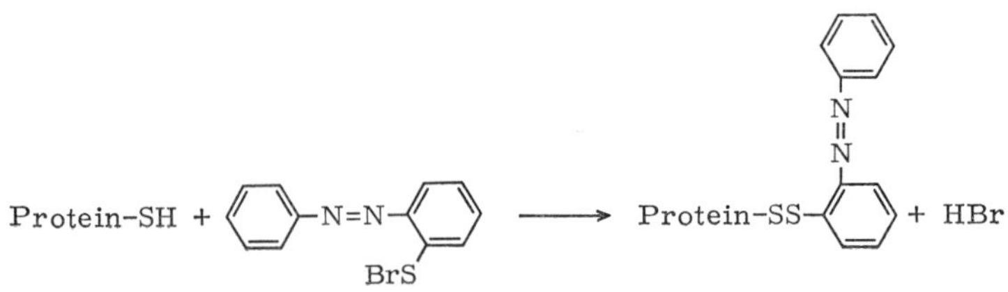

(13)

The lack of reactivity of the indole nucleus of tryptophan toward ABS–Br was explained in terms of the "sulfenamide nature" of the reagent, since its true structure is that of a 2-phenylbenzo-1-thio-2:3-diazolium ion (V) [169].

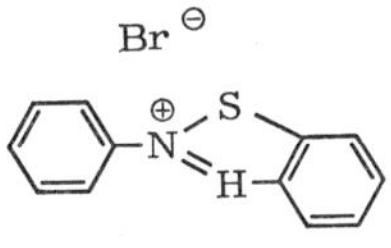

V

The reaction of ABS–Br with cysteine containing proteins is performed in aqueous media (buffer, pH 5) and then the labeled protein analyzed spectrophotometrically for ABS-content. Since the chromophore was shown to be influenced by the pH of the solution and by the protein structure, absorption measurements must be made in 5 to 8 M urea solution of pH 1 (λmax 353 mμ, $\mathscr{E} = 16{,}700$) [317].

The application of these procedures with different SH-containing proteins is under investigation [1089].

V. Estimation of Disulfides

The disulfide bonds cannot be determined directly in the intact protein and are most conveniently estimated by prior conversion to thiols followed by their quantitative determination. Thus, the cystine content of a protein is determined indirectly from the amount of sulfhydryl groups formed upon reduction.

Methods are available for estimation of the total amount of cysteine and cystine and, therefore, if the cysteine content is known the determination of cystine can be found by difference.

We must now consider the methods available for splitting disulfide bonds.

A. Reaction with Sulfite

Disulfides are cleaved by sulfite to S-sulfoacids and sulfide [Eq. (14)] [204, 428, 429, 631, 695].

$$\text{R–S–S–R} + SO_3^{2-} \longrightarrow RS^- + \text{RS–}SO_3^- \qquad (14)$$

The reaction was extensively studied [186, 666, 1018, 1019] and found to be reversible, so that the concentration of sulfite must be kept high.

Since the thiols formed (one SH group per S–S) during the reaction have to be determined in the presence of a large excess of sulfite, alkylating reagents cannot be used, because reaction takes place rapidly with sulfite too. Potentiometric titration with silver nitrate [185] or with mercuric chloride [183] was found to be a feasible procedure, as is amperometric titration with silver nitrate in an ammonia buffer [562] or mercuric chloride in borate buffer [1020].

The sulfitolysis step was also carried out in 8 M urea in order to unfold the protein molecule and to expose the disulfide bonds to the action of sulfite at 37 °C [182].

By mild oxidation (oxygen, sodium tetrathionate, iodosobenzoate) [49] RS^- is reconverted to R–S–S–R so that the reaction of sulfite progresses to the quantitative conversion to $RS–SO_3^-$. The same complete reaction may be achieved carrying out the sulfitolysis step in the presence of Cu^{2+} –ion [561, 1026]. In this case concomitant to the reduction of cupric ion all the disulfides are converted to the S-sulfonate.

The equivalence point of the reaction was determined amperometrically using the rotating platinum electrode. The method was used with proteins for analytical purposes as well as for a total cleavage of disulfide bridges [45, 246, 778, 1026].

The reaction with sulfite was used in some instances in the primary structure determination of proteins (*e.g.*, with trypsinogen and chymotrypsinogen [778]) in order to avoid the insoluble derivatives of the proteins obtained upon reduction and carboxymethylation. However a limitation in the use of the S-sulfoacids of proteins in sequence studies lies in the great instability of these derivatives in acidic as well as in alkaline media.

In similar way to the action of sulfite, the opening of the disulfide bonds in proteins by phosphorothionate was achieved [731, 733, 734]. It was shown in a recent study [732] that the reaction is a nucleophilic heterolytic scission, which proceeds according to the following (Eq. (15):

$$R–S–S–R + S–PO_3^{3-} \longrightarrow R–S–S–PO_3^{2-} + R–S^- \quad (15)$$

B. Reductive Cleavage

Reduction of disulfide bridges has some advantage with respect to the sulfite reaction since two SH groups per mole of disulfide are produced.

Older methods of reductive cleavage (zinc dust [1147], magnesium [879], sodium amalgam [562]) are no longer employed. Electrolytic reduction, although promising, has been rarely used [248].

Presently, the most widely employed reagents are β-mercaptoethanol [40, 1053], β-mercaptoethylamine [646] and dithiothreitol [$HS–CH_2–(CHOH)_2–CH_2–SH$] [206]. The reactions are carried out in 8 M urea [40] or in 4 to 6 M guanidine hydrochloride solutions at room temperature, or, for particularly resistant proteins, at higher temperatures.

The sulfhydryl groups formed by reduction can be estimated by using the methods described earlier (cfr. Section IV).

A general procedure of reduction of S–S bonds in a protein (ribonuclease) and their estimation by the carboxymethylation procedure was reported by Anfinsen and Haber [40], White [1126] and Crestfield, Moore and Stein [218]. The latter procedure is given below.

Procedure [218]. To 5 to 100 mg of protein in a 12 ml screw-cap vial (Kimble 609102) maintained under a nitrogen barrier [217], add 3.61 g of deionized, crystalline urea [70], 0.30 ml of EDTA solution (50 mg of disodium EDTA/ml), 30 ml of Tris buffer; pH 8.6 (5.33 g of Tris and 9 ml of 1 N HCl diluted to 30 ml with water), and finally 0.1 ml of β-mercaptoethanol. The solution is made up to 7.5 ml with water, and a solution 8 M in urea and 0.2% in EDTA is used to fill the vial completely. A disk of polyethylene is slid over the top in order to exclude air and is held in place with the screw cap.

After 4 h at 22 to 25 °C, the contents of the vial are transferred to a 25 ml beaker under the nitrogen barrier. A freshly prepared solution of 0.268 g of iodoacetic acid in 1 ml of 1 N NaOH is added to the reaction mixture. The iodoacetate added is slightly less on a molar basis than the amount of β-mercaptoethanol; the SH groups of cysteine residues react most rapidly. Three equivalents of Tris per equivalent of iodoacetate keep the solution more alkaline than pH 8.3.

Fifteen minutes after the addition of the iodoacetate, the sample may be applied to a 4 × 40 cm column of Sephadex G–75, medium particle size, equilibrated with 50% acetic acid. The sample is driven in and rinsed under air pressure to produce a column flow rate of 20 ml /h. Fractions of 8 ml are collected.

The protein in the effluent is located spectrophotometrically at 280 mμ and then recovered by lyophilization. A sample of the carboxymethyl-protein is hydrolyzed (HCl, 6 N, 110 °C, 24 h) and the content of S-carboxymethyl-cysteine determined by automatic amino acid analysis [207].

The cleavage of S–S bonds is a necessary pretreatment for proteins to be subjected to primary structure studies in order to facilitate enzymic hydrolysis of the protein chain. The new SH-groups formed upon reduction may reoxidize and must be stabilized by alkylation (iodoacetic acid or its amide [40], acrylonitrile [1106], bromoethylamine [624], ethyleneimine [441b]).

For sequence studies, reduction with sodium borohydride at pH 9 to 10 [441b, 698] is no longer used since it was found to produce cleavage of the peptide bond. There is a danger of "thiolation" of amino groups when reduction is attempted with impure thioglycolic acid containing reactive polythioglycolides and cyclic thiolactones [1126].

C. Oxidative Cleavage

The oxidation of a protein with performic acid (prepared by mixing 1 volume of 30% H_2O_2 with 9 volumes of formic acid [707]) quantitatively converts cysteine and cystine residues to the corresponding sulfoacids and methionine to its sulfone [892, 1052, 1061]. These oxidation products are stable to acid hydrolysis and can be separated by ion exchange chromatography after hydrolysis of the oxidized protein [441a, 892], allowing the determination of the total amount of cysteine and cystine. The yield of cysteic acid obtained from cysteine and cystine was over 90% of the theoretical value [295, 623, 698, 939, 1026].

At the end of the reaction, the performic acid is removed by lyophilization after addition of water, by evaporation *in vacuo*, or by precipitation of the protein with trichloroacetic acid or an organic solvent. A sample of the oxidized protein is subjected to acid hydrolysis and then analyzed for cysteic acid content by automatic amino acid analysis.

The reaction was first applied by Sanger *et al.* [856] in their sequential work on insulin and by Hirs on ribonuclease [441a]. A disadvantage presented by the method lies in the complete destruction of tryptophan [1061].

Another possibility for achieving the oxidative cleavage of disulfide bonds involves ozonization of the protein in 99% formic acid [803]. The yields of cysteic acid by this method, using free cystine, were 98%. The procedure is not selective for S–S bridges and other amino acids (methionine, tryptophan and tryosine) are modified.

VI. Estimation of Other Amino Acids

Little use is being made at the present time of methods for the direct determination of tyrosine, histidine, arginine in the intact protein. The determination of these amino acids can be accomplished easily and accurately by automatic amino acid analysis.

Some of these methods are more usefully employed to obtain an indication of the status of these amino acid residues in the protein. For a discussion concerning the use of selective reagents capable of discriminating between free and buried tyrosine, histidine and tryptophan residues see the review article of TIMASHEFF and GORBUNOFF [1058].

The tyrosine residue can be determined by the direct spectrophotometric method of GOODWIN and MORTON [261, 360] (see Section III.A). The Millon colorimetric method is unsatisfactory for following changes in the phenolic groups of the intact protein [105].

The formation of the phenoxide ion which absorbs at 290 to 295 mμ can be followed spectrophotometrically.

Not all the phenolic groups can be titrated with alkali at the same pH in the native protein as in the free amino acid [216, 1058]. Differences in reactivity of the hydroxyl group in the native and denaturated protein was often observed [1058].

Quite recently three new and promising tyrosine modifying reagents have been introduced which can serve as diagnostic reagents for free or buried tyrosine residues. N-Acetylimidazole [826], cyanuric fluoride [590] and tetranitromethane [824, 825, 980] are now extensively used in protein chemistry and they give clear indication of the status of tyrosine residues in proteins [1058].

Tyrosine and histidine react with freshly diazotized sulfanilic acid in sodium carbonate solution (Pauly reaction). When the reaction is performed in the Weiss-Szobolew modification [402, 448e, 1108] the color value obtained is equal to the sum of the contributions of histidine and tyrosine [448e]. If the tyrosine content is known, the determination of histidine can be found by difference. The use of diazo-1-H-tetrazole [465, 982] also has been used as a histidine- and tyrosine-modifying reagent capable of discriminating between free and buried residues.

Arginine can be estimated by the well known Sakaguchi reaction [847]. The guanidyl group reacts with α-naphthol and sodium hypochlorite in alkaline solution producing a red color. Since the color is unstable in presence of hypochlorite, urea is introduced in order to destroy hypochlorite after the reaction [1102]. The detailed procedure was described by KEYSER [539].

Since arginine linked in the protein molecule occasionally does not react completely with the Sakaguchi reagent, a procedure of prior acid hydrolysis of the protein sample has been suggested [488].

VII. Estimation of Amino and Amide Groups

Since the free amino groups of a protein can be readily calculated from the recovery of lysine by amino acid analysis of a protein hydrolyzate, its direct determination in the intact protein is no longer considered of importance. On the other hand, since the amide groups of glutamine and asparagine are hydrolyzed by acid or alkaline hydrolysis and the recovery of ammonia on the analyzer generally is not quantitative, the determination of the amide content in the intact protein is necessary.

A. Amino Groups

The total amino nitrogen has been determined by the ninhydrin reaction [744], formol titration [835] and the nitrous acid method of VAN SLYKE [1079, 1080]. The latter was shown to be the most useful procedure giving the best results. The reaction employed involves reaction of the primary amino groups with nitrous acid, formed by mixing sodium nitrite and cold acetic acid [(Eq. (16)].

$$\text{Protein-NH}_2 + \text{HONO} \longrightarrow \text{Protein-OH} + N_2 + H_2O \quad (16)$$

The nitrogen is measured by volumetric or manometric techniques. Accounts can be found in the original papers [1079, 1080] or, for a detailed description of the method, in the review article of HAMILTON [1058].

The 1-fluoro-2,4-dinitrobenzene (FDNP) introduced by SANGER [855] has the advantage that the dinitrophenyl-N bond is stable to acid hydrolysis so that the extent of reaction can be determined by colorimetric analysis of the $\mathscr{E}$-DNP-lysine after chromatographic separation [321, 797].

More recently the use of 2,4,6-trinitrobenzenesulfonic acid (TNBS) has been suggested [577, 757] as a reagent for the determination of the free amino groups in peptides and proteins. It was shown that TNBS reacts preferentially and quantitatively with $-NH_2$ under mild conditions to give the corresponding trinitrophenyl derivatives. No reaction occurs with imidazole nitrogen of histidine, the guanidinium group of arginine, or the hydroxyl group of tyrosine, serine and threonine. The trinitrophenyl derivative of amino groups can be quantitated spectrophotometrically.

B. Amide Groups

The amide groups of glutamine and asparagine release after acid or alkaline hydrolysis, ammonia which can be determined by automatic amino acid analysis [987]. However, the obtained values are only approximate since ammonia derived from other amino acids (in particular serine and threonine) and extraneous sources interferes. Therefore it is necessary to reach a compromise between the acid concentration, time and temperature of hydrolysis [974], and to measure the ammonia produced as a function of the time of hydrolysis. In this way the secondary deamination arising from partial destruction of amino acids can be evaluated and the ammonia content determined by extrapolating the ammonia produced to zero time.

An alternative method to the total hydrolysis of a protein and chromatographic separation of ammonia by automatic amino acid analysis involves the analysis of ammonia with a Conway microdiffusion vessel or Kjeldahl distillation apparatus by absorption at pH 9 into boric acid and successive determination by titration [402]. For this purpose detailed studies were carried out in order to find the best acid concentration to obtain complete hydrolysis of the amide linkages of the glutaminyl and asparaginyl residues.

Hydrolysis is usually performed in 2 N HCl at 100 °C for 3 to 7 h [196]. Alkaline conditions (1 N NaOH at 25 °C for 20 h) also can be employed [1002]. Gordon *et al.* [362] and Rees [818] satisfactorily used 12 N HCl at 37 °C for 10 days. Dilute acid (5.5 N HCl) and higher temperature (100 °C) lead to high values for the amide content.

The hydrolysis of a protein by enzymatic procedures offers the advantage that the glutaminyl and asparaginyl residues are preserved so that their direct determination is allowed by the use of chromatographic [987] or electrophoretic [178, 704] separation.

Hill and Schmidt [436b] employed a two-steps procedure, using papain first and then a mixture of leucine aminopeptidase and prolidase [228]. Several proteins of known composition were investigated and the values so obtained compared favorably with those obtained by acid hydrolysis. A single enzyme preparation, Viokase, a mixture of pancreatic enzymes, has been tried [1065]. Complete hydrolysis was achieved for peptides but for proteins the extent of cleavage was only about 50 to 80%.

On the other hand enzymic procedures are currently used in the determination of the presence of asparagine and glutamine in small peptides. A digest of the peptide with leucine aminopeptidase can be analyzed by paper electrophoresis [178, 704] or by the conventional quantitative amino acid analysis by column chromatography. However, since serine is eluted together with asparagine and glutamine, a modification of the temperature of column chromatography is need in order to lead to elution of serine in a different position [987].

Finally, the chemical method of Chibnall *et al.* [197, 199] for the estimation of asparagine and glutamine should be mentioned, though intrinsic difficulties limit its extensive use. The carboxyl groups of aspartic and glutamic acids as well as the carboxyl terminal residue are converted to esters by treatment of the protein with methanolic HCl at 25 °C for 24 h, while asparagine and glutamine remain unaffected. Reduction with $LiAlH_4$ in tetrahydrofuran converts the esterified groups to the corresponding alcohols of the aspartyl and glutamyl residues, and after acid hydrolysis γ-hydroxy-α-amino-butyric acid and δ-hydroxy-α-amino-valeric acid are obtained, respectively, while asparagine and glutamine yield aspartic and glutamic acid and the carboxyl terminal residues give β-aminoalcohols. After separation of these components, the relatives amounts are estimated by quantitative procedures.

VIII. Detection of Amino Acids and Peptides in Paper Chromatograms

Chromatographic methods have played a fundamental role in various areas of organic and biological chemistry. The most important applications of chromatography have been in the field of protein sequence determination. Since the introduction of automatic amino acid analysis of protein hydrolysates, less use has been made, in protein chemistry work, of the qualitative or quantitative separation of amino acids by paper chromatography. Of more interest on the other hand has been the role played in the fractionation of peptides by paper chromatography or electrophoresis, after the chemical or enzymic fragmentation of protein (Chapter 6).

In this respect the detection of the peptides in the paper or thin-layer chromatograms, by the use of non-specific reagents or by tests specifics for single amino acid residues, play an important role. Much of the usefulness of peptide mapping is derived from the variety of amino acid-specific reagents that may be used to stain the peptide map (Chapter 6).

Because of the wide application of this method and its adequate coverage in many books and review papers [106, 550, 600], a description of only a few procedures on paper or thin-layer chromatography will be discussed below.

A. Non-Specific Reagents

A number of reagents have been employed for detecting amino acids and peptides on chromatograms or for revealing the location of spots for subsequent elution.

For color development, dipping of the chromatogram in a solution of the appropriate reagent is recommended. Of course, for the reagent a solvent must be chosen in which the substances to be detected are insoluble. For the spraying of chromatograms, a glass atomizer may be used. Care must be taken not to overload the paper with the spray reagent, since the spots tend to diffuse, especially when aqueous solvents are employed.

A multiple procedure of spot detection can be done by successive color reaction, *e.g.*, ninhydrin, Ehrlich reagent, Sakaguchi, or diazo reagents (see further) leading to the possibility of revealing substances of different nature on the same chromatogram. In this way most of the amino acids may be identified on the same chromatogram by their position and characteristic color reactions.

1. Ninhydrin

The use of ninhydrin (triketohydrindene hydrate) for the detection of amino acids and peptides plays the most important role. However, the reagent is more sensitive with amino acids then with peptides. Several modifications of the reagent composition were proposed; some of them are listed below.

Method a. The chromatograms are sprayed with or dipped in 0.25% (w/v) ninhydrin in acetone [1062]. After drying, they are placed in a warm (50 to 70 °C) chamber for 1 to 2 h and kept in the dark overnight.

Method b. A 0.3% (w/v) ninhydrin solution in 95% ethanol is used [770, 854]. The color is developed at room temperature in the dark for 18 h.

Method c. When alkaline salts have been used to buffer the paper, the alkali must be neutralized by the incorporation of 2 to 4% (v/v) of acetic acid into the ninhydrin solution [210]).

Other procedures for the ninhydrin reaction are described. The use of ninhydrin in pyridine [831], in n-butanol containing 10% phenol [168], and in 10% (v/v) acetic acid [494] has been recommended.

2. Starch: Iodide

The developing solvent is very thoroughly removed at 60 °C. The chromatogram is then exposed in a chamber to Cl_2 gas for 10 min. When a Cl_2-bottle is not available, the gas can be produced by mixing 3 N HCl with a 10% $KMnO_4$ solution. The excess Cl_2 is carefully removed in the hood. The paper is then sprayed with 1% soluble starch: 1% KI solution [796, 843]. Blue spots appear. The procedure is very useful for acylated amino acids and for peptides. The reaction appears to be [Eq. (17)]:

$$\text{-CONH-} \xrightarrow{Cl_2} \text{-CONCl-} \xrightarrow{KI} \text{-CONH-} + KCl + \tfrac{1}{2} I_2 \qquad (17)$$

3. Indicator Dyes

Advantage may be taken of the buffering capacity of acidic, basic, or amphoteric substances by treating the chromatogram with a dilute solution of an indicator dye. After spraying with 0.1% solution of the sodium salt of bromothymol blue and drying, the chromatogram is dipped in 0.2% acetic acid [765]. The sheet turns yellow except for the blue spots indicating amino acids or peptides. A 0.04% solution of bromocresol green in ethanol [960] as a spraying reagent also was employed.

4. Isatin

The chromatogram is sprayed with 0.2% isatin in acetone and then heated for 10 min in a water-saturated oven at 70 °C [969]. Upon reaction the amino acids give rise to various colors: proline and hydroxyproline give blue colors; aspartic and glutamic acid dark purplish blue; serine, threonine and tyrosine light brown.

B. Specific Reagents

1. Tryptophan

Indole derivatives react with p-dimethylaminobenzaldehyde giving a purple color. Tryptophan, like 3-substituted indole derivatives, reacts with the reagent only in strong acidic solution. The reaction was successfully used in the Spies and Chambers method for the quantitative estimation of the tryptophan content in proteins (cf. Section III.B.3) as well as for detection of tryptophan derivatives in paper chromatograms. The reaction is specific; the only other amino acids reacting with the reagent are kynurenine and citrulline [318]. These give a pink yellow color but are rare in proteins. Prepare a fresh mixture of 1 g of p-dimethylaminobenzaldehyde, 90 ml of acetone and 10 ml of concentrated HCl [970]. After spraying, the color is developed at room temperature. Another feasible procedure use 1% p-dimethylaminobenzaldehyde in 1 N HCl or 1 ml of concentrated HCl plus 99 ml of ethanol followed by exposure to HCl fumes [1032]. The Ehrlich reaction may be stabilized by first spraying the chromatogram with 1% (w/v) methionine, drying in air and then spraying with p-dimethylaminobenzaldehyde in HCl-acetone [713].

2. Cysteine and Cystine

Since cysteine in paper chromatograms is oxidized to cystine after application to the filter paper [1063], the methods of detection of cystine are valuable also for cysteine-derivatives.

Sodium Nitroprusside [1062]. *Reagent 1*: Dissolve sodium nitroprusside (1.5 g) in 5 ml of 2 N H_2SO_4. Then 95 ml of methanol and 10 ml of 28% ammonia are added. The solution is filtered and stored in a refrigerator.

Reagent 2: Two grams of NaCN is dissolved in 5 ml of water and diluted to 100 ml with methanol. *Tests*: For cysteine use reagent 1; for cystine dip into reagent 1, dry slightly, and, while still damp, dip into reagent 2; for both cysteine and cystine prepare reagents at double strenghth and treat with equal mixture of 1 and 2.

Feigl's Sodium Azide: *Iodine Reaction*. The dry chromatograms are sprayed with 0.05 N iodine in 50% ethanol containing 1.5% of sodium azide. The spots are best seen under a UV light [1137]. A freshly prepared solution of 0.01 M iodine in 0.5 M KI plus 0.5 M NaN_3 can also be employed [953]. The sensitivity of the reaction is equivalent to 0.5 μg of methionine.

Platinic Iodide. Add in the following order 4 ml of 0.002 M $PtCl_6^{2-}$, 0.25 ml of 1 N KI, 0.4 ml of 2 N HCl, and 76 ml of acetone. The chromatograms are developed by dipping. Cystine, cysteine, methionine as well other reducing agents give a white spot on a red-purple background [1062]. If phenol, lutidine, etc. are the developing solvents, all traces must be removed with ether/acetone, petroleum ether, etc. *Palladous Chloride* in 0.1 N HCl may be used in place of chloroplatinic acid [1062].

3. Methionine

For methionine the platinic iodide and sodium azide: iodine reagents (see cystine above) can be usefully employed.

Sodium Nitroprusside. Dissolve 500 mg of sodium nitroprusside in 10 ml of water. Then add 500 mg of $NH_2OH \cdot HCl$, followed by 1 g of $NaHCO_3$. After the evolution of gas has stopped, add 2 drops of bromine. Remove the excess bromine by aeration, filter the dark green or black-brown solution and dilute to 25 ml.

The reagent solution is stable for 2 weeks. Allow to stand 24 h, filter again and store in the dark. Just before use, dilute the reagent with an equal volume of saturated Na_2CO_3 and filter. Spray and expose moist chromatograms to steam for a few minutes [510].

4. Arginine. Sakaguchi Reaction

The chromatograms are sprayed with a 0.1% solution of α-naphthol in 1 N NaOH. After drying, the paper is sprayed with NaOCl solution prepared from an equal mixture of ethanol and commercial NaOCl. Arginine and arginine-containing compounds appear as red spots. Another more sensitive procedure (0.2 μg of arginine) can also be employed [4]. Dissolve 0.01% α-naphthol in ethanol containing 5% urea. Add KOH to a total concentration of 5% just before spraying. After air drying for a few minutes, spray slightly with 0.7 ml of bromine in 100 ml of 5% KOH.

A modification of the Sakaguchi reaction consists in dipping the chromatogram into 0.1% 8-hydroxyquinoline in acetone. After the chromatogram has dried, it is dipped into a solution of 0.02 ml of bromine in 100 ml of 0.5 N NaOH. Arginine and other guanido compounds give orange-red spots [494].

5. Histidine

The imidazole group of histidine as well the phenol group of tyrosine are readily detected on paper chromatograms by means of Pauly's reaction.

Sulfanilic Acid. To a fresh mixture of 2 volumes of sulfanilic acid (9 g of the compound in 90 ml of concentrated HCl /l of water) plus 1 volume of 5% (w/v) $NaNO_2$, add 1 volume of 20% (w/v) NaOH. Spray the chromatogram and after air drying spray further with 10% Na_2CO_3 solution. Histidine appears as bright yellow spot with a red margin [638].

Several modifications of the procedure were suggested [106]. The use of sulfanilamide [136], *p*-bromoaniline [104] and *p*-anisidine [862] was also reported.

6. Tyrosine

This amino acid reacts with the Pauly reagent (see histidine above).

α-Nitroso-β-Naphthol. Spray the chromatogram with 0.1% α-nitroso-β-naphthol in 95% ethanol. After drying in warm air, spray with 10% aqueous HNO_3 and heat at 100 °C for 3 min [3]. The reaction is given by phenols and 3- and 5-substituted phenols, but not by ortho-substituted or 3,5-disubstituted phenols.

Chapter 8

Sequence Determination

PEHR EDMAN

I. Introduction

Anyone entering the field of protein structure determination will be faced with a bewildering choice between techniques for sequence determination. To sort the useful from the less useful, or even unworkable, could be difficult indeed. However, well-trodden tracks exist. These will be described here in some detail together with traps not to be fallen into, sidetracks not to be taken, and spooks not to be feared.

A certain bias cannot be avoided in giving preference and prominence to certain techniques. This is so because the author would not like to recommend a procedure unless he is himself thoroughly familiar with it. However, alternative procedures will be mentioned and sometimes critically discussed. Certain procedures will be described, not because they have been worked out in sufficient detail for an evaluation, but because they represent novel approaches which may in time lead to superior sequence techniques.

The use of mass spectroscopy in sequence determination has not been treated by the author (in fact, he was never asked) since it would fall outside his competence.

II. Isothiocyanate Degradation

This procedure was first described some 20 years ago by the present author [262, 264]. In the following years it has been used extensively, in one or another of its modifications, to establish amino acid sequences of peptides and proteins.

A. Reaction Mechanism

The following series of reactions forms the basis of the phenylisothiocyanate degradation [87, 265, 266]:

$$C_6H_5\cdot NCS + NH_2\cdot CHR\cdot CO\cdot NH\cdot X \longrightarrow \underset{\text{I}}{C_6H_5\cdot NH\cdot CS\cdot NH\cdot CHR\cdot CO\cdot NH\cdot X} \quad (1)$$

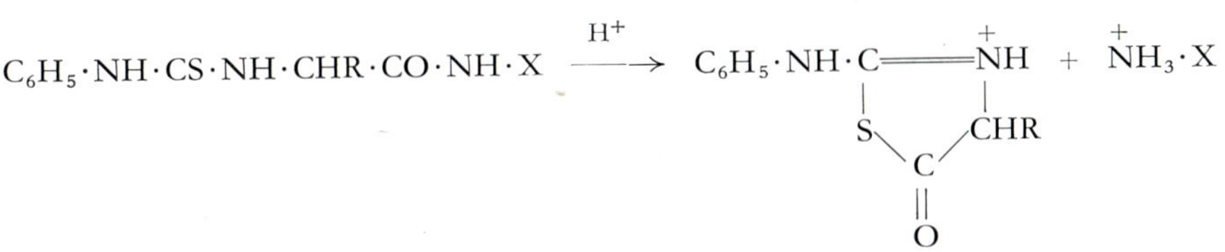

(2)

II

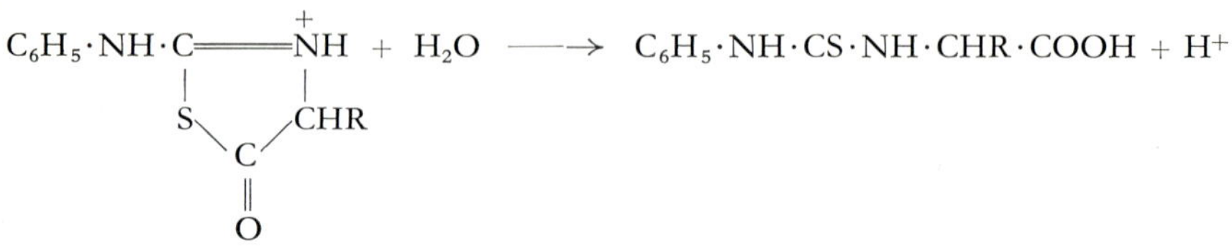

(3a)

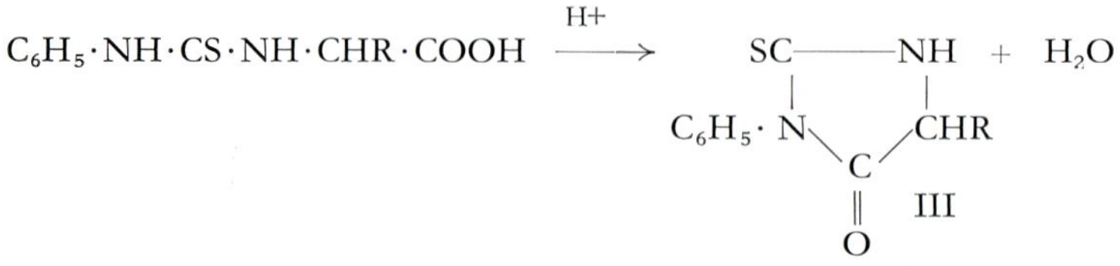

(3b)

The first reaction (1) is the formation of a phenylthiocarbamyl[1] derivative of the peptide (I) by coupling the free α-amino group with phenylisothiocyanate. This reaction will be referred to as the *coupling* reaction.

The next reaction (2) involves the cleavage of the PTC-peptide at the peptide bond nearest to the PTC substituent. This requires a strongly acid medium and leads to the formation of a 2-anilino-5-thiazolinone derivative (in the ammonium form) (II) and a peptide with one amino acid less than the original. This is termed the *cleavage* reaction.

The shortened peptide has a free α-amino group, and may therefore be subjected to a new cycle of the degradation.

The thiazolinone cleaved off is a derivative of the N-terminal amino acid, and could therefore in principle be used for its identification. In practice this is not possible because of the instability of this type of derivatives. However, the thiazolinone may be converted to the isomeric and stable 3-phenyl-2-thiohydantoin (III). This *conversion* reaction (3) takes place in an aqueous acid medium and consists in fact of two reactions, *i.e.* the hydrolysis of the thiazolinone to the PTC-amino acid (3a) and the cyclization of the latter to the corresponding PTH-amino acid (3b). The first reaction, however, is much faster than the second, and for practical purposes they may be treated as one reaction. It may be added that there exists also a direct route, which does not involve hydrolysis, from the thiazolinone to the phenylthiohydantoin [266]. This isomerization reaction, however, has not been studied in detail.

Needless to say, a full understanding of the nature of these reactions is essential for an intelligent use of them in sequence work. A detailed discussion is left to the sections dealing with the experimental conditions of the degradation. However, even at this stage the importance of separating the cleavage and the conversion reactions is emphasized. Since they are both acid catalysed it may seem logical to combine them. Yet, this does not take into account the fact that the conversion reaction requires an aqueous acid medium and is slow, whereas the cleavage reaction does not require water and is fast. A combination of the two reactions will therefore result in an unnecessary and prolonged exposure of the peptide to hydrolytic condi-

---

[1] Abbreviations used are: DFP, diisopropyl fluorophosphonate; PTC, phenylthiocarbamyl; PTH, 3-phenyl-2-thiohydantoin; TLC, thin-layer chromatograph; X (in formulas), rest of peptide.

tions. The better procedure is to extract the thiazolinone from the reaction mixture after cleavage, and to convert it in a separate operation.

It is illuminating to compare the mechanism of isothiocyanate degradation with those of some other reactions which have been proposed for the sequential degradation of peptides. In the degradation scheme of Léonis and Levy [601, 602] a dithiocarbamate of the peptide is initially formed, which under the influence of acid splits off the N-terminal amino acid in the form of a 2-thio-5-thiazolinone derivative (IV):

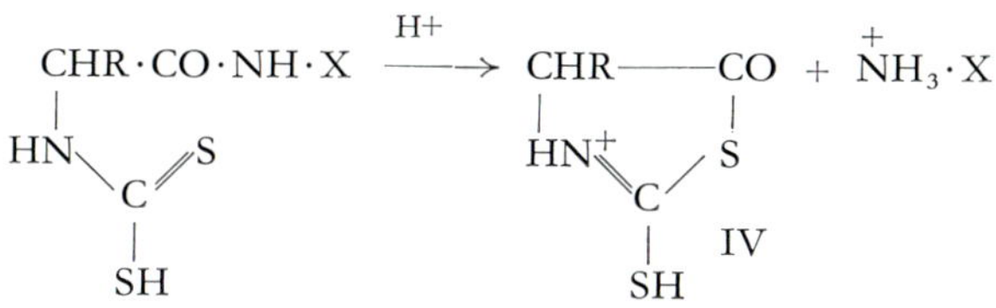

(4)

Similarly, in the procedure of Kenner and Khorana [535] an N-thioncarbalkoxy peptide in acid splits off a 2-alkoxy-5-thiazolinone derivative (V):

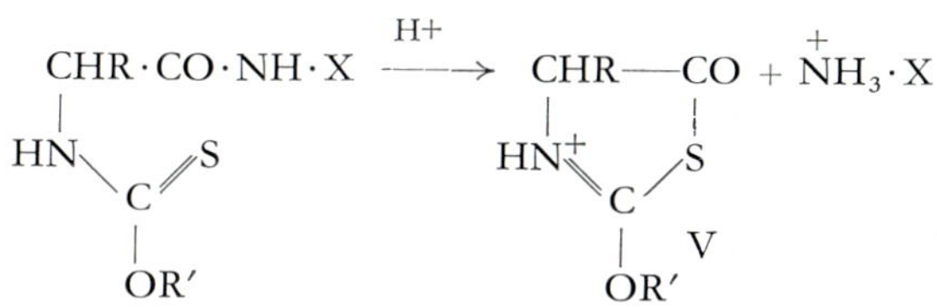

(5)

In the scheme of Elmore and Toseland [281, 282] a 2-acylamido-5-thiazolinone (VI) is formed by exposing an N-acylthiocarbamyl peptide to anhydrous acid (6):

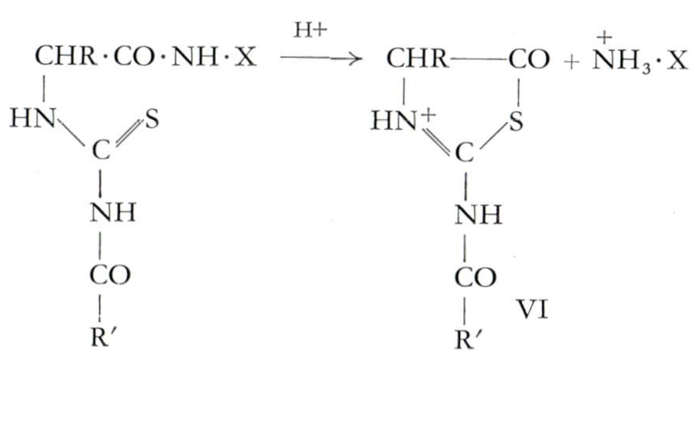

(6)

Barrett [58] has recently described a reaction (7) where N-thiobenzoylated peptides are easily cleaved in anhydrous acid forming 2-phenyl-5-thiazolinone derivatives (VII):

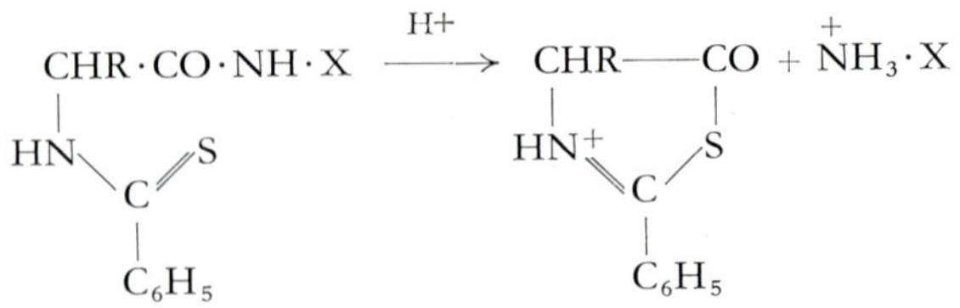

VII

(7)

In all these reactions the cleavage of the peptide bond is concomitant to the formation of a thiazolinone derivative. The explanation for the extreme facility of these reactions obviously lies here. On the other hand, the phenylisocyanate degradation of ABDERHALDEN and BROCKMANN [1] is, contrary to what is sometimes still claimed [47], unrelated to the reactions discussed, since obviously no thiazolinone can be formed here. The absence of a reactive thioketonic group makes it a very sluggish reaction.

It should be added that apart from the phenylisothiocyanate reaction none of the reactions mentioned here have been worked out into a generally applicable sequence procedure.

B. Preparation of Phenylthiohydantoins

The synthesis of a PTH-amino acid [152, 263, 271, 472, 473, 611][2] is usually carried out in two steps. First, the amino acid is coupled with phenylisothiocyanate in a weakly alkaline medium to produce the PTC-amino acid. The latter is then converted to the corresponding PTH-amino acid in acid medium, usually at elevated temperature.

1. Preparation of PTC-Amino Acids

The first reaction, *i.e.* the formation of the PTC-amino acid, is straightforward and the same for all amino acids. Phenylisothiocyanate reacts with the amino acid only in its amino form. The reaction rate is therefore dependent on the *pK* of the amino group, *i.e.* the higher the *pK* the lower the reaction rate. This applies particularly to lysine where the ε-amino groups ($pK = 10.53$) is noticeably slower in reacting. To ensure a complete reaction it is advisable to follow the coupling by the alkali consumption at a pH held constant at say 8.6. In practice one finds that the alkali consumption never ceases completely, but falls off to a constant rate due to the slow hydrolysis of the phenylisothiocyanate. After completion of the reaction excess reagents and by-products are removed by extraction, and subsequently the PTC-amino acid is precipitated from the aqueous solution by the addition of acid. No attempt is made to obtain an analytically pure preparation at this stage, since gradual conversion of the PTC-amino acid to the PTH-amino acid cannot be avoided.

The phenylisothiocyanate seems to react exclusively with primary and secondary (prolines) amino groups. Other functions, *e.g.* alcoholic, phenolic, imidazolic and guanidino groups, do not appear to react. At any rate these groups are unmodified in the end products, *i.e.* the PTH derivatives.

The amino acid (10 mmoles) is dissolved or suspended in pyridine-water (1:1; 50 ml) in a beaker. The pH is then adjusted to 8.6 (glass electrode) with 2 N NaOH, and the

[2] The PTH derivatives of S-allyl cysteine, S-ethyl cysteine and S-methyl cysteine may be prepared from the corresponding amino acids by the standard procedure described in this section (P. EDMAN, C. ROCHAT and H. ROCHAT, unpublished).

volume added is noted. The temperature is brought to 40 °C, preferably on an electric hot plate with provisions for magnetic stirring and phenylisothiocyanate (2.4 ml) is added. The solution is stirred and the pH is kept steady by the addition of 2 N NaOH from a burette. Usually the alkali consumption becomes slow and steady in less than 30 min, indicating the completion of the reaction. The total alkali consumption, including that of the initial pH adjustment, should be equivalent to the number of amino groups. Excess reagents and by-products are extracted with benzene (3 × 1 vol.). The remaining aqueous solution is cooled in ice and an amount of 2 N HCl equivalent to the total alkali consumption is added gradually and with stirring. By a further small addition of acid the pH is brought to about 1. The PTC-amino acid usually precipitates as a mass, which solidifies on storage in the cold for a few hours. The supernatant is discarded, and the preparation is then ready for conversion to the PTH-amino acid.

Some PTC derivatives do not precipitate or precipitate incompletely on acidification. *PTC-ariginine* and *PTC-histidine* may be precipitated isoelectrically at pH 7.5 and 3.5, respectively. The aqueous solution of *PTC-cysteic* acid is evaporated to dryness *in vacuo*.

In the case of *PTC-serine* and *PTC-threonine* the aqueous solution remaining after benzene extraction is made 0.5 N in HCl and the conversion is carried out in this medium as described below.

2. Conversion of PTC-Amino Acids

The conversion is carried out in aqueous acid usually at elevated temperature. Most PTH-amino acids are perfectly stable under these conditions and remain unaffected even if the conditions are more vigorous than is strictly necessary to bring about a complete conversion. However, certain PTH-amino acids tend to decompose on prolonged exposure to acid and heat. The chemical nature of these changes are discussed in the section "Properties of Phenylthiohydantoins", and here is only given the standard procedure applicable to most PTH-amino acids followed by special procedures for some PTH-amino acids.

The PTC-amino acid preparation is suspended or dissolved in 1 N HCl (25 ml). The temperature is quickly brought to 80 °C, kept there for 10 min, and then dropped to ambient. The reaction mixture is repeatedly extracted with ethyl acetate (3 × 1 vol.) and the extracts are combined and evaporated to dryness. The residue is taken up in a small volume of hot glacial acetic acid and made to crystallize by the gradual addition of water. Recrystallizations are made from ethanol-water (in the case of PTH-aspartic and PTH-glutamic acid ethanol should be avoided because of the risk of esterifying the carboxyl group).

PTH-arginine, *PTH-histidine* and *PTH-cysteic* acid are not extracted by ethyl acetate. The aqueous phase is evaporated to dryness *in vacuo*, water added, and the process repeated until excess HCl has been removed. PTH-arginine and PTH-histidine may then be recrystallized as their hydrochlorides from water and ethanol, respectively. The dry PTH-cysteic acid preparation is dissolved in a small volume of water. The solution is brought to pH 4.2 with 1 N KOH and evaporated to dryness *in vacuo*. The dry product is extracted with acetone-water (4:1), which leaves the inorganic salts behind. Most of the solvent is evaporated *in vacuo*, and to the resulting viscous solution is added acetone to the point of incipient crystallization, which is completed in the cold. Recrystallizations are made from water-ethanol. The resulting preparation is the potassium salt of PTH-cysteic acid.

The conversion of PTC-serine to *PTH-serine* is carried out in 0.5 N HCl at room temperature for 24 h. The precipitated product is filtered off and dried *in vacuo*. The dry product is extracted exhaustively with absolute ethanol and any insoluble residue filtered off and discarded. The ethanolic solution is concentrated at room temperature until incipient crystallization, which is completed by the addition of heptane. Recrystallization is made from ethanol. The procedure for *PTH-threonine* is the same as for PTH-serine, but the product is considerably less soluble in ethanol, and crystallization is therefore made from hot ethanol.

PTH-asparagine and *PTH-glutamine* are prepared from the corresponding PTC-derivatives by crystallization from boiling glacial acetic acid.

The over-all yield of PTH-amino acid calculated on the amino acid is generally about 80%. The asymmetric centre at C-5 of the thiohydantoin ring is completely or almost completely racemized during synthesis, and the configuration at the α-carbon of the starting material is therefore of little consequence.

Apart from elementary analysis and melting point determination the identity and purity of a PTH-amino acid preparation may be established by other criteria, *e.g.* chromatography (cf. "Identification of Amino Acids") and ultraviolet absorption (Table 1).

A wide range of PTH-amino acids are now commercially available.

C. Properties of Phenylthiohydantoins

PTH-amino acids are generally easily crystallizable compounds with high melting points. Their chemical stability is with few exceptions excellent. The less stable compounds are those with an OH or an SH group on the β-carbon of the side chain. These compounds show to a varying degree a tendency for β-elimination. So, for example, PTH-threonine tends to undergo the following reaction [263]:

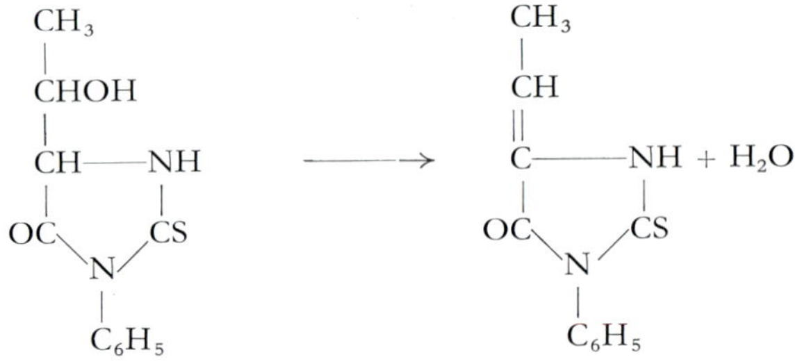

(8)

This tendency is stronger in PTH-serine. Here the formation of a highly reactive methylene group at C-5 of the thiohydantoin ring leads to further reactions, *e.g.* polymerization. In PTH-cystine and PTH-cysteine the tendency for β-elimination is so strong that the derivatives are not useful for identification purposes. On the other hand, the S-alkylated PTH derivatives of cysteine are more stable as is also PTH-cysteic acid. The tendency for β-elimination becomes important in the identification of the susceptible compounds, and this problem is discussed in the section on "Conversion". It seems likely that a tendency for β-elimination is also present in the corresponding thiazolinones.

The PTH derivatives of asparagine and glutamine are unstable to the extent that the amide group is hydrolyzed by acids and alkalis. In sequence determination these PTH-amino acids are found to be contaminated by the corresponding acids. PTH-tryptophan is also susceptible to the action of strong acids, presumably due to decomposition of the indole nucleus. However, the acid conditions of the degradation procedure do not seem sufficiently vigorous to affect appreciably PTH-tryptophan.

Solutions of PTH-amino acids show a tendency for photodecomposition, which becomes apparent after prolonged exposure to daylight or ultraviolet light. Solutions should therefore be stored away from light.

The PTH-amino acids are, with few exceptions, poorly soluble in water, and easily soluble in organic solvents. The exceptions are PTH-arginine, PTH-histidine and PTH-cysteic acid. This is important in the conversion procedure (cf. "Conversion"). After conversion the PTH-amino acids are extracted from the aqueous acid medium with ethyl acetate, leaving only the PTH derivatives of arginine, histidine and cysteic acid behind in the aqueous phase.

The PTH-amino acids may be hydrolyzed by acids or alkalis to regenerate the amino acid. However, this requires vigorous conditions. Certain labile amino acids are destroyed in the process. The identification of the amino acid after hydrolysis of

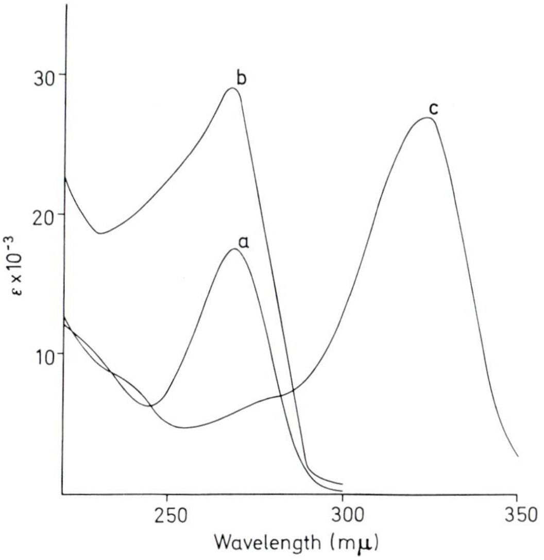

Fig. 8-1. Ultraviolet absorption spetra of PTH-leucine (*a*), PTH-N$^{\varepsilon}$-PTC-lysine (*b*) and PTH-dehydrothreonine (*c*) in ethanol

the PTH has been used in sequence work and this approach is discussed in the section "Identification of Amino Acids".

All PTH-amino acids show a strong absorption in the ultraviolet with a maximum around 268 mμ and a minimum around 245 mμ (Fig. 8-1). The molar absorptivity at 269 mμ[3] is usually around 16,000 (Table 1). The ratio A_{245}/A_{269} lies around 0.4. This ratio is a useful index in assessing the purity of a PTH-amino acid preparation since impurities tend to increase the absorption at 245 mμ more than at 269 mμ. PTH-N$^{\varepsilon}$-PTC-lysine is exceptional in that it has a much higher molar absorptivity at 269 mμ, and also a higher ratio A_{245}/A_{269} (Table 1). This is due to the additional absorption caused by the N$^{\varepsilon}$-PTC-group. The prolines also show a high ratio. The β-elimination product of PTH-threonine has, because of the conjugation of the

[3] This wavelength is used throughout for quantitation of PTH-amino acids by spectrophotometry although it does not always coincide exactly with the wavelength of maximum absorption.

newly formed double bond with the thiohydantoin ring, an entirely different spectrum with a maximum at 323 mμ (Fig. 8-1). PTH-dehydroserine has a similar spectrum, but the compound is unstable, and is rapidly transformed into other products with less characteristic absorption. The strong ultraviolet absorption of PTH-amino acids may be used for their location in chromatograms and in quantitation.

The individual PTH-amino acids have characteristic infrared spectra, which may be used for their identification as shown by RAMACHANDRAN *et al.* [811].

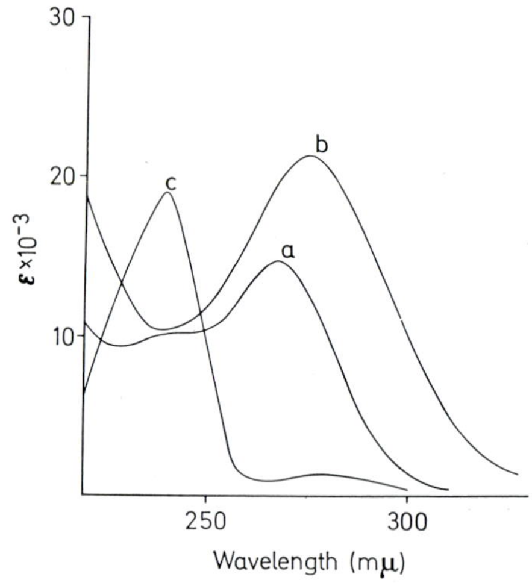

Fig. 8-2. Ultraviolet absorption spectra of PTC-glycine in ethanol (*a*) and in 0.1 N hydrochloric acid (*b*)

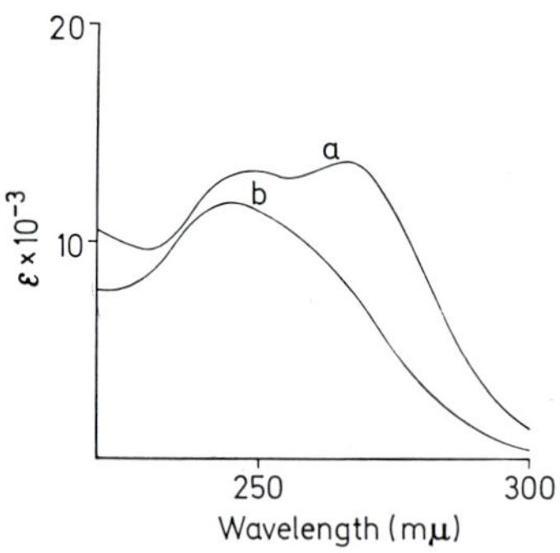

Fig. 8-3. Ultraviolet absorption spectra of phenylthiourea (*a*), *s*-diphenylthiourea (*b*) and phenylcarbamyl leucine (*c*) in ethanol

The ease of racemization at the asymmetric C-5 of the thiohydantoin ring is a tendency shared by other hydantoins [138, 224]. The optical rotary dispersion of phenylthiohydantoins have been reported on by DJERASSI *et al.* [247].

D. Identification of Amino Acids

Two different principles are commonly used in the identification of the amino acid residue removed in a degradation cycle. The *direct* method relies on the positive identification of the residue split off. The *indirect* method depends for the identification on a difference in analysis of the peptide before and after a degradation cycle.

PTH-amino acids may be separated and identified by chromatography. For this purpose paper chromatography [594, 953], thin-layer chromatography [191] and partition chromatography, either in liquid-liquid [837] or gas-liquid [790] systems, have been proposed. Paper and thin-layer chromatography techniques have been most commonly used, and have been worked out in all detail. These techniques are not well suited for the identification of some PTH-amino acids with strongly ionized groups in the side chain, *i.e.* PTH-arginine, PTH-histidine and PTH-cysteic acid, but here spot reations and paper electrophoretic techniques may be used.

The paper chromatographic procedures preceded the TLC techniques in time, but have now been replaced by the latter, which will therefore be treated first. However, only the paper chromatographic techniques are suitable for quantitation, and therefore still have an important role.

Table 1. *Molar absorptivities and melting points of phenylthiohydantoins. Light absorptions are measured in ethanolic solution*

PTH-Amino acid (m.w.)	ε_{245}[a]	ε_{269}[a]	$\frac{\varepsilon_{245}}{\varepsilon_{269}}$	M. p.[b] °C
PTH-alanine (206.3)	6,300	16,000	0.39	185
PTH-arginine-HCl (327.8)	6,000	15,900	0.38	189
PTH-asparagine (249.3)	6,500	17,200	0.38	235—237
PTH-aspartic acid (250.3)	6,500	16,100	0.40	229
PTH-cysteic acid, K salt (324.4)	5,900	16,600	0.36	193—197 (d.)
PTH-glutamic acid (264.3)	6,200	15,900	0.39	166—167
PTH-glutamine (263.3)	6,500	17,000	0.38	201—211 (d.)
PTH-glycine (192.2)	6,300	14,900	0.42	245—248 (d.)
PTH-histidine-HCl (308.8)	6,400	15,500	0.41	200—206 (d.)
PTH-isoleucine (248.3)	6,500	17,000	0.38	173—175
PTH-leucine (248.3)	6,000	16,700	0.36	178
PTH-n$^{\varepsilon}$-PTC-lysine (398.5)	21,000	29,000	0.72	162—164
PTH-methionine (266.4)	6,900	17,100	0.40	132
PTH-methionine sulphone (298.4)	6,400	15,300	0.42	210
PTH-phenylalanine (282.4)	5,700	15,500	0.37	187
PTH-proline (232.3)	9,600	14,300	0.67	179
PTH-s-allylcysteine (278.3)	7,300	17,600	0.41	134
PTH-s-carboxymethylcysteine (296.5)	7,100	16,600	0.43	164
PTH-s-ethylcysteine (266.3)	6,600	16,300	0.40	128
PTH-s-methylcysteine (252.3)	6,500	16,400	0.40	171
PTH-serine (222.3)	5,500	15,500	0.35	176—178 (d.)
PTH-threonine (236.3)	7.200	15,600	0.46	194—198 (d.)
PTH-tryptophan (321.4)	8,000	19,700	0.40	177
PTH-tyrosine (298.4)	6,600	15,600	0.42	216
PTH-valine (234.3)	6,200	16,500	0.38	206

[a] The data are mostly from reference [951].
[b] The data are mostly from references [263, 271, 472 and 473].

1. Thin-Layer Chromatography

The systems described here are not necessarily superior to others that have been proposed. However, they have stood up to the test of time, they permit unambiguous identification of all amino acids normally encountered in proteins and some others in addition, and, not least important, they invariably give satisfactory results in the hands of a competent technician.

The room where the chromatography is performed should be thermostatted at 20 °C ± 2. Chromatography is ascending, and is carried out in a glass tank with inner dimensions 22 × 14 cm and 22 cm in height. The glass lid should be tight fitting with one or two holes bored in it, to allow the introduction of the solvent into the tank without lifting the lid. Glass throughs to hold the developing solvent sit on the bottom of the tank. A sheet of filter paper covers the walls of the tank on the inside (systems E and H), and is held in place by loops of stainless steel wire.

The thin layer consists of silica gel to which have been added a fluorescent dye and calcium sulphate as a binder (Camag, Muttenz, Switzerland, type DF-5). If the iodine-azide reaction will be used later to locate the PTH-amino acids, starch is also incorporated in the layer. This may be done routinely, since the starch does not interfere in any way. A slurry of silica gel (50 g) in 0.5% aqueous starch solution (110 ml) is spread in a 0.3 mm thick layer on glass plates, 20 × 20 × 0.4 cm, using an applicator (*e.g.* Camag). The plate is allowed to dry in a horizontal position overnight at room temperature. Lines are then scored in the layer to separate the sample lanes. The lines are drawn parallel to each other and to the direction in which the layer has been spread. The lines are 0.9 cm apart, beginning 2.5 cm from what will be the lower edge of the plate during chromatography, and ending 7.5 cm from the top edge of the plate. Where the lines end, a transverse line is scored across the plate. The lanes ensure that the chromatogram is not distorted should the migration of the solvent not be even over the whole plate. The plates are suitably stored over P_2O_5 in a Perspex box.

Three solvent systems are used:

a) Xylene/Formamide (System D) [272]

The plate is dipped into a solution of formamide in acetone (3:7) for 1 min, and then placed horizontally to dry in the air for a few minutes. The formamide-acetone stock solution should be protected from evaporation, but after some use it tends to become concentrated, and a fresh solution should then be made up. The references and samples in ethylene chloride solution are applied at the start of the lanes, avoiding the two extreme lanes, where chromatography is often disturbed. When many samples are to be handled, it is convenient to use an applicator manifold (Fig. 8-6) which allows ten samples to be applied simultaneously. The plate is placed in the trough at the bottom of the tank, xylene is added to the through, and chromatography is started without prior equilibration. It is allowed to proceed until the solvent front in all lanes has reached the line scored across the plate. This usually takes about 40 min. The plate is then dried in an oven at about 100 °C for 15 min. The formamide should be evaporated as it absorbs the ultraviolet light used in locating spots.

b) n-Butyl Acetate/Propionic Acid/Formamide (System E) [272]

The layer is impregnated with formamide as described for system D. The butyl acetate (60 ml) is saturated with water (1.0 ml), and propionic acid (1.8 ml) added. This mixture is saturated with formamide (8 ml). Only the upper phase is used. The filter paper lining of the tank is moistened with the solvent mixture. The plate is placed in the empty through, and allowed to equilibrate for 1/2 h, and chromatography is then started by filling the through with the solvent mixture through the hole in the lid. The chromatography takes about 35 min.

d) Ethylene Chloride/Acetic Acid (System H)[4]

To ethylene chloride (60 ml) is added glacial acetic acid (14 ml). The paper lining of the tank is moistened with this mixture, which is also filled in the through. No equilibration is required. The development takes about 50 min.

The following specifications or purification procedures are recommended for the solvents used in TLC and paper chromatography:

Xylene is shaken three times with conc. H_2SO_4 (0.1 vol.) for 1/2 h and with cooling to below 30 °C. The organic phase is washed with H_2O, dried over pellets of KOH overnight, and finally distilled.

---

[4] P. Edman and I. Kluh, unpublished.

n-Butyl acetate is distilled, and the fraction boiling at 126–127 °C collected.

Propionic acid is refluxed for 4 h with CrO_3 (10 g/750 ml), distilled on a Widmer column, and the fraction boiling at 140–141 °C collected.

n-Heptane is shaken three times with concentrated H_2SO_4 (0.1 vol.), and then washed successively with H_2O, 10% aqueous NaOH and H_2O. It is finally dried over anhydrous Na_2SO_4 and distilled. B.p. 98 °C.

Ethylene chloride and formic acid are *purissimum* grades (Fluka A.G., Buchs, Switzerland).

Acetic acid is a *purissimum* (distilled from CrO_3) grade (Fluka A.G.).

Formamide is a *purissimum* grade (Fluka A.G.). It is important for successful chromatography that the formamide is pure. Some reagent grade formamide preparations have been found to be heavily contaminated with ammonium formate, and are then unsuitable.

The samples and the PTH-amino acid references are dissolved in ethylene chloride. Many other solvents tend to dissolve out the formamide at the application point, and thus disturb the chromatogram. However, some PTH-amino acids are poorly soluble in ethylene chloride, notably the PTH derivatives of aspartic acid, asparagine, glutamic acid and glutamine, and it may therefore be necessary briefly to warm the solution before use. The solutions are best stored in the cold (− 15 °C) and away from light. They are then stable for 1 to 2 months, but after that time foreign spots may appear in the chromatograms. It is convenient to use three reference mixture solutions consisting of the PTH derivatives of (I) *asparagine, serine methionine sulfone, glycine, alanine, S-methyl cysteine, valine* and *proline*, (II) *glutamine, threonine, tyrosine, lysine, methionine, S-allyl cysteine* and *leucine*, and (III) *aspartic acid, glutamic acid, tryptophan, S-ethyl cysteine, phenylalanine* and *isoleucine*, isoleucine, respectively. With this arrangement the relevant reference spots are well separated from each other. A suitable concentration of each PTH-amino acid for samples and reference solutions is 5 μmoles/ml, and of these solutions 2 μl (or 0.01 μmole) is applied to the plate. *PTH*-lysine produces a more intense spot, and *PTH-proline* a weaker spot, and are therefore preferably used in half and double this concentration, respectively.

The spots may be located by either or both of two independent methods. The first method [297] depends on the fact that divalent sulphur compounds catalyse the reaction between free iodine and azide ions. The plate, which has been pretreated with starch, is sprayed with a solution of iodine-potassium iodide and sodium azide. The layer takes on a blue color, and against this background appear bleached aeras where the PTH-amino acids are located. The iodine-azide reaction is very sensitive, and it is possible to detect less than 0.1 μg of PTH.

The spray reagent is prepared by mixing equal volumes of (*a*) an aqueous solution 0.01 M in I_2 and 0.5 M in KI, and (*b*) an aqueous solution 0.5 M in NaN_3. The reagent is applied in a fine spray. The spraying should be done in a hood with a good draft because of the toxic nature of the reagents. To prevent the plate from drying before the spots are sufficiently bleached it is placed under a Perspex cover together with a dish of hot water or a hot solution of dilute ammonia. This helps to bring out the spot of PTH-proline, which otherwise is slow to appear.

The iodine-azide reaction has properties that make it less than ideal for its purpose. Firstly, it is not sufficiently specific, since any compound that reacts with iodine will give a positive reaction. Secondly, the intensity of a spot is not proportional to the amount of PTH. Given enough time to develop, even extremely small amounts of PTH (or impurities) will give rise to positive reaction. Therefore, this author has come to rely more on another way of locating PTH spots. This method is based on the strong ultraviolet absorption of PTH derivatives. A fluorescent indicator with excitation in the ultraviolet and emission in the visible is incorporated in the layer as already described.

The ultraviolet light source is a battery of tube-shaped low pressure mercury lamps (Phillips Electrical Ltd., type TUV 15 W) which have been arranged in parallel to form a square. When the plate is held with the layer towards the light source, and viewed from the other side, the PTH-amino acids appear as dark spots against a luminous background. The contrast is considerably increased if the plate is viewed through a Wratten No. 45 filter. A permanent record of the chromatogram may be obtained by photographing the plate through the same filter. Kodak Micro-file film, Type 5453, is suitable for this purpose.

The method is sensitive to about 0.2 μg of PTH in a spot. It has the advantage over the iodine-azide reaction that the intensity (and area) of the spot is roughly proportional to the amount of PTH. The specificity is of course low, since any material absorbing light at 254 mμ would give a positive reaction. However, in practice, false reactions are rarely seen. The iodine-azide reaction may be applied to the plate afterwards. However, it is then advisable to disregard faintly positive reactions unless they tally with the findings of the fluorescence extinction method.

It should be added that some PTH-amino acids already show a characteristic color reaction when the plate is heated during drying. Thus, *PTH-glycine* productes a brownish-red spot, and *PTH-tyrosine* and *PTH-tryptophan* may be recognized by a lemon-yellow coloration. Some other PTH-amino acids also give rise to colors under these circumstances, but the reactions are too faint to be of use in the identification.

A chart showing the positions of the PTH-amino acids in the various systems is shown in Fig. 8-4. To this the following comments may be added. Generally, the D system resolves the non-polar PTH-amino acids, and the E and H systems resolve the polar PTH-amino acids. However, neither of the latter systems alone will allow the identification of all polar PTH-amino acids, but will do so when used together. The normal practice is to use only the D and H systems, and to resort to the E system only when one wants to differentiate between *PTH-tyrosine* and *PTH-glutamic* acid, or between *PTH-methionine sulfone* and *PTH-aspartic* acid. The PTH derivative of *S-carboxymethyl cysteine* (not shown in the chart) falls between PTH-aspartic and PTH-glutamic acid in system H, and coincides with PTH-aspartic acid in system E, and is difficult to differentiate from these PTH-amino acids. Here the resolution may be increased to a sufficient degree in the following way. Chromatography is performed in the H system, the plate is allowed to dry in the air, and is then irrigated in the same system. The present author prefers other S-alkyl, *e.g.* S-methyl, S-ethyl or S-allyl, derivatives of cysteine in degradation work, since *PTH-S-methyl cysteine*, *PTH-S-ethyl-cysteine* or *PTH-S-allyl cysteine* are more conveniently identified (Fig. 8-4). In the a system *PTH-leucine* and *PTH-isoleucine* are close together, but their relative positions are very reproducible, and may be relied on in the differentiation. However, it may be necessary to rerun the chromatogram, *e.g.* if the solvent front is skewed. In the rerun one places the unknown in the middle and the appropriate reference mixtures (II and III) in the neighboring lanes. A similar rerun may also be indicated for other less well resolved pairs of PTH-amino acids.

Samples obtained in a degradation may show spots due to compounds other than the expected PTH-derivatives. *PTH-serine,* due to partial decomposition during conversion (cf. "Conversion"), is always contaminated by other products, which are seen as multiple spots between PTH-serine and the front in solvent systems E and H (because of the lower yield of PTH-serine the sample should be applied in double strength). The positive identification of serine requires that a spot of PTH-serine is clearly present in the chromatogram. *PTH-threonine* is always accompanied by

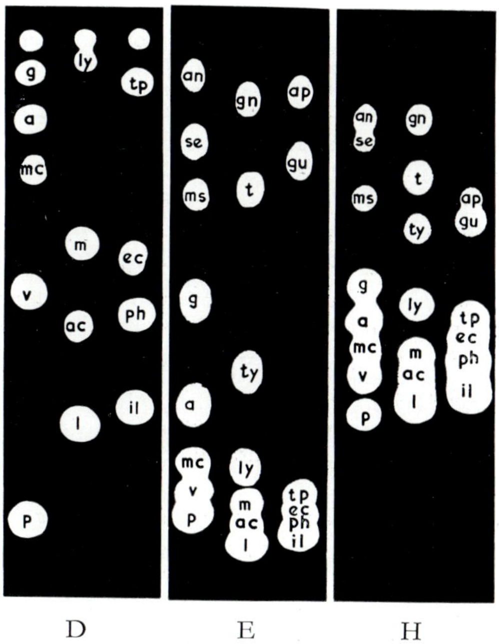

Fig. 8-4. Chart showing the TLC of PTH-amino acids in systems D, E, and H, (see text). PTH derivatives of asparagine (*an*), serine (*se*), methionine sulfone (*ms*), glycine (*g*), alanine (*a*), S-methyl cysteine (*mc*), valine (*v*), proline (*p*), glutamine (*gn*), threonine (*t*), tyrosine (*ty*), lysine (*ly*), methionine (*m*), S-allyl cysteine (*ac*), leucine (*l*), aspartic acid (*ap*), glutamic acid (*gu*), tryptophan (*tp*), S-ethyl cysteine (*ec*), phenylalanine (*ph*) and isoleucine (*il*)

some PTH-dehydrothreonine, which may be seen as a faint spot between PTH-alanine and PTH-methionine in system D. The spots of *PTH-asparagine* and *PTH-glutamine* are always accompanied by strong spots of PTH-aspartic and PTH-glutamic acid respectively. This helps greatly in the identification, since the latter are better separated than the former (Fig. 8-4). A spot at the origin becomes more apparent as the degradation progresses. It is particularly noticeable in system D, and is due to the accumulation of background material. *Phenylthiourea* and *s-diphenyl-thiourea* normally are not seen in the chromatograms, since they are removed in the extractions following coupling, but may appear if the extractions have not been thorough enough. Phenylthiourea remains at the origin in system D, in system E it appears between PTH-threonine and PTH-glycine, and in system H it coincides with PTH-glycine. Diphenylthiourea appears in system D between PTH-methionine and PTH-valine, and in systems E and H among the fastest moving PTH-amino acids.

2. Identification of PTH-Arginine, PTH-Histidine and PTH-Cysteic Acid

These PTH-amino acids may be identified by chromatography, but their positions on the chromatogram are not reproducible, and other methods are therefore preferred.

PTH-arginine is identified by a spot reaction on filter paper using a modified Sakaguchi reaction [3].

Two spray reagents are used: Reagent A consists of a solution of α-naphthol (5 mg) and urea (2.5 g) in ethanol (50 ml), to which is added immediately before use a solution of NaOH (3 g) in water (5 ml). Reagent B is a solution of Br_2 (0.70 ml) in ice cold aqueous NaOH (5 g NaOH in 100 ml). This solution should be freshly prepared. The sample (approximately 0.1 μmole) is deposited as a small spot on a strip of Whatman No. 1 filter paper. The aqueous phase of the preceding degradation cycle (if this does not contain PTH-arginine) serves as a blank. The strip is sprayed with reagent A until moist, allowed to dry in the air, and then sprayed with reagent B. A red color signifies a positive reaction.

PTH-histidine is identified by a spot reaction with a modified Pauly reagent [260].

Three solutions are used: Reagent A is a solution of sulphanilic acid (0.45 g) and HCl (0.5 ml conc. HCl) in water (50 ml). Reagent B is a solution of $NaNO_2$ (1.25 g) in water (25 ml). This reagent should be freshly prepared. Reagent C is a solution of Na_2CO_3 (5 g) in water (50 ml). All reagents are chilled in an ice bath. Reagents A and B are mixed immediately before use. The sample (approximately 0.1 μmole) solution is applied as a small spot on a strip of Whatman No. 1 filter paper and dried. It is then sprayed first with the mixed A and B reagents and, after drying, with the C reagent. An orange color marks a positive reaction. A blank should be used.

PTH-arginine, PTH-histidine and *PTH-cysteic acid* may be identified by their electrophoretic mobilities on filter paper[5].

A conventional apparatus for paper electrophoresis is used. The buffer is a 0.025 M potassium phosphate buffer, pH 6.20, the paper is Whatman No. 1, and the voltage approximately 10 V/cm. Reference PTH-amino acids are run alongside the sample. After 2 to 3 h the current is discontinued, the paper dried, and the spots located in the usual manner. At pH 6.20 PTH-arginine moves cathodically, PTH-histidine remains stationary (except for the electroendosmotic displacement), and PTH-cysteic acid moves anodically.

Other systems for the identification of PTH-amino acids using TLC have been described by Cherbuliez *et al.* [191], Brenner *et al.* [155], Pataki [769], Wang *et al.* [1098] and Jeppsson and Sjöquist [493]. It is difficult, and perhaps pointless, to give preference to any particular system for the TLC of PTH-amino acids. More important is probably that the operator is fully conversant with the system he has chosen.

The sensitivity with which PTH-amino acids can be detected may be considerably increased by suitable radioactive labeling. Cherbuliez *et al.* [192] and Laver [598] have utilized S^{35} labeled phenylisothiocyanate to prepare labeled PTH-amino acids. It is then possible to locate and estimate in a paper chromatogram as little as 0.1 mμ-mole of PTH-amino acid [598].

[5] In an earlier publication [270] is described the use of thin-layer electrophoresis on silica gel for this purpose. This technique has later been found insufficiently reproducible to be recommended.

3. Quantitative Paper Chromatography

In the identification of PTH-amino acids the paper chromatographic procedures have been superseded by the TLC techniques. For quantitative determinations, however, paper chromatography still plays an important role. In sequence work, the quantitation of PTH-amino acids poses no problem, since normally only one PTH-amino acid is produced at a time, and this may be determined by direct spectrophotometry. On the other hand, the quantitative N-terminal determination of a mixture of proteins, or of a single protein with several different N-terminal amino acids, necessitates the separation of the PTH-amino acids prior to quantitation. This may be done by paper chromatography, since the separated PTH-amino acids can be eluted from the paper, and determined spectrophotometrically. The technique described [272, 739] has been used successfully in the author's laboratory for many years. It was originally described, perhaps somewhat too modestly, as semiquantitative. In fact, the results for most PTH-amino acids are accurate to $\pm$ 5%, and the method may therefore qualify as quantitative. The technique is in large part similar to or identical with that already described for the TLC of PTH-amino acids, and an abbreviated account should therefore be adequate.

The chromatography room should be thermostatted at 20 $\pm$ 2 °C. Chromatography is descending, and is carried out in a glass tank, 20 $\times$ 30 cm and 60 cm in height. The glass lid should be well fitting with one or two holes bored in it to allow the introduction of the solvent withouth lifting the lid. A sheet of filter paper covers the inside walls of the tank, and is held in position by loops of stainless steel wire. The filter paper used in the chromatography is Whatman No. 1.

Three solvent systems are used:

a) Xylene/Formamide (System D)

The system and the procedure are the same as described for TLC. It is convenient to use for the formamide solution a small measuring cylinder, into which the rolled paper is dipped. The soaked paper is immediately placed between sheets of filter paper to remove excess liquid, and suspended in the air for a few minutes to allow the acetone to evaporate. After completion of the chromatography the paper is suspended and dried in a stream of warm air (*e.g.* from a hair dryer) until it feels dry to the fingers.

b) n-Butyl Acetate/Propionic Acid/Formamide (System E)

The system and the procedure are the same as described for TLC of PTH-amino acids, and the impregnation of the paper with formamide and the final drying of the paper are the same as described for system D.

c) n-Heptane/Ethylene Chloride/Formic Acid (System F)

The upper phase obtained after mixing *n*-heptane (1 vol.), 75% aqueous formic acid (2 vol.) and ethylene chloride (2 vol.) is used for developing the chromatogram, and the lower phase is used to soak the filter paper lining of the tank. An equilibration time of $^3/_4$ h is required. It is often found, when a new tank is set up, that the first runs are unsatisfactory with a tendency for phase separation on the paper. This trouble usually disappears after the tank has been in use for a day or two. The liquid accumulating in the bottom of the tank should not be removed until necessary, since this seems to upset the equilibration. The final drying of the paper is the same as described for system D.

To locate the PTH spots a fluorescent screen is used [272]. The screen is made of a sheet of translucent material, and one side is covered with a fluorescent material which is excited by ultraviolet light, and emits visible light. The paper is stretched over the fluorescent surface of the screen, and viewed through the screen, while illuminated by an ultraviolet light source. The spots appear as dark areas on an illuminated background. The light source described in the preceding section on TLC is also suitable here. The fluorescent indicator recommended has its secondary emission in the red, and outside the range of visible light emitted by the light source. A considerable increase in contrast may therefore be had by viewing the chromatogram through red colored glasses. The screen is made in the following way.

A sheet of polyethylene, 40 × 30 × 0.15 cm, is coated on one side with a layer of fluorescent cadmium borate (AB Lumalampan, Stockholm, Sweden). Finely cut polyethylene is dissolved in boiling heptane (0.25 g/100 ml), and a fine suspension of cadmium borate in this medium is prepared. The suspension, while still hot, is sprayed on to the polyethylene sheet. After drying in the air the polyethylene sheet is placed on a glass plate with the coated surface upwards, and then left in an oven at 100 °C to allow the coating to sinter. The spraying and sintering processes are repeated until an even opaque layer of fluorescent material is obtained.

The purification of solvents is described in the section "Thin-layer Chromatography".

A preliminary qualitative chromatography is made to establish which PTH-amino acids are present in the mixture.

In the quantitative determination 0.03 to 0.06 μmoles of PTH are applied to the paper. The solution of the sample to be determined is applied to the paper alongside an accurately made solution of the reference PTH-amino acid. The reference solution is applied in two different concentration levels, one above and one below the expected concentration of the sample. An adjacent empty lane serves as a blank. All solutions are made up in ethylene chloride. They are applied to the paper 3 cm apart with constriction pipettes. Certain PTH-amino acids are poorly soluble in ethylene chloride, and it may be necessary to use a larger volume, and apply several times with drying in between.

After completion of the chromatography and drying of the paper the spots are marked out in pencil with the aid of the fluorescent screen and cut out with a margin of approximately 3 mm. An area corresponding in size and level is cut out from the blank lane. The cut out pieces, after being divided in smaller pieces with scissors, are extracted with 1.5 ml of 70% aqueous ethanol overnight at room temperature and in the dark.

The absorbances of the extracts are read at 269 mμ in micro quartz cells (1 cm light path) with a capacity of 1 ml. Two corrections are made, one for background absorption due to the paper and one for losses of PTH during chromatography. The first is made by subtracting the absorbance of the paper blank from the other absorbances. The second correction is made by multiplying the sample absorbance with a correction factor. This factor is the ratio between the expected and the found absorbances of the reference samples, the expected absorbances being calculated from the molar absorptivities in Table 1. This correction factor varies depending on the PTH-amino acid, or more precisely on its position in the chromatogram. In system F the chromatographic losses are almost negligible, but may in system D be considerable, amounting to 25 to 35% for the spots closest to the solvent front. The amount of PTH-amino acid in the sample is calculated from the corrected sample absorbance using the values in Table 1.

Table 2 shows the R_f-values of the PTH-amino acids in the three systems. Most of the PTH derivatives are sufficiently well separated in one or the other system to allow elution without contamination from adjacent PTH-amino acids. However, in certain combinations of PTH-amino acids there is a partial overlap, and the estimation is then less accurate.

PTH-arginine, *PTH-histidine* and *PTH-cysteic acid* are estimated after electrophoresis on paper (cf. "Identification of PTH-Arginine, PTH-Histidine and PTH-Cysteic Acid").

The combination of chromatography and ultraviolet spectrophotometry for the estimation of PTH-amino acids occurs in many variants. In this area SJÖQUIST [954 to 956] has been particularly prolific. He has shown that most PTH-amino acids

Table 2. *R_f-values of phenylthiohydantoins in paper chromatography systems D, E and F*

PTH-amino acid	R_f-value in system		
	D	E	F
PTH-asparagine	0	0.31	0.80
PTH-serine	0	0.47	0.14
PTH-glutamine	0	0.41	0.15
PTH-aspartic acid	0	0.38	0.16
PTH-glutamic acid	0	0.56	0.27
PTH-threonine	0	0.55	0.34
PTH-tyrosine	0	0.75	0.41
PTH-glycine	0.09	0.63	0.62
PTH-alanine	0.18	0.77	0.78
PTH-N^{ε}-PTC-lysine	0.05	0.85	0.81
PTH-tryptophan	0.12	0.86	0.82
PTH-methionine	0.45	0.89	0.88
PTH-valine	0.53	0.87	0.89
PTH-proline	0.83	0.90	0.89
PTH-phenylalanine	0.57	0.90	0.91
PTH-isoleucine	0.65	0.92	0.91
PTH-leucine	0.67	0.92	0.92

may be separated on a partition column, and estimated with a considerable degree of accuracy [955]. The incorporation of a light absorption flow cell in the effluent line and the use of a recording spectrophotometer make the estimation particularly rapid and convenient. In conjunction with a microprocedure for the quantitative synthesis of PTH-amino acids from amino acids [957] the method may also be used for amino acid analysis. SJÖQUIST has also described several paper chromatographic systems suitable for quantitation of mixtures of PTH-amino acids [956].

Gas chromatography has also been suggested for the separation and identification of PTH-amino acids. PISANO *et al.* [790] have demonstrated the feasibility of a very rapid identification of a great number of PTH-amino acids. Difficulties were originally experienced with the PTH derivatives of serine, threonine, asparagine, glutamine and the basic amino acids. Most of these difficulties now seem to have been overcome by trimethylsilylation of the PTH-amino acid [409, 738, 789], and gas chromatography now offers an alternative method in the identification of PTH-amino acids, and be particularly valuable as an adjunct in the rapid automated sequence determination. The general techniques for gas-liquid chromatography are described in Chapter 5.

The amino acid may be regenerated from the PTH-amino acid by hydrolysis, and identified as such. This was in fact the procedure used in the first account of the phenylisothiocyanate degradation [264]. The PTH-amino acid was hydrolyzed in 0.25 N barium hydroxide at 140 °C for 48 h. In most cases the parent amino acid was recovered, but expectedly PTH-arginine gave rise to ornithine, PTH-asparagine produced aspartic acid (PTH-glutamine, PTH-serine and PTH-threonine were not available), and the yields were generally low. Levy [608] advocated the use of acid hydrolysis although the PTH derivatives of serine, threonine and tryptophan (and presumably those of asparagine and glutamine) did not produce the parent amino acid. Africa and Carpenter [8] have studied the hydrolysis more thoroughly, and have found that the best results are obtained by alkaline hydrolysis with rigorous exclusion of oxygen. Under these conditions the yields were generally high and often quantitative. However, the general shortcoming of the hydrolytic procedure was again apparent in that the PTH derivatives of serine, threonine, arginine, asparagine and glutamine did not produce the parent amino acids.

In the *indirect* method the amino acid residue is identified by a difference in analytical composition of the peptide before and after a degradation cycle. Two techniques are commonly in use.

The first is due to Hirs *et al.* [445c]. Here quantitative amino acid analyses [987] are performed on the peptide before and after a degradation cycle. The reduction of a particular amino acid by one mole indicates the residue split off (in practice the reduction is often considerably less than one mole). The technique was used extensively in the now classical elucidation of the first enzyme structure [445c, 974, 975], and has since been widely used in structural studies. Its chief, and perhaps unique, merit lies in the fact that the experimental findings are easily documented. This is done by presenting the results of the amino acid analyses.

The second technique was devised by Gray and Hartley [369]. Its essential feature is that only the N-terminal amino acid is determined before and after a degradation cycle. The new N-terminal amino acid created by a degradation cycle is the one next in sequence. This has the advantage over the preceding technique that the identification is less ambiguous. In the technique of Hirs *et al.* [445c], it is not infrequently seen that more than one amino acid is reduced after a degradation cycle. The uncertainty which this leads to is avoided by the identification of the new N-terminal acid. For identification the sensitive dansyl technique is used, which permits sequence determination on very small amounts of peptide. This technique is not described here, since it is treated in another chapter of this book (cf. "End Group Determinations"). For the following discussion, however, it is important to know that the identification by the dansyl technique requires complete hydrolysis of the peptide.

It should be clear from what has been said earlier in this chapter that the present author favors the direct identification method, and it only remains to give the reasons.

The chief criticism of the indirect method derives from the fact that it requires a complete hydrolysis of the peptide. Therefore the difference in composition of the peptide before and after a degradation cycle relates only to the *hydrolysate* of the peptide, but not necessarily to the peptide itself. Thus, an amino acid residue may be so modified by the hydrolysis that it cannot be identified by the indirect method. This applies to asparagine, glutamine and tryptophan. Here one is forced to rely on even more indirect and circumstantial evidence for the identification, such as the electrophoretic migration rate of the peptide before and after removal of the amino acid (asparagine and glutamine), or the action of proteolytic enzymes (tryptophan). Furthermore, many proteins contain modified amino acid residues, *e.g.* by a covalently bound carbohydrate moiety. If these groups are removed by hydrolysis in concentrated acids, and they usually are, they cannot be located by the indirect method. The awareness of their presence may not even be aroused, because there is nothing in the method to indicate their presence. The weaknesses enumerated here are inherent in the indirect method, and cannot be removed by any improvement in technique.

The value of a sequence technique depends largely on the length of sequence which may be determined. The important factor here is a high repetitive yield, *i.e.* the yield of amino acid calculated from one degradation cycle to the next. Losses of only a few per cent severely limit the length of the degradation (cf. "Sequential Degradation"). The indirect method requires that in each degradation cycle an aliquot of the peptide is removed for analysis. This removal in effect reduces the repetitive yield, and therefore prevents extended degradation.

The rapid automated technique for sequence determination calls also for rapid identification of amino acids. Here the indirect method cannot be used for reasons which should be obvious.

The indirect method was conceived at a time when reliable techniques for the positive identification of amino acid residues removed in a degradation were not available. Today several techniques for this purpose exist. They embrace all amino acids normally occurring in proteins, permit an unambiguous identification, and require little experimental time and labor. On the other hand, the indirect techniques are based on an analytically less sound principle, are not comprehensive, are more cumbersome, and are not suited for automated sequence techniques. For these reasons the indirect method should no longer be used.

E. Sequential Degradation[6]

The value of any sequence method depends largely on the length of the sequence that can be elucidated by it. A large, if not the larger, part of the time spent in determining the primary structure of a protein is taken up by the preparation of (*a*) fragments sufficiently small to be determined by the available sequence techniques, and (*b*) fragments furnishing overlapping sequences. Obviously, the magnitude of this arduous task would be substantially reduced, if the sequence method is capable of determining larger structures. Ideally, it should be possible to dispense with the fragmentation altogether. However, present techniques for various reasons fall short of this ideal, although the isothiocyanate degradation is closest to it.

---

[6] This section describes many observations and practices in the author's laboratory which previously have not been published by the author. They have, however, been published by DOOLITTLE [249] without permission from the present author.

The chemical yield becomes a major factor in the determination of long sequences, which may be illustrated by a simple calculation. Let it be arbitrarily assumed that a sequence determination ceases to give useful information when the over-all yield, *i.e.* the yield over *n* degradation cycles, has fallen to 30%. It may then be calculated that an average repetitive yield, *i.e.* from one cycle to the next, of 97, 98 and 99% would allow 40, 60 and 120 cycles, respectively. It is then obvious that even minor side reactions become significant.

It has already been mentioned (cf. "Reaction Mechanism") that each step in the degradation consists of three operations, *i.e. coupling, cleavage* and *conversion.* Only the two first-mentioned reactions are of importance for the repetitive yield, and the associated side reactions and their elimination will be discussed in some detail.

1. Coupling [2, 52, 53]

The reaction is:

$$C_6H_5{\cdot}NCS + NH_2{\cdot}CHR{\cdot}CO{\cdot}NH{\cdot}X \longrightarrow C_6H_5{\cdot}NH{\cdot}CS{\cdot}NH{\cdot}CHR{\cdot}CO{\cdot}NH{\cdot}X \qquad (9)$$

Amino groups react with phenylisothiocyanate only in their *amino* form, and the reaction rate is therefore dependent on the pK of the amino group. The higher pK of the ε-amino group of lysine must therefore also be taken into consideration. A reaction pH of 9 and a temperature of 40 °C give a convenient rate of reaction. Because hydrogen ions are produced during the reaction a buffer is required, and this should preferably be volatile for easy removal after completion of the reaction. The amine used here, N-dimethylallylamine, is suitable because, (*a*) it is a tertiary amine and therefore unreactive to phenylisothiocyanate, (b) it has $pK{\sim}9$, and (*c*) it is volatile. To raise the solubility of the non-polar phenylisothiocyanate it is necessary to use a reaction medium with a high content of an organic solvent miscible with water. A mixture of pyridine and water (1:1) is suitable, since it is also a good solvent for peptides and proteins.

Several side reactions have to be considered. Most important of these is the oxidative desulphuration of the phenylthiocarbamyl group observed by Ilse and Edman [472]. The oxygen dissolved in the medium is sufficient to bring about this reaction. The resulting phenylcarbamyl group acts as a blockage to the degradation, since obviously no thiazolinone can be formed. This side reaction may be eliminated by performing the coupling reaction in an inert atmosphere, *e.g.* nitrogen [472]. In the manual degradation it is not practicable to exclude oxygen completely. This is probably the main reason for the lower repetitive yields compared to the sequenator procedure.

The amino group may be made unreactive by various blocking reagents. Important are aldehydes, which tend to form Schiff bases. Traces of aldehydes occur universally, and precautions should be taken to remove them from solvents and reagents. This applies particularly to the sequenator procedure.

It seems likely that heavy metal ions could interfere with the degradation. However, this is hypothetical, and based only on the well known fact that thiourea derivatives are strong chelating agents.

Reactions of phenylisothiocyanate with functional groups other than primary amino groups and the secondary amino groups of the prolines have not been observed by the present author. Yet, it has been suggested that the imidazole group of histidine and the phenolic group of tyrosine [565] become thiocarbamylated during the coupling. If so, these reactions are reversed at a later stage, since the derivatives do not appear among the products of the degradation or in the synthesis of PTH-amino acids (cf. "Preparation of Phenylthiohydantoins"). An interesting observation, referred to by KONIGSBERG [565] is the reaction of phenylisothiocyanate with free sulfhydryl groups of cysteine residues, a reaction which subsequently leads to scission of the peptide chain at the cysteine residues. In preparations used for degradation the sulfhydryl groups have normally been blocked beforehand, and this reaction should therefore not occur.

A side reaction which does not involve the peptide is the hydrolysis of phenylisothiocyanate:

$$C_6H_5 \cdot NCS + 2\,H_2O \xrightarrow{OH^-} C_6H_5 \cdot NH_2 + CO_2 + H_2S \qquad (10)$$

$$C_6H_5 \cdot NH_2 + C_6H_5 \cdot NCS \longrightarrow \underset{\text{VIII}}{C_6H_5 \cdot NH \cdot CS \cdot NH \cdot C_6H_5} \qquad (11)$$

The *s*-diphenylthiourea (VIII) may simulate a PTH-amino acid in the chromatographic identification (cf. "Thin-layer Chromatography"), and should be removed (together with phenylthiourea formed from ammonia). This is done in the extractions following coupling.

The reaction is carried out in a thick-walled centrifuge tube (inner diam. 1.4 cm) with a capacity of 9 ml, and provided with a C 14 stopper. The weight of the dry tube is determined. The preparation to be degraded remains in this tube during all subsequent operations.

The peptide or protein preparation (0.1 to 5.0 μmoles) is weighed into the tube and dissolved or finely suspended in pyridine-water (1:1; 1 ml) made 0.4 M with respect to N-dimethylallylamine. The pH is adjusted to 9 with dilute aqueous trifluoroacetic acid. Phenylisothiocyanate (50 μl) is added, the tube flushed with N_2, stoppered, and left at 40 °C for 1 h with occasional shakings to keep the solution saturated with the isothiocyanate. The solution is then repeatedly extracted with benzene (5 × 2 ml) using centrifugation to separate the phases. The aqueous phase is subsequently freeze-dried. The sticky residue is repeatedly extracted with ethyl acetate (3 × 0.5 ml) leaving behind a powdery material. The residual ethyl acetate is removed first in a gentle stream of N_2 and finally *in vacuo*. The dry preparation is weighed at this stage in each degradation cycle.

To this description the following comments are added. The amount of material used in the degradation depends on the nature of the preparation and on the purpose of the degradation. If the peptide is short, and the intention is to carry the degradation through to the C-terminal end, it is advisable to use a large amount. The manipulative losses increase as the peptides become shorter, mainly due to an increased solubility of the peptide in the organic solvents used for washing. A warning of increasing losses is given by a more rapid drop in the weight of the PTC-peptide. The main losses in the coupling operation occur during the extraction with ethyl acetate. In the last degradation cycles one may leave out the ethyl acetate extractions, but should then expect spots caused by impurities, *e.g.* diphenylthiourea, to appear in the chromatograms. On the other hand, if it is only a matter of identifying the N-terminal amino acid in a protein, 0.1 μmole is ample. For an accurate quantitative estimation somewhat more, 0.2 to 0.3 μmoles, are required. In preparations with several different N-terminal amino acids, the quantitative determination involves separation by paper chromatography (cf. "Quantitative Paper Chromatography"), and the material requirements are higher, say 0.3 μmoles of the component occurring in the lowest concentration.

The solubility of the preparation in the coupling medium usually poses no problem. If the preparation is soluble in aqueous media, its solubility in the coupling medium is sufficient for the reaction with phenylisothiocyanate to be quantitative. Some proteins form a fine suspension rather than a clear solution in the coupling medium, but this does not seem to interfere with the reaction. Where it is necessary to handle larger amounts of protein, *e.g.* proteins of high molecular weight, it is advisable to scale up all volumes, as a very concentrated protein solution is difficult to extract properly. It is sometimes seen that a protein becomes insoluble after a few degradation cycles, and forms a gummy material. This material is difficult or impossible to handle in the degradation procedure. It may be mentioned already here, that this problem is much less troublesome in the sequenator procedure, because the insoluble protein is still reactive when spread in a thin film.

2. Cleavage [108, 267, 269]

The reaction taking place is:

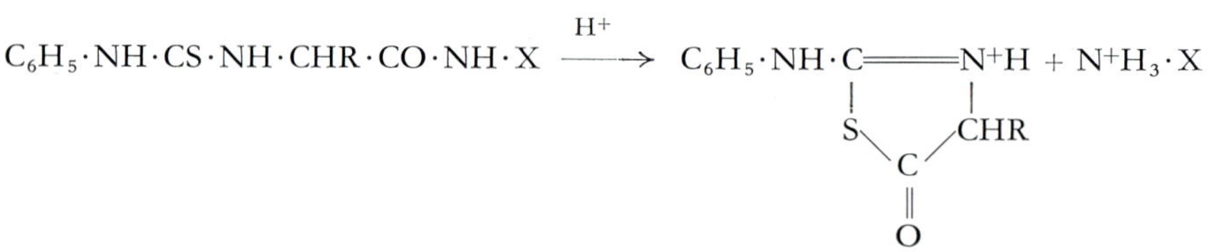
(12)

The two most important facts about this reaction is that it is non-hydrolytic, and that it is rapid. The cleavage may thus be brought about by the action of anhydrous acid for a short time. The peptide bonds are therefore not exposed to hydrolytic conditions. The rate of cleavage varies somewhat depending on the amino acid residue. PTC-leucylglycine in acetic acid saturated with hydrogen chloride is cleaved in less than 3 min at 37 °C [268]. Some other amino acid residues react more slowly, but kinetic data are not available. It may be surmised that the relative rate constants for the various amino acids residues are similar to those observed in the better studied conversion reaction [472].

If suitable conditions are used for the cleavage, few side reactions are observed. However, if the conditions are too severe, *e.g.* glacial acetic acid saturated with hydrogen chloride at 100 °C for 1 h [445c], side reactions become prominent. The reactions described involve specific amino acid side chains. Thus, it is reported [975] that, when the degradation exposes a glutamine residue in N-terminal position, this cyclizes to form a pyroglutamic acid residue (IX):

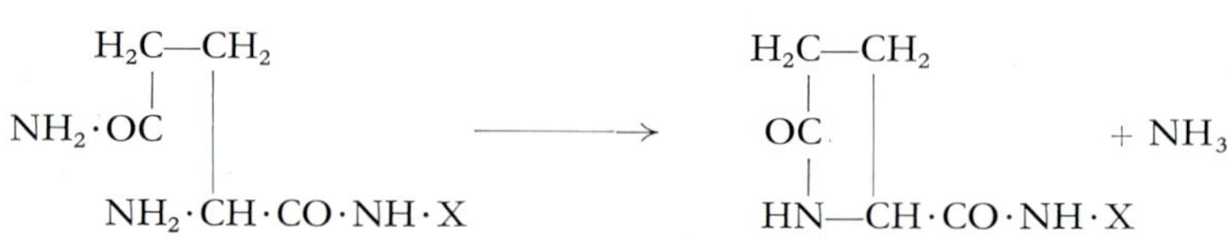
(13)

IX

As the α-amino group is now blocked, the degradation comes to a halt. Reports conflict [375, 565, 975] whether this reaction takes place under the milder cleavage conditions suggested by KONIGSBERG and HILL [567], *i.e.* anhydrous trifluoroacetic acid at 25 °C for 1 h. Under the conditions recommended here, anhydrous trifluoroacetic acid at 40 °C for 15 min, this reaction has not been observed. It is relevant that the side reaction follows the cleavage reaction. Measures should therefore be taken to protect the shortened peptide from the conditions causing pyroglutamyl formation. This is done by removing the trifluoroacetic acid and by storing the shortened peptide in the cold until the next cycle is started.

Under the conditions of HIRS *et al.* [445 c], serine and threonine residues become acetylated at their hydroxyl groups, and when the adjacent amino group is released in the course of the degradation, an O → N acetyl migration takes place, blocking the α-amino group [975]. An $\alpha \rightarrow \beta$ peptide shift at aspartic acid and asparagine residues has also been reported to take place under the above conditions [975]. It should be stressed, that all these reactions are caused by too severe conditions, and are not unavoidable concomitants of the cleavage reaction. Nevertheless, they serve to emphasize the importance of not exposing the peptide unnecessarily to the cleavage medium.

In the sequenator procedure the protein goes through a great number of degradation cycles, and side reactions which are not noticed in shorter degradations become apparent [270]. What is seen is the cumulative effect of minor side reactions. It is difficult to establish with certainty the nature of these reactions, since in each degradation cycle they may only implicate a few thousands of a micromole of the protein. The mechanisms proposed are therefore no more than tentative.

The first reaction to be mentioned is probably not a side reaction in a strict sense. It makes itself noticeable in a gradually increasing overlap between consecutive cycles in the degradation. This overlap is not reduced, if the time allowed for cleavage is increased. On the other hand, if the thiazolinone is extracted, and the cleavage operation repeated, the overlap is substantially reduced or disappears altogether. The suggested explanation [270] is that the end point of the cleavage reaction is an equilibrium rather than a quantitative reaction, although the equilibrium is close to a quantitative reaction. This interpretation is supported by observations made on model substances[7].

The other side reaction noticeable in extended degradations is the appearance in the chromatogram of an increasing general background of other PTH-amino acids. This effect is apparent only if large structures, say with more than 60 amino acid residues, are being degraded, and is the more pronounced the larger the structure. The cause of the background is not known with certainty. It may be significant that the dominating amino acids in the background are also those which occur most frequently in the structure. A small degree of non-specific cleavage along the peptide chain would produce this result. A hydrolytic action of the acid seems unlikely, since precautions have been taken to perform the cleavage reaction under anhydrous conditions. An alternative explanation is acidolysis caused by the organic fluoro acid used in the cleavage. A precedent is the observation that anhydrous trifluoroacetic acid may cause acidolysis of peptide bonds [571].

[7] P. EDMAN, unpublished.

The amide groups of asparagine and glutamine residues are not appreciably hydrolysed even after many degradation cycles. However, a small degree of hydrolysis would be difficult to detect, since it would be obscured by the partial hydrolysis of the amide group that occurs during the subsequent conversion reaction (cf. "Conversion"). At any rate, no problem arises in the identification of asparagine and glutamine residues. The situation is not as clear in regard to tryptophan residues. The repeated exposure to anhydrous acid expectedly would lead to some destruction of the indole nucleus, and this has also been reported in the phenylisothiocyanate degradation of gramicidin C [864]. On the other hand, when apomyoglobin was degraded by the sequenator procedure, a tryptophan in position 14 [270] could be identified in the normal way (Fig. 8-11). The stability of tryptophan residues during extended degradations needs further investigation.

Another side reaction, involving histidine residues, has been reported [61, 62]. It is claimed that when the degradation reaches a histidine residue, not only this residue but also the one following it in the sequence is released. Both appear as PTH derivatives in the identification procedure. Without questioning the correctness of these observations, the author can only state that he has never encountered this side reaction. Examples are the three histidine residues in position 24, 36 and 48, respectively, of the apomyoglobin structure [270], where the histidine residues behaved normally (Fig. 8-11).

As has been pointed out on several occasions, the cleavage reaction takes place under anhydrous acid conditions. The medium first proposed, nitromethane saturated with hydrogen chloride [264], is suitable for short peptides, but not for larger peptides because of their poor solubility. A better solvent is glacial acetic acid saturated with hydrogen chloride, in which the cleavage reaction proceeds with equal ease [268]. A medium in which, in addition, proteins also are soluble, trifluoroacetic acid, was originally suggested by Elmore and Toseland [283], and first used in the isothiocyanate degradation by the present author [267]. Konigsberg and Hill [567] have studied more thoroughly the use of trifluoroacetic acid for this purpose. They recommended anhydrous trifluoroacetic acid at 25 °C for 1 h as a standard set of conditions applying generally to cleavage operations and causing little or no side reactions. The sequenator procedure [270] called for a medium with a lower vapor pressure, and for this reason heptafluorobutyric acid was introduced. The lower homologs of the organic perfluoro acids are in many ways ideal media for the cleavage reaction. They are (*a*) excellent solvents for peptides and proteins [528], (*b*) sufficiently strong acids to bring about a rapid cleavage, (*c*) sufficiently volatile to allow easy removal after completion of the reaction, and (*d*) easy to prepare in a high degree of purity. On the debit side is tendency to cause acidolysis of peptide bonds (cf. above). From their known properties [1120] one would expect them to cause O-acylation of serine and threonine residues. This may well occur, but, if so, does not lead to the expected complication, *i.e.* and O → N acyl migration and consequent blockage of the amino group, when released in the course of the degradation [975].

The dry PTC preparation is dissolved in trifluoroacetic acid (50 to 100 μl), the tube stoppered and placed in a water bath at 40 °C for 15 min. The shortened peptide is precipitated with ethylene chloride (3 ml). The precipitate, which is often difficult to see because of a refractive index close to that of ethylene chloride, is spun down in a centrifuge, and the ethylene chloride drawn off with a Pasteur pipette. To the sticky precipitate is added a fresh portion of ethylene chloride (1 ml), and the precipitate is then thoroughly macerated with a glass rod. The ethylene chloride extracts, which contain the thiazolinone, are combined and transferred to a glass-stoppered centrifuge tube (inner diameter 1 cm, capacity 5 ml). The tube containing the precipitated peptide is placed without delay in a precooled desiccator together with a dish containing pellets of KOH. The desiccator is evacuated on the oil pump and put in the cold room over night. This preparation is used for the next coupling. The combined ethylene chloride extracts are evaporated to dryness in a stream of N_2, and the residue is then ready for conversion.

Particular instances may warrant small variations in this procedure. Thus, if larger amounts of protein are handled, it may be necessary to increase the volume of trifluoroacetic acid. When short peptides are being degraded, considerable losses may occur in the ethylene chloride extraction. These may be reduced by (*a*) evaporating off most of the trifluoroacetic acid subsequent to the cleavage, and (*b*) reducing the volume of ethylene chloride used in the extraction.

The rapid and careful removal of excess of trifluoroacetic acid from the residual peptide and the storage of the latter in the cold serve to prevent side reactions., *e.g.* pyroglutamyl formation from an N-terminal glutamine residue (cf. above).

3. Conversion [108, 267, 270, 472]

This comprises, strictly speaking, two reactions:

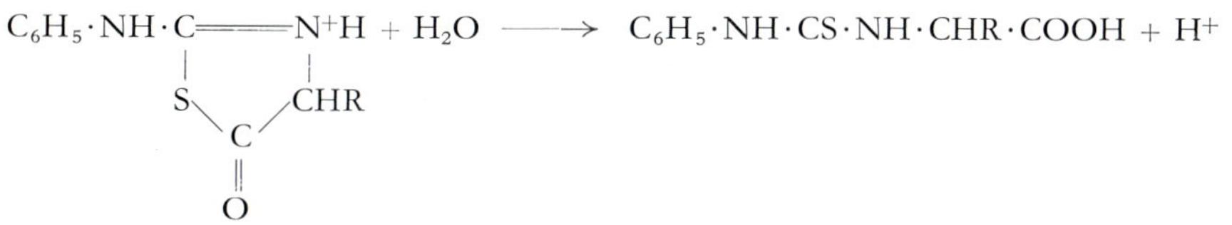

(14)

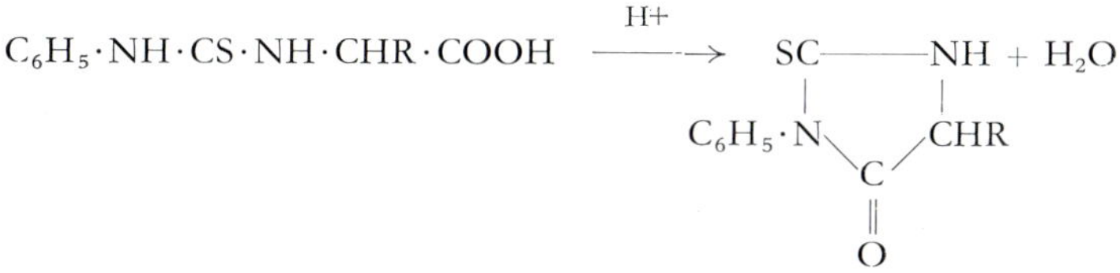

(15)

However, the first reaction is much faster than the second, and for practical purposes they may be treated as one reaction.

The conversion is carried out in an aqueous medium. The kinetics of the cyclization reaction has been studied in some detail [472]. The reaction follows first order kinetics, and, expectedly, the rate increases with the hydrogen ion concentration. The rate varies depending on the amino acid. Thus, the rate constants ($k_1 \times \text{min}^{-1}$) at pH 1.0 and 80 °C for the conversion of PTC-valine and PTC-glycine, are 0.13 and 0.03, respectively.

Several side reactions are of importance. As has already been mentioned (cf. "Coupling"), the phenylthiocarbamyl group is easily desulfurized by oxidation, *e.g.* through the action of oxygen dissolved in the medium. The conversion is therefore carried out in an atmosphere of nitrogen.

Certain PTH-amino acids tend to decompose during the conversion. Most rapidly decomposed is *PTH-serine*, which initially undergoes a β-elimination reaction (cf. "Properties of Phenylthiohydantoins") followed by other reactions, presumably polymerizations. The β-elimination reaction is slower at higher hydrogen ion concentrations [472], and it is therefore possible to select a pH for the conversion reaction which minimizes the tendency for β-elimination. Under the conditions described below the PTH-serine is obtained in about 40% yield. Several of the decomposition products later appear in the chromatograms as multiple spots. *PTH-threonine* shows much less tendency for β-elimination, and is recovered in near quantitative yield. The decomposition product, PTH-dehydrothreonine, is, unlike that of PTH-serine, rather stable, and reveals its presence by an ultraviolet absorption peak at 323 mμ (Fig. 8-1) and by a spot in the TLC (cf. "Thin-layer Chromatography"). The amide groups of *PTH-asparagine* and *PTH-glutamine* are partially hydrolysed, and PTH-aspartic and PTH-glutamic acid are formed, respectively. This fact is taken advantage of in the chromatographic differentiation of PTH-asparagine and PTH-glutamine, since the acids are better separated than the amides. For quantitation by paper chromatography (cf. "Quantitative Paper Chromatography") it is preferable to hydrolyze the amides completely to the acids.

In the final extraction of the conversion mixture with ethyl acetate, *PTH-arginine*, *PTH-histidine* and *PTH-cysteic acid* remain in the aqueous phase.

To the dry sample of the thiazolinone is added 1 N HCl (0.2 ml). The tube is flushed with a jet of N_2, care being taken also to stir the solution. The tube is stoppered, heated at 80 °C for exactly 10 min, and then cooled under the tap. The solution is extracted repeatedly with ethyl acetate (3 × 1 ml), and the extracts combined. If many samples are to be extracted, it is convenient to use a manifold as shown in Fig. 8-8. The organic phase is evaporated to dryness in a jet of N_2. Here again one may use a manifold (Fig. 8-7). The residue is dissolved in a suitable volume of ethylene chloride (approximately 200 μl/1 μmole of PTH). The aqueous phase is frozen for a later identification (cf. "Identification of PTH-Arginine, PTH-Histidine and PTH-Cysteic Acid").

All solutions of PTHs should be stored at — 15 °C and away from light.

4. Quantitation and Identification

The amount of PTH in the sample is determined from the ultraviolet absorption at 269 mμ. The position of the actual absorption maximum varies from one PTH-amino acid to another, and may be as low as 266 mμ. The shape of the absorption curve provides valuable information about the optical purity of the sample and in certain cases about its identity. The absorption measurements should therefore cover the whole range 340 to 220 mμ. This makes a recording ultraviolet spectrophotometer almost indispensable.

Typically (cf. "Properties of Phenylthiohydantoins"), a sample obtained in a degradation should have no absorption at wave lengths longer than 300 mμ, an absorption maximum around 268 mμ, a minimum around 245 mμ and a gradually increasing end absorption at shorter wave lengths. The ratio $\varepsilon_{245}/\varepsilon_{269}$ is a sensitive index of the purity of the product. This ratio should normally be around 0.4 (Table 1). The notable exceptions are the PTH derivatives of N^{ε}-PTC lysine and PTH-proline. Due to decomposition products formed during the conversion (cf. "Conversion") samples of PTH-serine and PTH-threonine show atypical spectra. The absorption curve of PTH-serine samples have a lower absorption at the maximum, and the peak is broader. Samples of PTH-threonine show, in addition to the typical absorption, a smaller peak with a maximum at 323 mμ, which is caused by PTH-dehydrothreonine (Fig. 8-1).

The sample solution in its stoppered tube is heated briefly in hot water to insure complete solution of the sample, and immediately cooled. A measured volume (approximately 10 μl) is added to a light absorption cell (1 cm path length) containing ethanol (3.0 ml). The solution is thoroughly mixed (ethylene chloride is much heavier than ethanol) in the cell with the aid of a Pasteur pipette. The absorbance is read against the same ethanol. The total quantity of PTH in the sample is calculated from the formula:

$$\mu\text{moles PTH} = \frac{a}{b} \times K \times A_{269}$$

where a is the total sample volume, b the volume used for spectrophotometry, K a factor and A_{269} the absorbance at 269 mμ. The magnitude of the factor K depends on the particular PTH-amino acid and is found in Table 3.

Table 3. *Factors (K) for calculating yields of phenylthiohydantoins from ultraviolet absorption at 269 mμ*

PTH-amino acid	K	PTH-amino acid	K
PTH-alanine	0.188	PTH-methionine sulphone	0.196
PTH-arginine	0.189	PTH-phenylalanine	0.193
PTH-asparagine	0.174	PTH-proline	0.210
PTH-aspartic acid	0.186	PTH-S-allyl cysteine	0.171
PTH-cysteic acid	0.181	PTH-S-carboxymethyl cysteine	0.181
PTH-glutamic acid	0.189	PTH-S-ethyl cysteine	0.184
PTH-glutamine	0.176	PTH-S-methyl cysteine	0.183
PTH-glycine	0.201	PTH-serine	0.194
PTH-histidine	0.194	PTH-threonine	0.192
PTH-isoleucine	0.177	PTH-tryptophan	0.152
PTH-leucine	0.180	PTH-tyrosine	0.192
PTH-N^{ε}-PTC-lysine	0.104	PTH-valine	0.182
PTH-methionine	0.175		

The identification of PTH-amino acids has already been described in detail (cf. "Thin-layer Chromatography", and "Identification of PTH-Arginine, PTH-Histidine and PTH-Cysteic Acid").

5. Reagents and Solvents

Much emphasis is laid on the purity of reagents and solvents.

Phenylisothiocyanate (*purissimum* grade, Fluka A.G.) is distilled *in vacuo* at 1 mm of Hg. The reagent is somewhat unstable on storage even in the cold, and is therefore distilled only as required.

N-dimethyallylamine is synthesized as follows [214]. N-dimethylammonium chloride (50 g) is dissolved in water (38 ml) in a thick-walled glass bottle (capacity 250 ml) with a tight fitting glass stopper. Allyl chloride (41 ml) is added, and the mixture cooled in an ice bath. Pellets of NaOH (30 g) are added in portions over 15 min, with the stopper only lightly fitted to allow gas to escape. The reaction is started by gentle shaking of the reaction mixture, and the rate is controlled by cooling. After 45 min the reaction mixture is brought to room temperature, and the reaction completed at this temperature. When the reaction has subsided sufficiently, the stopper is tied down and the bottle shaken mechanically overnight. The oily amine layer is separated, and to the aqueous phase is added solid KOH (1 g/ml of aqueous phase), whereupon a new layer of amine is formed, which is added to the bulk. The amine is distilled off on a Widmer column. The distillation is interrupted when the temperature rises and water starts to distill. The distillate is refluxed with phthalic anhydride

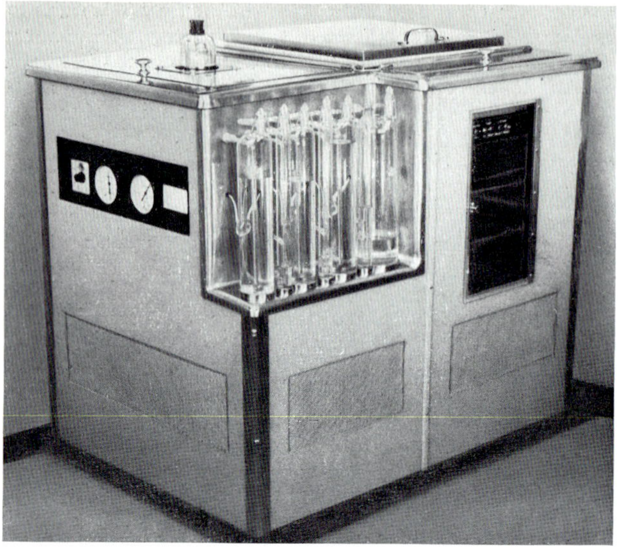

Fig. 8-5. Protein sequenator. Reprint. from Ref. [270], p. 86, by courtesy of Europ. J. Biochem.

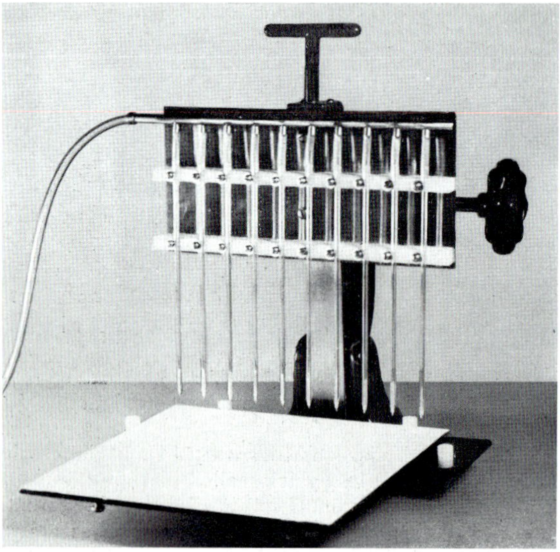

Fig. 8-6. Applicator manifold

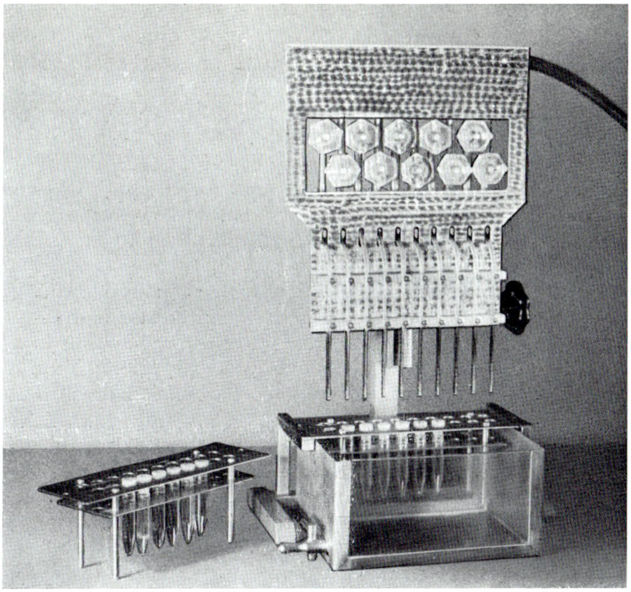

Fig. 8-7. Evaporator manifold

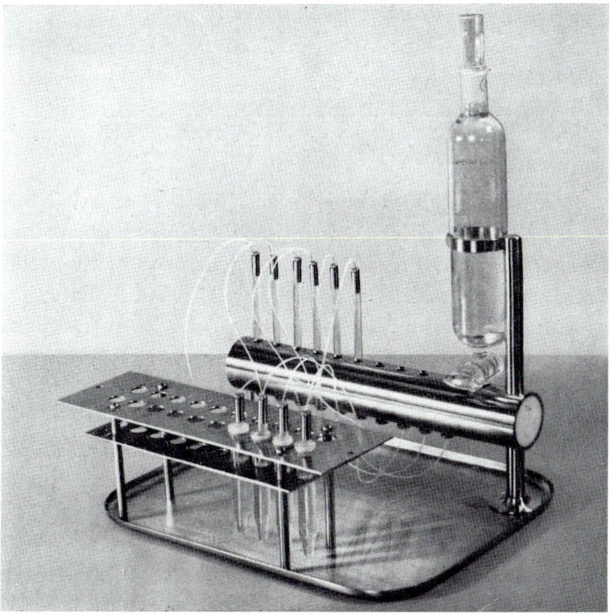

Fig. 8-8. Extractor manifold

(5 g/100 ml) for 3 h, and then distilled on a short column of glass beads. The free amine should be protected against atmospheric CO_2 during all operations. B.p. 59 to 60 °C. (All operations involving allyl chloride should be carried out in the hood and with the use of rubber gloves.)

Pyridine (L.R. grade, British Drug Houses Ltd., Poole, England) is refluxed over solid KOH for 3 h and then distilled. B.p. 115 to 116 °C.

Benzene (practical grade) is stirred over concentrated H_2SO_4 with several changes of acid, and then washed with distilled water. It is shaken in a mechanical shaker overnight with an aqueous $KMnO_4$ solution (30 g/l). The $KMnO_4$ is removed by washing with distilled water and the preparation dried over KOH. It is finally distilled on a Widmer column. The fraction boiling at 80 to 81 °C is collected.

Ethyl acetate (practical grade) is stirred, first with aqueous Na_2CO_3 (50 g/l) and then with a saturated solution of $CaCl_2$ in water. The organic phase is drawn off, shaken with solid $KMnO_4$ (1 g/l) in a mechanical shaker overnight, and then washed with distilled water until colorless. It is then dried, first over anhydrous $CaCl_2$ (200 g/l) for 24 h and then for the same period over anhydrous $CaSO_4$ (200 g/l), from which it is distilled using a splashhead B.p. 77 °C.

Trifluoroacetic acid (Fluka A.G.) is refluxed with solid CrO_3 until the color does not appear to change. The acid is then distilled off and again refluxed with another portion of CrO_3 to ensure that the oxidation has been exhaustive. It is then dried over anhydrous $CaSO_4$, distilled, and the fraction boiling at 72 to 73 °C collected.

Ethylene chloride (*purum* grade, Fluka A.G.) is stirred twice with concentrated H_2SO_4 (0.1 vol.), then washed in succession with distilled water, dilute Na_2CO_3 solution and distilled water. It is then shaken in a mechanical shaker overnight with an aqueous solution of $KMnO_4$ (30 g/l). The $KMnO_4$ is removed by repeated washings with distilled water, and the preparation then dried over anhydrous $CaSO_4$, and finally distilled. B.p. 83 to 84 °C.

Ethanol should have negligible light absorption at wave lengths longer than 240 mμ.

6. Protein Sequenator (Fig. 8-5)

This instrument (Fig. 8-5) was designed to speed up the process of primary structure determination.

The design and operation of the sequenator has recently been described in full detail [270]. Since then no major changes have been made, and description in outline will therefore suffice here.

The operation of the instrument is based on the principle that solutions of reactants and extracting solvents are spread in thin films inside a rotating cyclindrical glass cup. This principle makes it simple to perform the many mechanically diverse operations in the degradation. Thus, a spinning film is ideally suited for extraction by another film of an immiscible solvent sliding over its surface, for centrifugation and also for drying under reduced pressure, because of the large surface and the stabilizing centrifugal force.

The sequenator performs only the coupling and cleavage operations. The conversion reaction is carried out in the conventional way. The omission of the latter operation simplifies the design problem, but does not reduce the speed of the analysis, since the conversion may conveniently be carried out on a large number of samples simultaneously. The evaporation (Fig. 8-7), extraction (Fig. 8-8) and application (Fig. 8-6) manifolds make the conversion and the TLC identification of many samples no more time comsuming than for a single sample.

The general design of the sequenator is shown in Fig. 8-9. The reaction vessel is a cyclindrical glass cup (*A*) mounted on the shaft of an electric motor (*B*). The cup spins continuously, and solutions and solvents entering the cup are therefore spread as thin films on the wall of the cup. Reagents and solvents enter through the feed line

(*R*) at the bottom of the cup. Extracting solvents climb to the top of the cup, where they are scooped off, and leave through the effluent line (*S*). The cup is enclosed in a bell jar (*Q*), and the system can be evacuated by means of a vacuum pump (*P*). The reaction chamber is thermostatted at 50 °C.

Reagents and solvents are stored in reservoirs (*C*) and are admitted to the cup through an assembly of valves (*D*). The reservoirs are under a constant low pressure of nitrogen supplied by a nitrogen cyclinder (*H*) and pressure regulators (*K*). The contents of the bell jar are likewise held at a fixed although lower pressure of nitrogen through a similar arrangement. The pressure differential between the reservoirs and

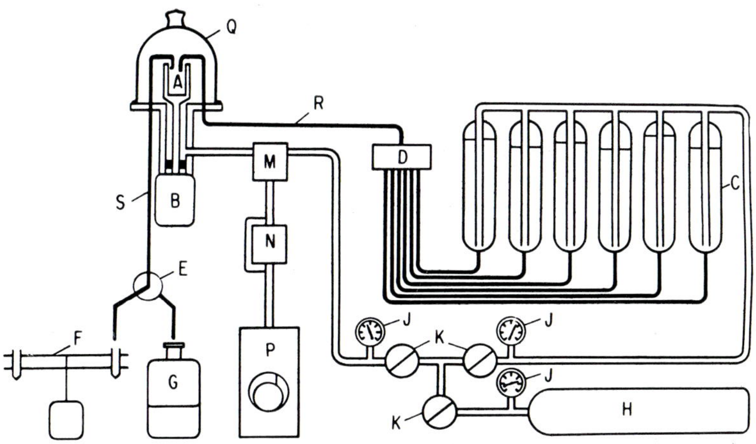

Fig. 8-9. Diagram of sequenator. *A* spinning cup; *B* electric motor; *C* reagent (solvent) reservoir; *D* valve assembly; *E* outlet stopcock assembly; *F* fraction collector; *G* waste container; *H* nitrogen cylinder; *J* pressure gauges; *K* pressure regulators; *M* 3-way valve; *N* 2-way valve with bypass; *P* rotary vacuum pump; *Q* bell jar; *R* feed line; *S* effluent line. Gas lines are doubly contoured, and liquid lines are filled. Reprinted from Ref. [270], p. 81, by courtesy of Europ. J. Biochem.

the bell jar is constant, and the volume of reagent or solvent admitted to the cup is therefore determined by the time a valve is kept open.

The effluent line leads, *via* a 3-way stopcock (*E*), either to a fraction collector (*F*) or to a waste container (*G*). The valves in the assembly (*D*) and the gas valves (*M* and *N*) are operated by solenoids and the 3-way outlet stopcock (*E*) and the fraction collector (*F*) by electric motors. All these functions are governed by an electronic programming unit.

A detailed drawing of the reaction chamber is presented in Fig. 8-10.

The design of the sequenator requires modification of certain reagents and solvents as compared to the manual procedure. Thus, the large volume of the bell jar necessitates the replacement of more volatile reagents with reagents of lower vapour pressure in order to reduce losses due to evaporation. N,N,N′,N′,-tetrakis-(2-hydroxypropyl)-ethylenediamine (Quadrol) and *n*-heptafluorobutyric acid have been substituted for N-dimethylallylamine and trifluoroacetic acid, respectively.

Another deviation from the manual procedure is that the cleavage operation is repeated in each cycle. The reason for this is discussed under "Cleavage". The following reagents and solvents are used (cf. Table 4):

Reagent 1 is a solution of phenylisothiocyanate.
Reagent 2 is a buffer of QUADROL-trifluoroacetic acid (pH 9.0).
Reagent 3 is heptafluorobutyric acid.
Solvent 1 is benzene.
Solvent 2 is ethyl acetate.
Solvent 3 is 1-chlorobutane.

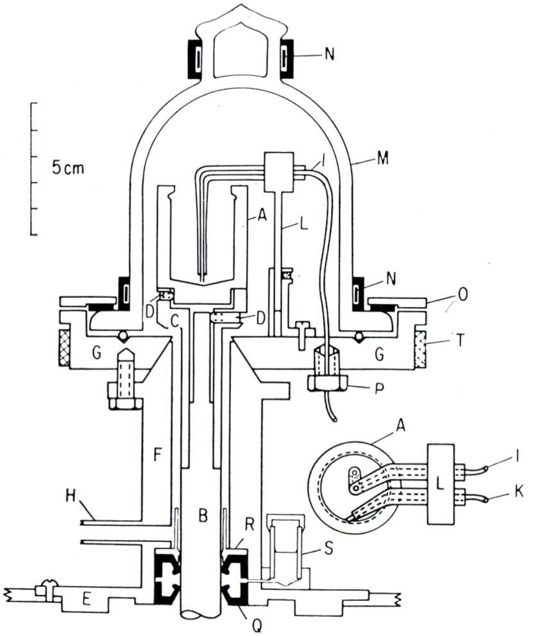

Fig. 8-10. Reaction chamber. *A* glass cup; *B* motor shaft extension; *C* cup support; *D* grub screws; *E* motor support; *F* column; *G* base plate; *H* side tube; *I* feed line; *K* effluent line; *L* adjustable stand; *M* bell jar; *N* electrodes; *O* rubber padded ring; *P* sealing bolt; *Q* oil seal; *R* teflon sleeve; *S* oil reservoir; *T* band heater. A top view of the cup, the feed line and the effluent line is shown in the lower right hand corner. Reprinted from Ref. [270], p. 82, by courtesy of Europ. J. Biochem.

The programming of the degradation cycle requires that it be divided in 30 stages. The stages, their duration and associated operations are shown in Table 4. The time required for one degradation cycle is 93.6 min, which is equivalent to 15.4 cycles in 24 h.

The repetitive yield is about 98%. This high yield makes it possible to carry out much longer degradations with the sequenator than with the manual procedure. With apomyoglobin D of the humpback whale it was possible to perform 60 degradation cycles, and establish the N-terminal structure [270]:

Val.Leu.Ser.Asp.Ala.Glu.Trp.Gln.Leu.Val.Leu.Asn.Ile.Trp.Ala.Lys.Val.Glu.Ala.Asp.Val.
1 2 3 4 5 6 7 8 9 10 11 12 13 14 15 16 17 18 19 20 21

Ala.Gly.His.Gly.Gln.Asp.Ile.Leu.Ile.Arg.Leu.Phe.Lys.Gly.His.Pro.Glu.Thr.Leu.Glu.Lys.
22 23 24 25 26 27 28 29 30 31 32 33 34 35 36 37 38 39 40 41 42

Phe.Asp.Lys.Phe.Lys.His.Leu.Lys.Thr.Glu.Ala.Glu.Met.Lys.Ala.Ser.Glu.Asp.
43 44 45 46 47 48 49 50 51 52 53 54 55 56 57 58 59 60

Table 4. *Operations in a cycle of the sequenator*[a]

Reagents 1, 2 and 3 are in reservoirs I, II and III, respectively, and solvents 1, 2 and 3 occupy reservoirs IV, V and VI, respectively. The valves are numbered correspondingly. Other functions are the 3-way vacuum-pressure valve (M), the 2-way vacuum valve (N), the 3-way outlet stopcock (E) and the motor driving the fraction collector (F). A + or a − sign means that the operating solenoid or motor is energized or deenergized, respectively, at the beginning of the stage. The + sign also means that a valve is open, in the case of valve M to the vacuum line. The three positions of the outlet stopcock are indicated by the letters c (collect), v (vacuum) and w (waste). The duration of stages refers to a reaction temperature of 50 °C.

Stage	Duration min	(Volume) (ml)	Function I	II	III	IV	V	VI	M	N	E	F
1. Reagent 1	0.15	(0.40)	+								+w	+
2. Reagent 2	1.75	(0.40)	−	+								−
3. Reaction	30.00			−							+v	
4. Restric. vacuum	3.00								+			
5. Vacuum	6.00									+		
6. Delay	0.05								−	−		
7. Solvent 1	5.00	(11.5)				+					+w	
8. Solvent 2	8.00	(23.5)				−	+					
9. Delay	1.00						−					
10. Restric. vacuum	3.00								+		+v	
11. Vacuum	6.00									+		
12. Delay	0.05								−	−		
13. Reagent 3	0.70	(0.23)			+						+w	
14. Reaction	3.00				−						+v	
15. Vacuum	1.50								+	+		
16. Delay	0.05								−	−		
17. Solvent 3	1.75	(5.5)						+			+c	
18. Delay	1.00							−				
19. Restric. vacuum	3.00								+		+v	
20. Vacuum	1.00									+		
21. Delay	0.05								−	−		
22. Reagent 3	0.70	(0.23)			+						+w	
23. Reaction	3.00				−						+v	
24. Vacuum	1.50								+	+		
25. Delay	0.05								−	−		
26. Solvent 3	2.25	(7.0)						+			+w	
27. Delay	1.00							−				
28. Restric. vacuum	3.00								+		+v	
29. Vacuum	6.00									+		
30. Delay	0.05								−	−		

[a] Reprinted from Ref. [270], p. 87, by courtesy of Europ. J. Biochem.

The identification of the PTH-amino acids in solvent system D are shown in Fig. 8-11. The termination of the degradation was determined by two factors, the cumulative effect of a 2% loss of yield in each cycle, and the gradual appearance in the chromatograms of an increasing general background of other PTH-amino acids, which eventually made the identification impossible. The origin of this background is discussed under "Cleavage". In the degradation of very long peptide chains (more than 60 to 70 residues) the background is a limiting factor. On the other hand, the degradation of very short peptides in the sequenator is subject to other limitations. The difficulty here is the small difference in solubility between the thiazolinone and the short peptide, making a clean extraction of the former impossible (cf. "Cleavage"). Therefore considerable losses of material occur in the last few cycles of the degradation. These losses may be reduced by leaving out the repetition of the cleavage operation. In this way it was possible to use the sequenator for the complete sequencing of the human fibrinopeptide A, a hexadecapeptide [108][8].

The application of the manual degradation procedure to proteins is often restricted by the fact that proteins become completely insoluble after a few degradation cycles. This limitation is much less severe in the sequenator procedure. It appears that the protein, when spread in a thin layer, retains much of its reactivity even after it has become insoluble.

7. General Comments

Most problems likely to be encountered in the course of a degradation have already been commented upon and it only remains to discuss the scope of the degradation technique.

The repetitive yield in the manual version of the phenylisothiocyanate degradation is 90 to 95%. The loss in yield is probably mostly due to oxidative desulphuration of the phenylthiocarbamyl group. This in its turn is caused by small amounts of oxygen, which are difficult to exclude in the manual technique. Higher losses are to be

[8] P. Edman, unpublished.

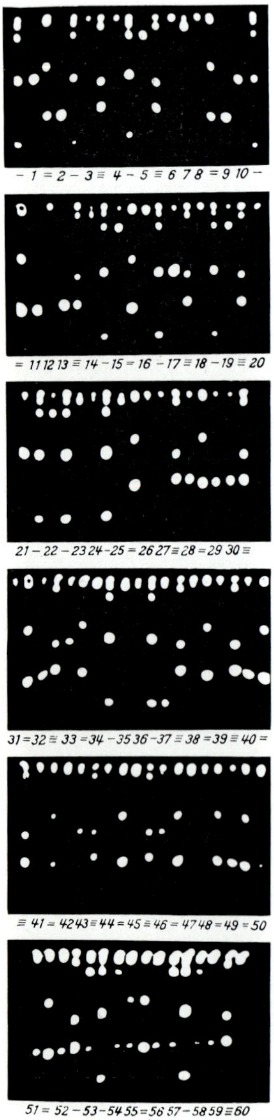

Fig. 8-11. Degradation of apomyoglobin. TLC identification using solvent system D of the PTH-amino acids obtained in first 60 degradation cycles. Reference mixtures are indicated by horizontal bars, and show in order from bottom to top the migration of (—) PTH-proline, PTH-valine, PTH-alanine, PTH-glycine; (=) PTH-leucine, PTH-methionine; (≡) PTH isoleucine, PTH-phenylalanine and PTH-tryptophan. Reprinted from Ref. [270], p. 89, by courtesy of Europ. J. Biochem.

expected when the degradation approaches the C-terminal end of the peptide. The causes of these losses and ways to minimize them are discussed in the sections "Coupling" and "Cleavage". The degradation of a protein is sometimes prevented by the fact that the preparation becomes insoluble after a few degradation cycles. Sometimes this problem may be overcome by oxidizing the cystine residues to cysteic acid residues [269]. Another limitation in the degradation of proteins is a gradual increase in background amino acids.

In not unfavorable cases one may expect the manual technique to produce 15 to 20 degradation cycles with clearly interpretable results. To perform one degradation cycle requires the best part of a working day. However, it is quite possible for a trained person to run in the same time several parallel degradations.

A blocked N-terminal is not infrequently encountered in peptides and proteins. This may be due to the amino group being acetylated, *e.g.* cytochromec [642], to a Schiff base, *e.g.* hemoglobin A_{Ic} [460] or to an N-terminal pyroglutamic acid residue, *e.g.* human fibrino-peptide B [107, 108]. The last may represent the native configuration, but may also be an artifact produced during isolation [974], since N-terminal glutaminyl residues are notorious for their great tendency to cyclise to pyroglutamyl residues [672] (cf. "Cleavage"). Regardless of the mode of blocking, the isothiocyanate degradation obviously cannot be applied. The way out has been to create a new N-terminal amino acid, preferably close to the blocked group, by the use of some specific cleavage method, *e.g.* proteolysis. The structure of the fragment with the blocked N-terminal has to be determined by other techniques [108]. A promising approach to the problem of N-terminal pyroglutamic acid has recently been described by DOOLITTLE and ARMENTROUT [250]. They have succeeded in isolating from a strain of *Pseudomonas fluorescens* a new enzyme, pyrrolidonyl peptidase, which is claimed to split off selectively N-terminal pyroglutamic acid residues [1073]. An enzyme with similar specificity has also been reported for *Bacillus subtilis* [1031].

The great tendency for an $\alpha \to \beta$ peptide shift at aspartyl and asparaginyl peptide bonds is well documented [496, 721, 975, 1024]. This rearrangement may occur even under mild conditions, *e.g.* during the isolation of a peptide [396]. A β-aspartyl peptide bond is not cleaved in the isothiocyanate procedure and further degradation is therefore blocked [375, 1104].

A blocking reaction involving N-terminal S-carboxymethyl cysteine residues has recently been described [976]:

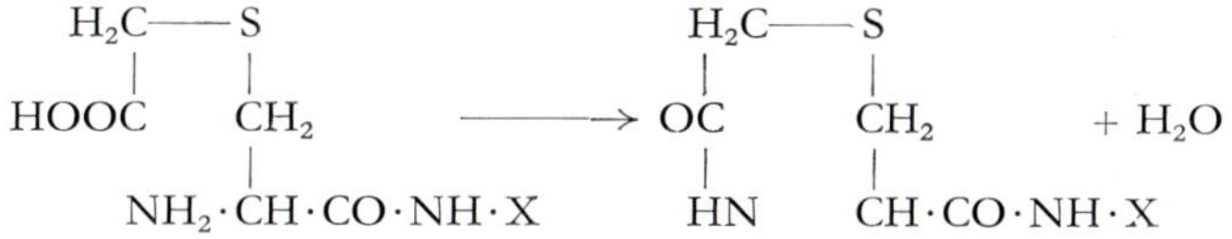

(16)

The reaction takes place even under mildly acid conditions. This may be another reason for using other S-alkyl derivatives of cysteine, *e.g.* S-methyl or S-ethyl cysteine (cf. "Identification of Amino Acids"), in sequence determinations.

Sequential degradation is normally not carried out on peptides with intact cystine residues. In a few instances where this has been done, it was found that a half cystine residue does not block the degradation. Expectedly, no PTH derivative is obtained in the corresponding degradation cycle.

The possibility that an amino acid residue is modified, *e.g.* carries a prosthetic group, has generally been disregarded in sequence determinations. The wide use of the method of indirect identification (cf. "Identification of Amino Acids") is conducive to this practice. The presence of a modified amino acid residue may, in fact, have been more frequent than is apparent from degradation protocols. The attachment of a carbohydrate moiety through an N-glycosidic linkage to an asparagine residue or through an O-glycosidic linkage to residues of serine and threonine have been established [729]. Almost nothing is known about the behavior of such modified residues in the isothiocyanate degradation. However, FLETCHER *et al.* [307] have shown that an asparagine-carbohydrate complex from ovalbumin with an N-glycosidic linkage between the β-amide group and the carbohydrate forms what appears to be a normal PTH derivative.

The value of quantitative N-terminal determinations in the characterization of proteins has not been realized generally. This is surprising, since evidence of this kind is even more informative than, and often complementary to, ultracentrifugation data. Thus, for a pure protein preparation a quantitative N-terminal determination will give direct information about the minimum molecular weight. This figure, it should be noted, is independent of the quarternary structure of the protein as contrasted to physico-chemical data. In addition, N-terminal analysis may, in the absence of blocked N-terminals, provide information about the number of peptide chains in the smallest unit. It may also reveal, by the occurence of N-terminal amino acids in fractional molar quantities, the cleavage of peptide bonds, *e.g.* by the accidental action of proteolytic enzymes during the preparation [251], changes which are not necessarily revealed by physico-chemical measurements. For these reasons it seems appropriate to require, when new protein preparations claimed to be homogeneous are described, that the data should also include a quantitative N-terminal determination.

In many other cases N-terminal determination may be used to great advantage, and a few will be mentioned here. One instance is the characterization of protein mixtures, *e.g.* plasma proteins [739], by the quantitative composition of their N-terminal amino acids. Another example is monitoring the progress of a protein purification. The first phase in a structural determination is usually a fragmentation of the structure, and here an N-terminal analysis of the initial digest may serve as a check that fragments are not lost in the later isolation procedure. Also, a quantitative N-terminal determination on an isolated fragment will define its size.

F. Related Procedures

The isothiocyanate degradation exists in numerous modifications. Most of these concern less essential features, and the advantages, if any, would be so marginal that they will not be described here. However, certain modifications, because they are widely in use, or because they represent new approaches will be mentioned.

The use of organic isothiocyanates other than phenylisothiocyanate has been investigated. Thus, 4-dimethylamino-3,5-dinitrophenylisothiocyanate [820] and *p*-azobenzeneisothiocyanate [812] have been suggested because they give rise to strongly colored thiohydantoins, which are easy to locate on chromatograms and to quantitate. Naphthylisothiocyanate has reactivities similar to phenylisothiocyanate, but naphthylthiohydantoins have a much stronger absorption than phenylthiohydantoins in the ultraviolet [526]. This advantage is at least partly offset by the fact that the absorption maximum is at 222 mμ, a region of the spectrum which is inconvenient to use. Methylisothiocyanate has also been used successfully [822, 597]. The higher volatility of the methylthiohydantoin derivatives are said to make them more suitable for identification by mass spectroscopy.

The isothiocyanate degradation of proteins is limited, at least in the manual version of the technique, by the fact that the protein sometimes becomes insoluble and unreactive after only a few degradation cycles. In the often used paper strip method of FRAENKEL-CONRAT [321] the protein is absorbed on a piece of filter paper, and remains there during all subsequent operations, *i.e.* couplings, cleavages and conversions. Since the protein is prevented from forming a dense coagulum, it retains at least partly its reactivity and it is therefore possible to carry through several successful degradation cycles. SCHROEDER [902] has later modified the paper strip method to allow the degradation also of peptides.

In the paper strip method the cleavage and conversion operations are combined. As has already been pointed out (cf. "Reaction Mechanism") the former reaction is rapid and the latter is slow. As a result the peptide is unnecessarily exposed to the hydrolytic and otherwise detrimental (*e.g.* destruction of tryptophan) action of a strong aqueous acid.

LAURSEN [597], DIJKSTRA *et al.* [245] and SCHELLENBERGER *et al.* [878] have all reported on the successful isothiocyanate degradation of short peptides covalently bound to resins. The C-terminal carboxyl group is linked either by an amide or an ester bond to a suitably modified polystyrene resin. The chief advantage of attaching the peptide to a solid support is that all operations in coupling and cleavage may be done by filtration. This would simplify an eventual automation of the procedure. For smaller peptides there is the additional advantage that losses due to extractions are eliminated.

In establishing the linkage between the peptide and the resin the bridge head on the peptide side poses some problems. Ideally, the bridge head should be exclusively the C-terminal carboxyl group. However, in practice it could be difficult to restrict the reaction to this group, since the carboxyl groups of aspartic and glutamic acid residues would have similar reactivities. For short peptides this may not matter so much, but for long peptides and proteins with many side chain carboxyl groups it would be a problem. When the degradation passes the attached amino acid residue, this as well as the anchorage point would be lost. The development of methods which would selectively engage the C-terminal carboxyl group therefore seems needed.

The method of STARK [997] uses the principle of solid support in reverse, since here the isothiocyanate group is carried by the solid support. The peptide in solution is coupled to the isothiocyanate groups of the resin. In a subsequent cleavage step the shortened peptide is released from the resin leaving the removed amino acid residue attached to the resin. The procedure is then repeated with a fresh batch of resin. In its present stage of development the technique gives rather low repetitive yields. The identification of the amino acid removed depends on the indirect method with its known limitations (cf. "Identification of Amino Acids").

Several other degradation reactions related to the isothiocyanate degradation have already been discussed (cf. "Reaction Mechanism").

III. Other Chemical Degradation Procedures

In all the schemes for sequential degradation discussed so far the essential feature of the mechanism has been the formation of a thiazolinone. However, other schemes with different mechanisms exist. These have generally not been worked out in such detail that they may be applied routinely to sequence determinations, and therefore only the principles will be presented.

A. From N-Terminus

In the method of Holley and Holley [459], the driving force in the cleavage of the peptide is the formation of a lactam. The terminal amino group is first reacted with 4-carbomethoxy-2-nitrofluorobenzene. The nitro group is then catalytically hydrogenated to an amino group (17). The lactamization and the concomitant peptide bond scission (17) produces a 7-carbomethoxy-3,4-dihydro-2-quinoxalone derivative (X) containing the N-terminal amino acid residue. The reaction takes place in aqueous solution under rather mild conditions, 5 h at ambient temperature or 15 min at 70 °C:

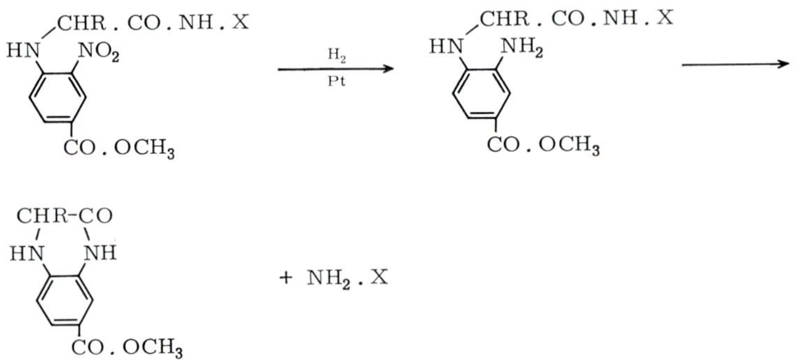

X

(17)

The procedure has been tried out on short peptides with non-polar amino acid residues. The yield was 84% on the average. A variant of the procedure [474b, 507] uses instead the 2,4-dinitrophenyl derivative of the peptide, which is hydrogenated to the 2,4-diaminophenyl derivative, and made to form a lactam in analogous fashion. The general applicability of this type of procedure remains to be demonstrated.

A remarkable degradation reaction has recently been described by Buckingham, Collman and coworkers [164, 165, 208, 209]. A tetradentate Co(III) complex, β-$[Co(trien)OH(H_2O)]^{2+}$ (trien = triethylenetetramine), reacts with a peptide (or an amino acid ester) as follows:

$$\begin{aligned}&\beta\text{-}[Co(trien)OH(H_2O)]^{2+} + NH_2\cdot CHR\cdot CO\cdot NH\cdot X \longrightarrow\\&\beta\text{-}[Co(trien)NH_2\cdot CHR\cdot CO_2]^{2+} + NH_2\cdot X\end{aligned} \qquad (18)$$

The reaction between the cobalt complex and the peptide is stoichiometric and an excess of the former leads to a scission also of the peptide bond next in sequence. The reaction conditions used were rather mild, aqueous solution around pH 7, 65 °C and a few hours. The amino acid-cobalt complex split off is strongly colored and may be identified by paper or thin-layer chromatography. So far only short peptides with non-polar amino acids have been studied.

B. From C-Terminus

The first method for the sequential degradation of peptides by chemical means is due to Schlack and Kumpf [880]. Their idea, inspired by the studies of Johnson and Nicolet [499] and Komatsu [564] on 1-acyl-2-thiohydantoins, was first to convert the C-terminal amino acid residue into a 2-thiohydantoin substituted in the 1-position by the rest of the peptide (XII) and subsequently to hydrolyze off the acyl group (20). A new C-terminal residue would then be created, on which the procedure could be repeated. The formation of the 2-thiohydantoin derivative was achieved by treating the peptide at 100 °C for $^1/_2$ h with ammonium rhodanide and acetic anhydride. This led to the formation of an acylisothiocyanate (XI) which rearranged spontaneously (19) to the 1-acyl-2-thiohydantoin derivative (XII):

(19)

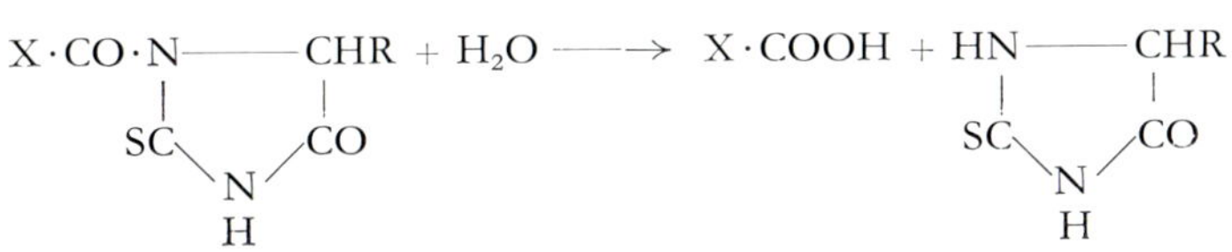

(20)

The exact role of the acetic anhydride in the formation of the acylisothiocyanate is not clear. It may either induce the C-terminal residue to form an oxazolinone, or form a mixed anhydride with the C-terminal carboxyl group. Either of these compounds would then react with the thiocyanic acid to form the acylisothiocyanate.

The formation of the 1-acyl-2-thiohydantoin derivative causes difficulties with some amino acids. Thus, proline is, by its structure, incapable of forming a derivative of this kind. Furthermore, in a thorough study Swan [1025] was unable to form the 1-acyl-2-thiohydantoin derivatives of arginine, aspartic acid, glutamic acid, serine and threonine. In N-acylated aspartic and glutamic acids the acetic anhydride induced the formation of cyclic anhydrides, and these were found to be unreactive towards thiocyanic acid. The reason for the failure with the hydroxyamino acids is not clear.

Kenner *et al.* [531] have suggested a milder way of preparing the acylisothiocyanate. For this purpose they used the reagent diphenyl phosphoroisothiocyanatidate (XIII) which produces the 1-acyl-2-thiohydantoin in excellent yield.

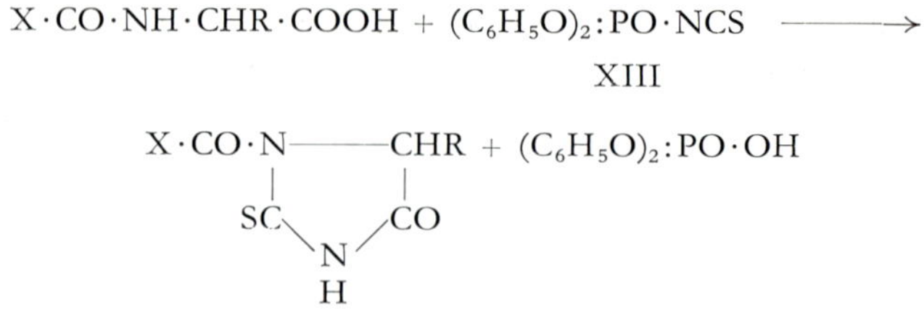

(21)

This reaction has only been applied to small peptides containing amino acid residues with unreactive side chains, and nothing is therefore known about possible side reactions.

The hydrolysis of the 1-acyl-2-thiohydantoin (20) may be done by acids or alkalis. SCHLACK and KUMPF [880] used 1 N sodium hydroxide for 3 h at room temperature, but WALEY and WATSON [1096] and KJAER and ERIKSEN [554] showed that these conditions were unnecessarily vigorous, 0.01 N sodium hydroxide at room temperature for 1 h being sufficient to bring about complete hydrolysis.

The application of the reaction, even solely as a C-terminal method, to larger peptides and proteins [53, 1057, 1070, 1096] has produces less than satisfactory results. Perhaps more thorough studies on model peptides are required. Bearing in mind the very reactive nature of acylisothiocyanates, a search for reactions competing with the isomerization reaction (19) could be worthwhile.

A notable extension in the applicability of the method has recently been reported by STARK [996]. Milder conditions for the formation of the acylisothiocyanate (XI) as well as for the deacylation step (20), *e.g.* 0.1 M acethydroxamic acid in 50% aqueous pyridine for 2 h at 50 °C, have permitted the inclusion of all amino acids except aspartic acid and proline in the scheme. Even with large peptides it was usually possible to carry through a sequential degradation for at least two cycles and quite often for several more.

KHORANA [541] has suggested the use of carbodiimides for the removal of C-terminal residues. The peptide is first made to react under very mild conditions with *p*-tolylcarbodiimide (R′·N:C:N·R′; R′ = *p*-$CH_3 \cdot C_6H_4$) to form an acylurea (XIV). The latter derivative is then degraded at room temperature with 0.01 N sodium hydroxide in aqueous ethanol (22) when the C-terminal amino acid is split off as an N-tolylcarbamyl toluidide derivative (XV). However, in a quantitatively

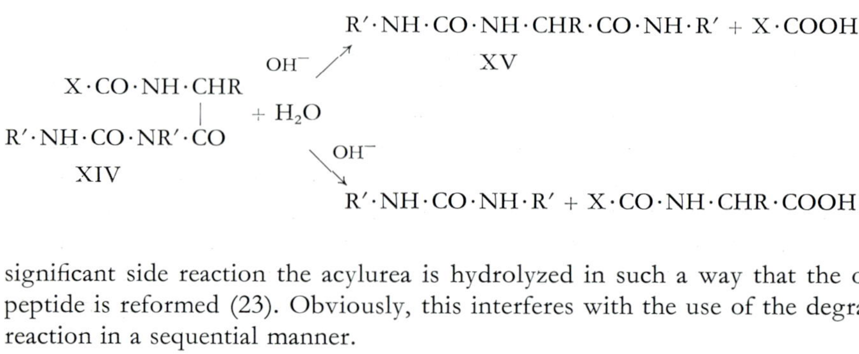

(22)

(23)

significant side reaction the acylurea is hydrolyzed in such a way that the original peptide is reformed (23). Obviously, this interferes with the use of the degradation reaction in a sequential manner.

In the degradation scheme of BAILEY [48] the peptide is first made into an ester, and the ester is then split reductively with a metal hydride, *e.g.* lithium borohydride (24). The resulting β-hydroxy amide (XVI) is then induced, *e.g.* by the action of phosphorus oxychloride, to undergo an N$\rightarrow$O acyl shift (25) to form a β-amino

$$X\cdot CO\cdot NH\cdot CHR\cdot CO\cdot OCH_3 \xrightarrow{LiBH_4} \underset{XVI}{X\cdot CO\cdot NH\cdot CHR\cdot CH_2OH} + CH_3OH \quad (24)$$

$$X\cdot CO\cdot NH\cdot CHR\cdot CH_2OH \xrightarrow{POCl_3} \underset{XVII}{X\cdot CO\cdot O\cdot CH_2\cdot CHR\cdot NH_2} \quad (25)$$

$$X\cdot CO\cdot O\cdot CH_2\cdot CHR\cdot NH_2 \xrightarrow{LiBH_4} X\cdot CH_2OH + HOCH_2\cdot CHR\cdot NH_2 \quad (26)$$

ester (XVII). The latter may be cleaved reductively (26) to produce a new β-hydroxy amide, on which the degradation cycle may be repeated. The amino alcohol derivatives removed are identified, *e.g.* by paper chromatography [500]. Little is known about the scope and limitations of the method. Clearly it could not deal with peptides having internal serine and threonine residues, since cleavage of the peptide would occur at these residues.

IV. Enzymatic Degradation Procedures

The use of aminopeptidases and carboxypeptidases for sequence determination may seem attractive because of technical simplicity. However, the simplicity is only apparent, and for theoretical or practical reasons the scope of the approach is limited. Nevertheless, much valuable and perhaps otherwise unobtainable information has been gained in this way. All enzymatic degradation procedures are to a varying degree subject to the same limitations, and these will therefore first be discussed in a general way.

In the chemical methods *all* molecules of a peptide are made to go through a degradation cycle (at least ideally) before a new cycle is started. This is not so in an enzymatic degradation, and here lies its main limitation. The order of the amino acid residues is not determined in a stepwise fashion, but rather from the rate at which the amino acids appear in the digest, *i.e.* an amino acid appearing faster than another presumably precedes it in sequence. The differences in rate of release tend to become smaller as the degradation proceeds, and ambiguity then ensues. Although the digestion technique in itself is simple, the rate measurements necessitate a supplementary technique for accurate amino acid analysis.

Another limitation lies in the specificity of the enzyme. So for example are proline and arginine residues in C-terminal position not released by carboxypeptidase A, and the degradation will therefore come to a halt at these residues. Furthermore, the rate at which a peptide bond is split may vary greatly depending on the particular amino acid residue occupying the terminal position. If a rapidly released amino acid follows in sequence a slowly released amino acid, the two amino acids will appear in the digest at similar rates.

A serious practical difficulty is for obvious reasons the contamination of the enzyme preparation with other proteolytic enzymes. This must be guarded against by the use of properly purified enzyme preparations and by the addition of specific inhibitors. Short peptides are less likely to be affected by traces of endopeptidases, but the likelihood of a susceptible peptide bond being present increases with the size of the substrate molecule.

The enzymatic degradation techniques have an advantage over the chemical techniques in that they provide information about the optical configuration of the amino acid released. Only L-forms are released, the occurence of an amino acid residue of the D-series constituting an absolute blockage to further degradation. This property of the enzymes is made use of in checking synthetic peptides for absence of racemization [451].

Procedures for the preparation and the activity measurement of the enzymes are described in another chapter (cf. "End Group Determinations"), and these descriptions will not be repeated here.

A. From N-Terminus

The only aminopeptidase used in the degradation of peptides from the N-terminal end is the leucine aminopeptidase from hog kidney (EC 3.4.1.1) described by SPACKMAN *et al.* [986]. The properties of the enzyme have been reviewed by SMITH and HILL [965]. Leucine aminopeptidase has not been prepared in a chemically homogeneous form. Less pure preparations have been found to display other proteolytic activities, *e.g.* carboxypeptidase [974] and various endopeptidase [326, 375] activities and also glutaminase activity [449]. The carboxypeptidase activity may be eliminated by further purification, but even the most pure preparations retain some endopeptidase activity. This can be considerably suppressed, but not entirely eliminated, by treating the enzyme preparation with DFP and iodoacetic acid [326]. Therefore the enzyme cannot at the present stage be used with confidence in determining N-terminal sequences of proteins and large peptides. On the other hand, shorter peptides, which are less likely to contain bonds susceptible to endopeptidase activity, may be degraded solely in accordance with the specificity of leucine aminopeptidase. However, when possible the sequence deduced should be corroborated by independent evidence.

Leucine aminopeptidase splits the peptide bond adjacent to the free α-amino group of the peptide. This peptide bond must be of the amide type. If the N-terminal residue is linked to a proline residue (imide bond) no cleavage takes place. All other α, α'-peptide bonds are split, but the rate varies greatly depending on the amino acid side chain. Amino acids with large non-polar side chains, *e.g.* leucine, are released most readily, whereas amino acids with polar side chains and proline are slow to come off [968].

The enzyme is a metallo-enzyme and requires for its activity and stability Mg^{2+} or Mn^{2+} ions. It loses its activity outside the pH range 7 to 9. A solution of the enzyme preparation, 0.005 M in Tris (pH 8.0) and 0.005 M in magnesium chloride, may be stored in the frozen state for long periods without loss of activity [437a].

A DFP-treated preparation is commercially available. It is supplied as a suspension in 75% saturated ammonium sulfate solution, 0.005 M in Tris (pH 8.0) and in magnesium chloride.

Before use the enzyme preparation is dialyzed against a 0.005 M Tris — 0.005 M $MgCl_2$ buffer (pH 8.5) in the cold. Any insoluble material is centrifuged off. The approximate concentration of the protein solution is determined spectrophotometrically using the formula

$$A_{280} \times 1.2 = \text{mg/ml}$$

where A_{280} is the absorbance at 280 mμ.

The correct amount of enzyme to be used in a digestion depends on several factors, *e.g.* the nature of the terminal amino acid residue and the amount of peptide, and is found by trial in small scale experiments. However, the range of 3 to 30 enzyme units/μmole of peptide may serve as a guide. For the digestion the following procedure is suggested.

The peptide is dissolved (1 μmole/ml) in 0.05 M Tris — 0.0025 M $MgCl_2$ buffer (pH 8.5) in a stoppered test tube. The solution is incubated at 40 °C, and a suitable aliquot of the solution of the activated enzyme is added. Samples (0.1 ml) are withdrawn at the start and at intervals during the next 10 to 20 h. A blank without substrate should be included.

The withdrawn sample is transferred to a glass-stoppered centrifuge tube containing the sample dilution buffer (2.1 ml) used in the amino acid analysis according to Spackman *et al.* [987]. The solution is quickly frozen and stored for later analysis. If necessary, the solution is cleared by centrifugation before analysis.

In calculating the results of the amino acid analysis it should be realized that the analytical procedure does not separate serine, asparagine and glutamine. It should also be borne in mind that an undigested small peptide may simulate an amino acid in the analysis. This should be suspected if the height of a peak decreases with time, or if it is unduly broadened. The Tris buffer produces a peak in front of lysine on the short column [445c].

A plot of concentration *versus* digestion time is made for each amino acid. The relative rates of release indicate the order in which the amino acid residues occur in the sequence, as has already been discussed.

Finally, it may be mentioned that a second amino peptidase from hog kidney has recently been described [1092, 1093]. This enzyme is more stable, does not require metal ion activation, and splits the peptide bonds of the various amino acids at a more uniform rate. No reports are available on its use in sequence determination.

B. From C-Terminus

Three different carboxypeptidases have been used for the release of C-terminal amino acids. Two of these, carboxypeptidases A and B, are obtained from pancreatic glands [43, 309], and the third, carboxypeptidase C, is isolated from the peel of citrus fruits [1184]. The properties of carboxypeptidases of animal origin have been reviewed by Neurath [737]. Much less is known about the citrus carboxypeptidase(s), but some properties have been described by Zuber [1184].

Although all three carboxypeptidases split the peptide bond adjacent to the C-terminal α-carboxyl group, they differ in their specificities. Bovine carboxypeptidase A releases all amino acids except proline and arginine, but the rate differs greatly depending on the amino acid side chain. The amino acids with aromatic or large non-polar side chains are split off faster than those with short or polar side chains. Released at the slowest rate are glycine and the acidic amino acids. Lysine is released slowly. It has been reported that the rate of release of the acidic amino acids may be increased by lowering the pH of the digestion [371, 958], and conversely that lysine is split off more rapidly at a higher pH [759]. A C-terminal homoserine (resulting from the cleavage of a peptide chain at a methionine residue by cyanogen bromide) is readily released [22].

The specificity of carboxypeptidase B [350] has not been as well defined. Like carboxypeptidase A it is inert to C-terminal proline. The enzyme releases preferentially arginine and lysine. The activity towards other amino acids is not quite clear. The most highly purified carboxypeptidase B preparations from pork are said to be uncontaminated by carboxypeptidase A activity [310], whereas a pure preparation of the enzyme from beef shows a considerable and apparently intrinsic activity against typical substrates of carboxypeptidase A, *e.g.* carbobenzoxyglycyl-L-phenylalanine [1142].

Carboxypeptidase C seems to have a wider specificity than the other two enzymes [1184, 1185]. Thus, even proline is released, and the rate of release of the other amino acids does not vary greatly, excepting glycine, which is slow to come off.

With carboxypeptidase A it is usually not possible to determine sequences longer than 3 to 4 residues. The limitations here are those inherent in the enzymatic method itself. However, the fact that some amino acids are not released at all (proline and arginine) and others at a very slow rate (glycine and the acidic amino acids) also effectively limits the length of the degradation. The use of carboxypeptidase A in combination with carboxypeptidase B removes the obstacle presented by arginine residues, and speeds up the otherwise slow release of lysine.

The use of carboxypeptidase A and B for the elucidation of large structures (including proteins) has been more successful, and led to fewer false results than the use of leucine aminopeptidase for the same purpose. This is probably due to the fact that extraneous proteolytic activities in carboxypeptidase preparations are better known, and more easily suppressed. A protein substrate in its native conformation is often found to be resistant to the action of carboxypeptidases, whereas the unfolded structure is easily digested. This unfolding may be brought about by breaking the disulphide bonds and converting the cysteine residues to stable derivatives by established methods.

Preparations of bovine carboxypeptidase A (EC 3.4.2.1) and porcine carboxypeptidase B (EC 3.4.2.2) are commercially available (but not preparations of carboxypeptidase C). These preparations are contaminated by endopeptidases, mainly chymotrypsin and trypsin, and may not be used as such. The trypsin and chymotrypsin activities are inhibited by treatment with DFP, and preparations so treated are procurable. Also, it should not be relied on that carboxypeptidase A preparations are entirely free of carboxypeptidase B and *vice versa*. Both enzymes require zinc ions for their activity [310, 1074].

The crystalline DFP-treated carboxypeptidase A preparation is obtained from the manufacturer as an aqueous suspension to which toluene has been added as a preservative. The preparation may be stored in the cold for long periods without loss of activity. The method of solubilizing the enzyme and the condition of the digestion described here are essentially those of Hirs *et al.* [445c].

Before use the enzyme crystals are washed with distilled water to remove traces of free amino acids. The preparation is then suspended in 2 M NaCl (6 ml), and brought into solution by the addition with stirring of 0.01 N NaOH to pH 10. The pH is immediately reduced to 8.3 by the addition of 1 M sodium phosphate buffer (0.6 ml), pH 8.0. After standing for 2 h at 4 °C the solution is clarified by centrifugation. The concentration of the enzyme solution is determined spectrophotometrically using the formula

$$A_{278} \times 0.51 = \text{mg/ml}$$

where A_{278} is the absorbance at 278 mμ.

The suitable ratio between enzyme and substrate is found by trial. The range of 0.1 to 0.5 mg enzyme per μmole of substrate serves as guide.

The peptide is dissolved (1 μmole/ml) in 0.1 M Tris buffer at pH 8.3, and the solution is incubated at 40 °C in a stoppered test tube. A suitable aliquot of the enzyme is added. A blank without substrate should be included.

Sampling, treatment of samples and analysis are the same as for leucine aminopeptidase digestions (see above).

The commercial DFP-treated carboxypeptidase B preparation is supplied as a frozen solution in 0.1 M sodium chloride. The protein concentration is determined from the formula

$$A_{278} \times 0.47 = \text{mg/ml}$$

where A_{278} is the absorbance at 278 mμ.

A suitable ratio of enzyme to substrate is approximately 5 enzyme units/micromole of substrate. The conditions of the digestion, the sampling, the treatment of the samples and the analysis are the same as for carboxypeptidase A. As has already been pointed out, the two enzymes may be used in conjunction.

No technical details are available for the digestion with carboxypeptidase C.

V. Conclusion

The task of determining amino acid sequences is likely to remain with us for quite some time. It is perhaps worth contemplating that the amount of DNA in a mammalian germ cell is equivalent to 2 billion base triplets, which is then the maximum coding capacity for the sequential ordering of amino acid residues. One may contrast this figure with the estimate that the sequential ordering of no more than 2,000 amino acid residues is at present published annually. Perhaps (and hopefully) only a small fraction of the coding capacity is actually used in protein synthesis, the rest being held in reserve.

However, all speculations apart, it is obvious that the amount of structural work waiting to be done is great indeed. It is therefore important that new developments in technique should permit not only accurate but also rapid analysis of protein structures. Other merits of a method, *e.g.* great sensitivity, are here of secondary importance. In fact, if the advantage is bought at the price of a more unwieldy technique, the method would represent a retrograde step.

The author wants to thank his wife, Dr. Agnes Henschen, for help and criticism in the preparation of this review.

Chapter 9

Reconstruction of the Primary Sequence of a Protein from Peptides of Known Sequence

ALFRED D. GOLDSTONE and SAUL B. NEEDLEMAN

I. Introduction

The reconstruction of the complete amino acid sequence of a protein is not unlike assembling a jigsaw puzzle. The ease with which the task is accomplished depends upon the details known about the smaller units making up the total picture. Thus, in the jigsaw puzzle, one is aided by examining the shape, color, size and pictorial composition of each piece in the puzzle. Similarly, in reassembling the amino acid sequence of a protein, one must have comparable information regarding the total amino acid composition, the amino acid content of the subunits of the protein structure, *i.e.* the individual peptides, and, preferably, the exact sequence of each of these peptides.

The technique generally used in preparing the protein for sequence determination is to break up the long amino acid chain, chemically or enzymically, into smaller fragments of less complexity and then to reconstruct the original protein structure by a logical and combinational examination of overlapping fragments obtained from different methods of breakdown. The merits of the various proteolytic enzymes and chemical hydrolytic procedures used in producing the peptide digests of the protein have been discussed elsewhere in this volume. Particular emphasis has been placed, in each case, upon the limited range of reactivity, that is, the high degree of specificity, exhibited by these methods. It should be noted, however, that computer methods are available for constructing the amino acid sequence of a protein from less information than is required by the conventional manual operations. This has been accomplished with fragmentary peptides obtained through the random acid hydrolysis of the protein. Under such conditions, no information regarding the specificity of the hydrolytic sites is obvious and reconstruction is effected only on the basis of the amino acid composition of the peptide and of the composition and determination of the N-terminal residues of the peptide. The use of computer techniques greatly facilitates the orientation of the large number of peptides obtained through any of the hydrolysis techniques. The difficulty in readily applying such methods to the random peptide procedure lies in the inability to obtain in pure form the diversity of small peptides resulting from acid hydrolysis.

In that the soft- and hardware essential for the computer oriented methods for the reassembly of peptides is not yet standard equipment in most laboratories, it will be best to examine first and in detail, the operations involved in the manual reconstruction of the complete protein sequence. To accomplish this, one must determine both the amino and carboxyl terminal peptides of the protein and correctly align the intervening peptides. Detailed discussions of such procedures are found in Chapter 3 but are repeated here in abbreviated form only to indicate how alignment of the terminal peptides is achieved in sequence determination.

II. Determination of the Amino Terminal Peptide

Comparison of the amino terminal residue of a protein shown to be composed of a single polypeptide chain with the amino terminal residues of the various peptides isolated after proteolytic digestion is the method by which one identifies the amino terminal segment of the protein. This can be done only in proteins whose amino terminal residue has a free amino group. Several methods mentioned earlier have been utilized in identifying this peptide once the amino terminal residue of the protein has been determined by either the DNFB method of SANGER [857], the EDMAN degradation [272] or, in certain cases, leucine aminopeptidase hydrolysis or other methods. In studies on lysozyme, FRAENKEL-CONRAT [322] showed that reduction and alkylation of the protein did not significantly alter the molecular weight from that of native lysozyme indicating it to be present as a single polypeptide chain. SCHROEDER [901] and JOLLES [502], among others, showed by the DNFB method that the protein contained only one amino terminal residue, lysine. Upon partial hydrolysis of DNP-lysozyme with 6 N HCl, SCHROEDER isolated a DNP-tetrapeptide whose amino acid sequence was Lys–Val–Phe–Gly suggesting this to be the amino terminal segment of the protein. CANFIELD and coworkers [176a], after tryptic hydrolysis of lysozyme, isolated 18 peptides. Rapid tryptic hydrolysis yielded a peptide with the sequence Lys–Val–Phe–Gly–Arg while slow tryptic hydrolysis gave a peptide whose sequence was Val–Phe–Gly–Arg together with free lysine. These observations together with the results obtained by SCHROEDER and the absence of any other tryptic peptide in which lysine was the amino terminal residue, characterized Lys–Val–Phe–Gly–Arg as the amino terminal segment of lysozyme. From the specificity of trypsin, the isolation of a peptide with lysine as the amino terminal residue is good evidence for its being the amino terminal tryptic peptide.

Another example of unequivocal identification of the amino terminal peptide by dinitrophenylation alone can be found in the structural work by TANIUCHI and ANFINSEN [1040], on an extracellular nuclease from *Staphylococcus aureus*. DNFB treatment of the pure protein produced DNP-alanine with a yield of 80 to 90%, thereby characterizing it as the amino terminal residue. The protein, which has four methionine residues, was treated with cyanogen bromide by a modification of the method of GROSS and WITKOP [379]. DNFB treatment of the resulting five peptides, isolated by gel filtration on Sephadex G-50 gave only one peptide containing alanine as the amino terminal residue and it could therefore be designated the amino terminal peptide in the protein. In cases where more than one peptide has an amino terminal residue identical to that found in the protein, techniques such as peptide overlap, which will be discussed fully later, must be employed. It was shown by HELLER and SMITH [431], using the DNFB method, that glycine was the amino terminal residue in *Neurospora crassa* cytochrome *c*. From peptide overlap and homology all but 41 residues located at the amino terminus could be positioned. Of the five chymotryptic peptides which comprise this segment, two had glycine as their amino terminal residues. The sequences of these peptides were Gly–Phe [C-1] and Gly–Leu–Phe [C-5], respectively, leaving the amino terminus still in doubt. Tryptic digestion of the protein yielded six peptides which encompassed the amino terminal 41 residues. Of these, two had glycine as their amino terminal residues and their amino acid sequences were Gly–Phe–Ser–Ala–Gly–Asp–Ser–Lys [T-1] and Gly–Ala–Asn–Leu–Phe–Lys [T-3]. Since only C-1 and T-1 had a common sequence at the amino terminus, namely, Gly–Phe, the amino terminal segment of the protein could be established as being identical with the sequence of T-1.

In proteins such as cytochrome *c* where the free amino group of the amino terminal residue is blocked, the method of detecting the amino terminal peptide is somewhat different. None of the methods previously discussed are of use since all require a free amino group for their chemical reactivity. In fact the lack of reactivity towards these various reagents is used as a criterion for determining whether the amino terminal residue of a protein is blocked. The ε-amino group of lysine, the imidazole group of histidine and even the sulfhydryl function of cysteine and the tyrosyl hydroxyl will react with DNFB, dansyl chloride or some of the other N-terminal reagents, but comparison with proper standards will eliminate any ambiguity.

Lack of reactivity with the Edman reagent while usually a good indication of a blocked α-amino group may give an ambiguous answer if glutamine is the amino terminal residue. Formation of pyrrolidone carboxylic acid, unreactive with the Edman reagent, might ensue and lead to the erroneous conclusion that the α-amino group of the amino terminal residue is blocked. Leucine aminopeptidase hydrolysis in many instances can help circumvent this difficulty. If glutamine is the amino terminal residue of a protein it will be released upon proteolytic digestion with leucine aminopeptidase. The blocking of the α-amino group of the amino terminal residue on the other hand renders the peptide bond formed by the residues γ-carboxyl group nonsusceptible to proteolytic attack by leucine aminopeptidase. In this way the presence of glutamine as the amino terminal residue can be determined.

Once it has been determined that a protein has a blocked amino terminus, peptides isolated after proteolytic digestion are screened for the absence of a free α-amino group by the methods mentioned above. When the peptide which proved to be the amino terminal peptide of dogfish heart cytochrome *c* [356] was digested with leucine aminopeptidase for 24 h no release of any amino acid could be detected, indicating it had a blocked α-amino group. The problem of identifying the blocking group as well as the blocked amino terminal residue now arises. In both the studies on dogfish cytochrome *c* and wheat germ cytochrome *c* [1007], a similar procedure was employed. In the case of dogfish heart cytochrome *c*, the amino acid composition of the amino terminal peptide after chymotryptic digestion was [Gly_2,Asp,Val_2,Glu,Lys_3,Phe]. Hydrolysis of this peptide with trypsin yielded three peptides whose amino terminal residues had free α-amino groups and a pentapeptide whose α-amino group was blocked. This pentapeptide whose amino acid composition was [Gly,Asp,Val,Glu,Lys] could therefore be designated the amino terminal segment of the original chymotryptic decapeptide. From the specificity of trypsin, lysine could be designated the carboxyl-terminal residue of this pentapeptide. Partial acid hydrolysis with 0.3 N HCl for 16 h at 105 °C hydrolyze selectively peptide bonds formed by both the

α-amino and β-carboxyl groups of aspartic acid. Such treatment resulted in the formation of three components which could be separated by electrophoresis at pH 1.9. Two of these components were ninhydrin positive indicating a free α-amino group. One of these components was free aspartic acid and the other component, which was neutral, had an amino acid composition of [Val,Glu,Lys]. This indicated that glutamic acid and not glutamine was present. Referring to the specificity of trypsin, lysine could be designated the carboxyl terminal residue of this tripeptide which had to be the carboxyl terminal segment of the pentapeptide. One step of the Edman degradation showed valine to be the amino terminus and this peptide could be written as Val–Glu–Lys. The third component was ninhydrin negative indicating no free amino group and characterizing it as the amino terminus of the peptide and of the protein. It was detected on paper by the method of RYDON and SMITH [12] and can be written as follows:

$$\text{—CO—NH– } + Cl_2 \rightarrow \underset{\text{N-chloro-peptide}}{\text{CO—NCl—}} \xrightarrow{KI} \text{—CO—NH— } + KCl + I_2$$

An aliquot of a peptide is chromatographed on paper, placed in an atmosphere of chlorine, and after thorough aeration, sprayed with a 1% starch-KI solution. The iodine released by the N-chloro-peptide is detected as blue-black spot on a faint blue background. This component migrated identically with authentic N-acetylglycine when chromatographed in butanol [4]:acetic acid [1]:water [5] for 16 h characterizing the amino terminal residue of dogfish cytochrome *c* as N-acetylglycine. The amino terminal pentapeptide could therefore be written N-acetyl–Gly–Asp–Val–Glu–Lys.

The approach used by NEEDLEMAN [724] in establishing the N-terminus of rabbit heart cytochrome *c* made use of enzymic and chemical operations different from those just described. The first ten amino acid residues were obtained as the chymotryptic peptide. Enzyme specificity placed phenylalanine at the C-terminal position. Treatment of Cl-10 with trypsin yielded five fragments including the N-terminal pentapeptide, (Ac,Gly,Asp,Val,Glu)-Lys, Gly–Lys, Gly–Lys–Lys, Ile–Phe and free lysine. The tryptic peptide now was subjected to the action of elastase which in spite of its usually broad range of specificity [640], hydrolyzed only the peptide bond C-terminal to valine in the present example. The order of residues in the terminal dipeptide was fixed by the specificity of trypsin. The blocked N-terminal tripeptide was insensitive to ninhydrin and was located on paper chromatograms with the Rydon-Smith procedure previously mentioned. A fourth enzymic operation, requiring papain at pH 4.25 [640], released the blocked N-terminal dipeptide and free valine. The identical dipeptide was obtained from the tryptic dipeptide Tl-5 under similar conditions. The final identification was made on the basis of the reactivity of carboxamide bonds, but not the free carboxyl group, with anhydrous hydrazine [14] to form the acyl hydrazines. The reaction yielded acetyl hydrazide, glycyl hydrazide and free aspartic acid confirming that aspartic acid, but not glycine, was C-terminal and had a free carboxylic acid group.

III. Determination of the Carboxyl Terminal Peptide

Kinetic studies with carboxypeptidase A and/or carboxypeptidase B is the procedure most widely used in the determination of the carboxyl terminal residue of proteins and of the carboxyl terminal peptide. When denatured cytochrome *c* [1 μmole] from *Neurospora crassa* [355] was incubated with 0.01 μmole DFP-treated carboxylpeptidase A for time intervals of 2 h, 18 h and 40 h, the various amino acids were released in the amounts shown in Table 1. It is evident from these results that alanine is the carboxyl terminal residue of the protein. The exact sequence of the remainder of the carboxyl terminal segment can not be deduced from the above data, but the presence of threonine, glutamic acid, methionine, isoleucine and phenylalanine in this area is strongly suggested. In certain instances, more than the terminal amino acid can be ascertained. From the studies on α-corticotropin [421], a plot of

Table 1. *Residues released after carboxypeptidase A treatment of Neurospora crassa cytochrome c (Ref.* [431]*)*

Amino acid	μmoles liberated per mole cytochrome *c*		
	2 h	18 h	40 h
Alanine	0.07	0.37	0.48
Threonine	0.04	0.15	0.13
Glutamic acid	0.01	0.07	0.07
Methionine sulfoxide	0.02	0.14	0.16
Isoleucine	0.01	0.08	0.10
Phenylalanine	0.00	0.04	0.04

the rate of release of amino acids after treatment with carboxypeptidase A, shown in Fig. 9-1, established Leu–Glu–Phe–COOH as the carboxyl terminal segment.

Hydrolysis of a protein with cyanogen bromide is another method by which one can quickly determine the carboxyl terminal segment of the protein. Cyanogen bromide attacks the sulfur atom of methionine causing cleavage at the carboxyl end of the methionyl residue with conversion to homoserine. All but one of the peptides isolated from a cyanogen bromide digest of the parent protein have a homoserine residue as their carboxyl terminal group. The peptide lacking the homoserine residue would comprise the carboxyl terminal peptide of the original protein. As mentioned above, Taniuchi *et al.* [1040] found four methionine residues to be present in an extracellular nuclease of *S. aureus* that they had isolated. Cleavage of the protein with cyanogen bromide yielded five major peptides as would be expected from the specificity of cyanogen bromide. The amino acid composition of the cyanogen bromide segments obtained in this manner is shown in Table 2. From these results it can be concluded that cyanogen bromide fragment E which contains no homoserine residue is the carboxyl terminal segment of the protein. Evidence for the establishment of the carboxyl terminal peptide can also be obtained by examination of the amino acid composition of peptides obtained from digestion of the protein with proteases with relatively limited substrate specificity. Trypsin, for example, acts primarily on the bond formed by the carboxyl group of arginine and lysine. Accordingly all peptides except the one found at the carboxyl terminus should contain at least one residue of arginine or lysine. The carboxyl terminal peptide, on the other hand, should contain neither of these residues except on the rare occasion when lysine or arginine happens to be the carboxyl terminal residue of the original protein. If the amino acid composition of five peptides which were isolated after tryptic digestion of a hypothetical protein were as found in Table 3, Peptide 3 could readily be designated the carboxyl terminal peptide since it lacks both lysine and arginine residues.

Several other chemical methods have been utilized for the determination of carboxyl terminal residues of a protein. Among these, hydrazinolysis and reduction

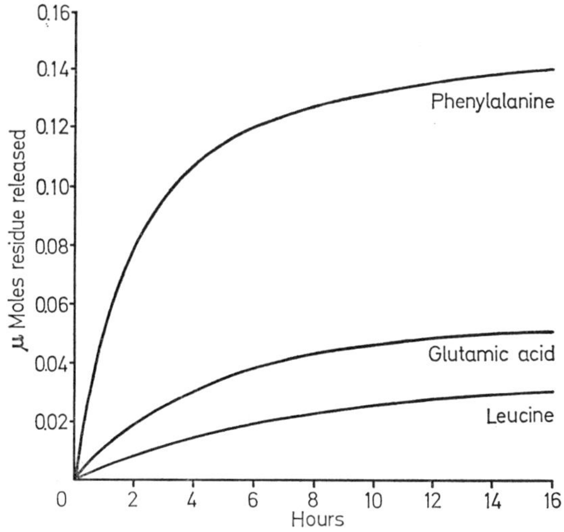

Fig. 9-1. Rate of release of amino acids from α-corticotropin by the action of carboxypeptidase A estimated as their DNP-derivatives (Ref. [421])

of C-terminal carboxyl group to amino alcohols are two of the more extensively studied methods. Hydrazinolysis, based on the work of Akabori [14], liberates the carboxyl terminal residue as a free amino acid while the remaining amino acids in peptide linkage are converted to the amino acid hydrazides. Further treatment with benzaldehyde transforms the amino acid hydrazides to water insoluble dibenzyl derivatives leaving the water soluble C-terminal amino acid in solution. The reactions are shown in Fig. 9-1. One serious disadvantage of this method is the low recovery of all C-terminal amino acids and the complete loss of the C-terminal amino acid if this residue is cystine, cysteine, glutamine or asparagine. The method of reduction of the C-terminal residue to an amino alcohol with agents such as lithium borohydride suffers from undesirable side reactions. These include partial reductive cleavage of some peptide bonds with subsequent production of amino alcohols making unequivocal determination of the C-terminal amino alcohol impossible in some cases.

Table 2. *Amino acid content in cyanogen bromide fragments (Ref.* [1040]*)*

Amino acid	Fragments				
	A	B	C	D	E
Lysine	5	1	5—6	4—5	5—6
Histidine	1		1		1
Arginine			1	1—2	2
Aspartic acid	2		1	3—4	6—7
Threonine	3—4		3	1	1
Serine	1		1		2—3
Glutamic acid	1	1	3	4	9
Proline	1	1	3		1
Glycine	1	1	2	4	2
Alanine	2—3		2	3	6
Cysteine					
Valine	1		2	4	4
Methionine					
Isoleucine	2		1—2	1	1
Leucine	3		2—3	5	5
Tyrosine		1	1	2	2
Phenylalanine			1—2	1	
Tryptophan					1
Homoserine	1	1	1	1	

Table 3. *Amino acid content of tryptic peptides from hypothetical protein*

Amino acid	Peptides				
	1	2	3	4	5
Lysine	1			1	
Histidine		1	1	2	
Arginine		1			1
Aspartic acid	2		2		
Threonine	1	2	3	1	4
Serine	3	1	2	3	1
Glutamic acid		3		2	
Proline	1			1	
Glycine	3	2			
Alanine	1	3	1		
Cysteine				1	
Valine	2	1	2		1
Methionine	1	2			
Isoleucine	2	1	3		3
Leucine	1			3	2
Tyrosine		2			
Phenylalanine	2	1	2		
Tryptophan				1	

IV. Alignment of Peptides by Analogy

In the case where there is a relatively large degree of homology between two proteins, and the amino acid sequence of one of the proteins is known, it is sometimes possible to correctly align the peptides which arise from the digestion of the unknown protein with a single proteolytic enzyme by comparing these peptides with those similarly obtained from the known protein. Examples of this can be found in the studies done on cytochrome *c*.

Since the amino acid sequence of horse heart cytochrome *c* was determined by SMITH and MARGOLIASH [641], the sequences of cytochrome *c* from over 30 other species have been elucidated. The homology between some of these proteins made it possible to properly orient along the polypeptide chain the chymotryptic digest peptides of an unknown cytochrome *c* by comparing them with the chymotryptic

Table 4. *Comparison of homologous peptides from the Pig and Cow with the Whale (Ref.* [355]*)*

Pig, Cow	N-Acetyl-Gly-Asp-Val-Glu-Lys-Gly-Lys-Lys-Ile-Phe^{10}
Whale	N-Acetyl-Gly-Asp-Val-Glu-Lys-Gly-Lys-Lys-Ile-Phe
	←——— C–1 ———→
Pig, Cow	Gly-Arg-Lys-Thr^{40}-Gly-Gln-Ala-Pro-Gly-Phe
Whale	Gly-Arg-Lys-Thr -Gly-Gln-Ala-*Val*-Gly-Phe
	←——— C–5 ———→ ←— C–6 —→
Pig, Cow	Lys-Gly-Glu^{50}-Arg-Glu-Asp-Leu
Whale	Lys-Gly-Glu -Arg-*Ala*-Asp-Leu
	←——— C–15 ———→
Pig, Cow	Lys-Lys^{100}-Ala-Thr-Asn-Glu-COOH^{104}
Whale	Lys-Lys -Ala-Thr-Asn-Glu-COOH
	←——— C–17 ———→

peptides of previously studied species. Immunological studies [140] indicated that *Cetacea* [whale] was more closely related to *Artiodactyla* [cow, pig] than to any other mammalian order. When the amino acid sequence of the 17 chymotryptic peptides of cytochrome *c* isolated from the California gray whale became known [355] only two peptides were found to differ from analogous peptides found in the pig and cow; it was therefore possible to deduce the complete sequence of the entire protein by analogy. Table 4 illustrates the extreme analogy between some of the chymotryptic peptides found in whale and in pig, sheep and cow. As can be seen, once the amino acid sequence of the porcine and bovine cytochromes *c* were known, one could unequivocally place the chymotryptic peptides isolated from the whale cytochrome *c* into their proper position along the polypeptide chain.

Another example of how comparison of analogous peptides can lead to the proper alignment of these peptides in the protein can be seen in the studies of primate cytochrome *c*. The amino acid sequence of human cytochrome *c* was completed by MATSUBARA *et al.* [653] in 1962. Recently NEEDLEMAN [723] has completed the sequences of peptides isolated after chymotryptic hydrolysis of several primate cytochromes *c*. The almost identical nature of these peptides when compared to chymotryptic peptides of human cytochrome *c*, permitted their proper alignment in the protein chain. In *Erythrocebus patas*, for example only one chymotryptic peptide shows any variance form the analogous peptide found in human cytochrome *c*, differing from it in one amino acid. An alanine residue in the human peptide corresponding to position number 50 in the protein is replaced by a glutamic acid residue in *E. patas*. Thus complete sequence of the protein often can be determined on the basis of a single proteolytic digest, saving both time and material, if the homology between the protein in question and a protein whose sequence is known is sufficiently close.

V. Alignment of Peptides by Peptide Overlap

A. Digestion of the Protein with Two Enzymes of Different Specificity

1. Order of Residues with Both Sets of Fragments Known

In the vast majority of cases involving determination of the primary sequence of a protein there is not enough similarity between the protein in question and proteins whose sequences are known to utilize the technique of peptide analogy. In these instances one must invoke the method of peptide overlap. Reconstruction of the primary sequence of a protein by this method involves the orientation of the set of peptides obtained from one enzymatic digest with respect to the set of peptides resulting from another method of breakdown in such a manner as to have peptides of either one set overlap the sites of hydrolysis in the other. A trivial example will illustrate the general procedure. In the simplest case, let us assume that one method of breakdown has produced the fragment peptides of known order as:

A–B–C D–E F–G H–I–J K Set 1

where A, B, C, D, E, F, G, H, I, J, K represent amino acids and these occur only once in the protein. Let a second method of breakdown yield the peptides:

A–B C–D–E–F G–H I J–K Set 2

It is then evident that peptide A–B–C– in Set 1 overlaps the peptides A–B and C–D–E–F of Set 2. Fragment F–G establishes the overlap between the protein thus far reconstructed and fragment G–H of Set 2. The reassembly continues with fragment H–I–J overlapping G–H and J–K thereby completing the sequence. The position of fragment I in Set 2 is fixed because of the known order of amino acids in peptide H–I–J in Set 1. The orientation of fragments from the two sets can be summarized as:

A–B–C D–E F–G H–I–J K Set 1
A–B C–D–E–F G–H I J–K Set 2

2. Order of Residues with One Set of Fragments Known

As a second example, consider the case more frequently encountered, where the amino acid composition of both sets of breakdown fragments is known but the order of the residues within the fragments is known only for one set. Let parenthesis indicate that the order of the residues they enclose is unknown. For example, (A,B,C) represents the six permutations of A,B and C; (D,E,) represents either D–E or E–D. Then (A,B,C) (D,E,) represents the $6 \times 2 = 12$ arrangements of amino acids resulting from the six permutations in (A,B,C) followed by D–E or E–D, etc. If the fragment peptides of Set 1 above represent the fragments of unknown order and the peptides of Set 2 those of known order, we can proceed with the reconstruction of the primary sequence.

Fragment (A,B,C) has possible the six permutations A–B–C, A–C–B, B–A–C, B–C–A, C–A–B and C–B–A. The known order of Set 2 fragment A–B reduces these possibilities to A–B–C and C–A–B both of which contain the residues A and B in the proper sequential order. Peptide fragment (D,E) may have the order D–E or E–D while fragment (F,G) may exist as F–G or G–F. Together with peptide A–B, the peptide of Set 2, C–D–E–F, reduces the permutations and combinations to the single possibility A–B–C–D–E–F–G. It should be noted that Set 1 fragment (D,E) is completely contained within Set 2 fragment C–D–E–F and only the known order of the amino acids in the Set 2 fragment establishes the order as D–E rather than E–D since no overlap peptides exist in the present example in which residue D is in one fragment and E in another.

3. Order of Residues with Neither Set of Fragments Known

In the final general example, let us consider the case where the amino acid components of the fragments are known but the order within the fragments is unknown. Again, let the parenthesis indicate that the order they enclose is unknown. Focusing our attention on the first two fragments of Set 1 and Set 2 we have (A,B,C) and (D,E) of Set 1 and (A,B) and (C,D) of Set 2. Clearly (C,D) of Set 2 overlaps (A,B,C) and (D,E) of Set 1, but (A,B) is contained within (A,B,C). Hence two possible reconstructions result, namely, (A,B) (C) (D) (E) and (E) (D) (C) (A,B) where, in each possibility, the order of (A,B) still remains unknown. Such partial reassemblies frequently occur but afford considerable advantage in that with only minimal expenditure of effort on sequencing (ordering) procedures one can pin-point those few areas of the molecule which will require further analytical manipulation.

B. Reconstruction of the Protein Sequence by Manual Operation

Unfortunately, the situations encountered in the laboratory are rarely as simple as the trivial examples just presented. In the cases actually encountered, between 100 to 250 amino acids are linked through peptide bonds and each of the different enzyme digests may contain 50 or more major peptide fragments which must be used in reconstructing the complete amino acid sequence of the protein molecule. Since the number of combinations of n things taken r at a time ($n > r > 1$) increases more rapidly than does n itself, it is to be expected that the difficulties in reassembling the fragments of a protein will increase proportionally faster than the number of amino acids in the protein. It is in the logic and combinational problems produced by the increasing complexity of protein structure subjected to analysis that the computer finds great application.

In turning to an actual example of the reconstruction of the peptide fragments of a protein, two new factors must be introduced; these are not encountered in the trivial examples above. The first which increases the complexity of the problem to a large extent relates to the fact that in natural proteins most of the component amino acids occur many times in the protein; hence it is to be expected that many breakdown peptide fragments will contain the same amino acid residue or even combinations of several residues. Without the feature of unique occurrence given in the trivial examples above, the arrangement of fragments becomes considerably more involved.

It is perhaps fortunate that the second new feature works in the direction opposite to that just described. From the table giving the amino acid composition of rabbit [724] and dogfish [356] cytochrome *c*, it can be noted that several of the amino acid residues appear only one, two or three times in the protein. Thus in the rabbit

Table 5. *Tryptic peptides of rabbit cytochrome c (Ref.* [724]*)*

Peptide	Amino acid composition
T 1- 5	Lys, Asp, Glu, Gly, Val
T 6- 8	Lys_2, Gly
T 9- 13	Lys, Glu, Val, Ile, Phe
T 14- 25	Lys_2, His, Thr, Glu_2, Gly_2, Ala, Val, Cys_2
T 23- 25	Lys, Gly_2
T 26- 27	Lys, His
T 28- 39	Lys, His, Arg, Asp, Thr, Pro, Gly_3, Leu_2, Phe
T 40- 53	Lys, Asp_2, Thr_2, Ser, Glu, Gly_2, Ala_2, Val, Phe, Tyr
T 54- 55	Lys, Asp
T 56- 72	Lys, Asp_2, Thr_2, Glu_3, Pro, Gly_2, Met, Ile, Leu_2, Tyr, Trp
T 56- 73	Lys_2, Asp_2, Thr_2, Glu_3, Pro, Gly_2, Met, Ile, Leu_2, Tyr, Trp
T 74- 79	Lys, Thr, Pro, Gly, Ile, Tyr
T 80- 86	Lys, Gly, Ala, Met, Ile_2, Phe
T 88- 91	Lys, Arg, Asp, Glu
T 92- 97	Asp, Ala_2, Ile, Leu, Tyr
T 98- 99	Lys, Leu
T 100-104	Lys, Asp, Thr, Glu, Ala
T 101-104	Asp, Thr, Glu, Ala

protein, tryptophan (Trp) and serine (Ser) occur only once, arginine (Arg) and methionine (Met) twice and histidine (His) and proline (Pro) three times. Dogfish cytochrome *c* has one arginyl residue more and one methionyl residues less than the rabbit protein molecule. Such residues can serve as marker sites from which to begin the reassembly of the peptide fragments; this is of considerable aid in simplifying the reconstruction procedure. A second group of amino acid residues serve as marker in another fashion. The C-terminal residue of peptides released during enzymic or chemical hydrolysis generally are of a limited group relating to the specificity of the enzyme or chemical procedure used in obtaining the peptides. The list of the major peptides obtained as the result of tryptic hydrolysis of rabbit heart cytochrome *c* are given in Table 5 and the peptides resulting from chymotrypsin action are in Table 6. For comparison, the peptides of the dogfish cytochrome *c* tryptic and chymotryptic digest are shown in Table 7.

One can begin the reconstruction from one end of the molecule and proceed in an orderly fashion across the length of the protein chain or one may prefer to make use

Table 6. *Amino acid sequence of chymotryptic peptides of rabbit heart cytochrome c (Ref.* [724]*)*

Peptide	Amino acid composition
C 1- 10	Acetyl-Gly-Asp-Val-Glu-Lys-Gly-Lys-Lys-Ile-Phe
C 11- 26	Val-Gln-Lys-Cys-Ala-Gln-Cys-His-Thr-Val-Glu-Lys-Gly-Gly-Lys-His
C 23- 26	Gly-Gly-Lys-His
C 27- 33	Lys-Thr-Gly-Pro-Asn-Leu-His
C 34- 36	Gly-Leu-Phe
C 37- 46	Gly-Arg-Lys-Thr-Gly-Gln-Ala-Val-Gly-Phe
C 47- 48	Ser-Tyr
C 49- 55	Thr-Asp-Ala-Asn-Lys-Asn-Lys
C 55- 59	Lys-Gly-Ile-Thr-Trp
C 60- 67	Gly-Glu-Asp-Thr-Leu-Met-Glu-Tyr
C 68- 74	Leu-Glu-Asn-Pro-Lys-Lys-Tyr
C 75- 80	Ile-Pro-Gly-Thr-Lys-Met
C 81- 82	Ile-Phe
C 83- 88	Ala-Gly-Ile-Lys-Lys-Lys
C 88- 94	Lys-Asp-Glu-Arg-Ala-Asp-Leu
C 95- 97	Ile-Ala-Tyr
C 98-104	Leu-Lys-Lys-Ala-Thr-Asn-Glu

Table 7. *Comparison of differences in the amino acid sequence of peptides from dogfish and rabbit cytochromes c (Refs.* [356 and 724]*)*

Species	Peptide	Region	Difference
Rabbit	C 1- 10	8- 10	Lys-Ile-Phe
Dogfish			Lys-Val-Phe
Rabbit	C 11- 26	21- 23	Glu-Lys-Gly
Dogfish			Glu-Asn-Gly
Rabbit	C 27- 36	32- 34	Leu-His-Gly
Dogfish			Leu-Ser-Gly
Rabbit	C 37- 46	43- 45	Ala-Val-Gly
Dogfish			Ala-Gln-Gly
Rabbit	C 50- 56	50- 52	Lys-Asn-Lys
Dogfish			Lys-Ser-Lys
Rabbit	C 57- 64	57- 64	Gly-Glu-Asp-Thr-Leu-Met-Glu-Tyr
Dogfish			Gln-Gln-Glu-Thr-Leu-Arg-Ile-Tyr
Rabbit	C 83- 85	84- 85	Gly-Ile
Dogfish			Gly-Leu
Rabbit	C 86- 94	88- 93	Lys-Asp-Glu-Arg-Ala-Asp
Dogfish			Lys-Ser-Glu-Arg-Glu-Asp
Rabbit	C 98-104	100-104	Lys-Ala-Thr-Asn-Glu
Dogfish			Lys-Thr-Ala-Ala-Ser

of the marker information and begin the reconstruction at several loci in the chain permitting the segments of increasing length to "grow together". The consecutive approach will be illustrated later in the discussion of computer procedures. The present example will make use of the marker residues.

Since tryptophan and serine each occur only once in the total protein sequence, we will begin the reassembly of peptide fragments with these markers. From the list of tryptic peptides we note that only peptides T 56-72 and T 56-73 contain the tryptophanyl residue and from the list of chymotryptic peptides we need only C 55-59. Since T 56-72 obviously is completely contained in and is coincident with T 56-73 we may ignore this fragment in the present discussion. Since the order of residues in the chymotryptic peptide is known

T 56–73 (Lys_2,Asp_2,Thr_2,Glu_3,Pro,Gly_2,Met,Ile,Leu_2,Tyr,Trp)
C 55–59 Lys-Gly-Ile-Thr-Trp

we may rewrite the T 56-73 fragment as:

Gly-Ile-Thr-Trp-(Asp_2,Thr,Glu_3,Pro,Gly,Met,Leu_2,Tyr)-Lys-Lys

The lysine in position 73 must be C-terminal in T 56-73 because of the enzyme specificity of trypsin while the position of the lysine in C 55-59 is known from sequencing studies on this peptide. The two lysine cannot be the same. Tryptophan, of course, must be C-terminal in C 55-59 because of chymotryptic specificity. We might now search for a peptide fragment containing both proline and methionine from the list of chymotryptic fragments. Only C 75-80 meets this requirement and it must be eliminated from consideration on the basis of the location of lysine in the peptide sequence. In T 56-73, both proline and methionine lie towards the N-terminal side of the lysine at position 73. If peptide C 75-80 were subjected to tryptic cleavage, the two fragments Ile–Pro–Gly–Thr–Lys and free Met would result, placing methionine C-terminal next to lysine contrary to what is thus far known. It is necessary then to find separate proline- and methionine-containing peptides which relate to this region of the sequence. These include:

C 27–33 Lys-Thr-Gly-Pro-Asn-Leu-His
C 68–74 Leu-Glu-Asn-Pro-Lys-Lys-Tyr
C 60–67 Gly-Glu-Asp-Thr-Leu-Met-Glu-Tyr

Having eliminated C 75-80 from consideration, C 60-67 remains as the only methionine-containing peptide available. Since no lysyl residue is contained within C 60-67 and methionine (and proline) must be N-terminal to Lys (73), the methionine residue must lie to the N-terminal side of proline and C 68-74 must be the proline-containing peptide needed to extend the sequence in this region. Thus by making use of the methionine peptide one may rewrite the sequence above as:

Lys-Gly-Ile-Thr-Trp-Gly-Glu-Asp-Thr-Leu-Met-Glu-Tyr-Leu-Glu-Asn-Pro-Lys-Lys-Tyr

Tyrosine is C-terminal in C 68-74 because of chymotrypsin specificity.

Only tyrosine remains without overlap in a tryptic peptide. It is highly undesirable to attempt to establish an overlap on the basis of a single residue or even of two amino acids unless these are of the unique types of residues which can serve as markers. Tyrosine occurs too frequently in cytochrome *c* to be considered a marker residue and it would be best to leave this sequence at this point and return to it later when other segments in the protein molecule have been completed.

Since serine is the only other residue occurring only once in the protein, we will list all of the tryptic and the chymotryptic peptides which contain this amino acid. Peptide C 47-48 is completely contained in T 40-53 leaving only a single peptide for consideration. At present we need not pursue this region further because too few marker residues are available here to aid in the reconstruction of the sequence.

Turning to methionine, as an example of a marker residue which occurs twice in the protein, it should be recalled that one of these already has been located in the reconstructed segment covering positions 55 to 74. Peptide C 75-80 contains the second methionine together with proline. In that T 80-86 is the sole methionine-containing tryptic peptide not already accounted for, it clearly must overlap C 75-80 by the single methionyl residue. Thus one obtains:

Ile-Pro-Gly-Thr-Lys-Met-(Gly, Ala, Ile_2, Phe)-Lys

Again lysine must be C-terminal because of enzyme specificity. Overlap is permissible on the basis of a single residue in this case because of the infrequent occurrence of methionine in the protein sequence. Peptide T 74-79 clearly overlaps the N-terminus of this peptide because it is the sole tryptic peptide containing both isoleucine and proline. This results in:

Tyr-Ile-Pro-Gly-Thr-Lys-Met-(Gly, Ala, Ile_2, Phe)-Lys

By similar logic, peptide C 81-82 must be completely contained within the unordered region of this segment, it being the only one of the two –Ile–Phe– containing chymotryptic peptides which would make sense at this point. The addition of this peptide leaves only three residues in unknown sequence, *i.e.* (Gly,Ile,Ala), but the arrangement of the two peptides, (Gly,Ile,Ala) and (Ile–Phe), also is unknown. Thus one

Tyr-Ile-Pro-Gly-Thr-Lys-Met-[(Gly,Ile,Phe), (Ile-Phe)]-Lys

-(Gly, Ile, Ala)-Ile-Phe-Lys or Ile-Phe-(Gly, Ile, Ala)-Lys

must find a chymotryptic peptide which contains part of the sequence Peptide C 83-88 is coincident with the latter and establishes the overlap sequence uniquely from position 74 to position 88.

Arginine is another marker residue which may be used in the reconstruction of the cytochrome *c* sequence. Referring to the list of chymotryptic and tryptic digest peptides given in Tables 5 and 6, one finds two chymotryptic peptides, C 37-46 and C 88-94, and two tryptic peptides, T 28-39 and T 88-91, which contain arginine. From the order of amino acids, it is apparent that the two chymotryptic peptides are unique and occupy different parts of the sequence. Further, it is obvious that T 28-38, but not T 88-91, must overlap C 37-46. This is established by the absence of a glycyl residue in T 88-91 though one is found just prior to arginine in C 37-46. It is also noted that the composition of T 88-91 is compatible with that of C 88-94. Thus the orientation of these peptides gives:

I (His,Asp,Thr,Gly_2,Pro,Leu_2,Phe)-Gly-Arg-Lys-Thr-Gly-Gln-Ala-Val-Gly-Phe

and

II Lys-Asp-Glu-Arg-Ala-Asp-Leu.

Both satisfy the specificities of the enzymes used in obtaining these peptides as well as the requisite amino acid composition. The peptide fragment, C 27-33, contains both proline and histidine and clearly overlaps the unordered portion of I above to yield:

Lys-Thr-Gly-Pro-Asn-Leu-His-(Gly,Leu,Phe)-Gly-Arg-Lys-Thr-Gly-Gln-Ala-Val-Gly-Phe.

The unordered tripeptide is completely coincident with C 34-36 thereby establishing the order of amino acid residues over the entire region 27 to 46.

Of the tryptic peptides remaining unassigned, only T 92-97 has an aspartic acid and leucine occurring together. This establishes the overlap necessary to extend II giving:

Lys-Asp-Glu-Arg-Ala-Asp-Leu-(Ala,Ile,Tyr)

The order of the final three amino acid residues is fixed by the chymotryptic peptide fragment C 95-99.

Tryptic peptide T 101-104 is unusual in being completely free of any residue subject to hydrolysis either by trypsin or by chymotrypsin. It is the only such peptide listed in Tables 5 and 6 and must, thereby, hold the unique position of being the C-terminus of the protein. The sequence of C 98-104 contains these amino acids and provides the needed overlap for peptide fragments T 98-99, T 100-104 and T 101-104 thereby completing the C-terminal sequence. That this is in fact the C-terminal sequence can be established too by use of the techniques described in Chapter 2.

A heme-containing peptide has been obtained from cytochrome *c* by means of chymotrypsin, trypsin and pepsin digestion. The peptides obtained in the latter two cases are identical and are contained within the structure of C 11-26, the chymotryptic digestion heme peptide. Both T 1-5 and T 9-13 contain valine, glutamic acid and lysine necessary for establishing the overlap with the N-terminal end of the chymotryptic heme peptide. The selection of one of these is simplified by the fact that cytochrome *c* has been shown to have an acetyl blocked N-terminal glycyl residue. Peptide T 1-5, but not T 9-13, contains glycine and, accordingly, must constitute the N-terminal fragment while T 9-13 extends the sequence of C 11-26 to:

III (Ile,Phe)-Val-Gln-Lys-Cys-Ala-Gln-Cys-His-Thr-Val-Glu-Lys-Gly-Gly-Lys-His.

This selection can further be verified by examining both peptide fragments for the presence of a blocked N-terminal glycine.

It remains necessary now only to turn to peptide C 1-10 to provide the overlap through –(Ile–Phe)– with segment III above and with T 1-5 and T 6-8 to complete the N-terminal segment of the protein.

Reference to the table of amino acid composition shows that all the histidine residues have been accounted for in the long segments reassembled from the various peptide fragments. At this point, we may make use of peptide fragment T 26-27 to establish the overlap with segments extending from residues 1 to 26 and 27 to 46. Further, we may now use peptide fragment T 40-53 to fix the overlap with C 37-46, C 47-48 and C 49-55 in that no other fragments remain unused and the composition of T 40-53 can be accounted for by these chymotryptic peptides in the absence of any obvious marker residues.

The C-terminal –Lys–Lys–Tyr sequence of peptide C 68–74 overlaps the C-terminal part of T 56-73 and the N-terminal part of T 74-79 while peptide fragment C 83-88 is sufficient to fix the order of amino acids over the region 80-94 containing fragments C 83-88, C 88-94, T 80-86 and T 88-91. The C-terminal peptide is

Table 8. *Amino acid composition of rabbit and dogfish heart cytochromes c (Refs.* [356, 724]*)*

Amino acid	Residues per molecule cytochrome *c*	
	Rabbit	Dogfish
Lysine	18	17
Histidine	3	2
Arginine	2	3
Aspartic acid	10	7
Threonine	8	8
Serine	1	5
Glutamic acid	10	12
Proline	3	3
Glycine	13	12
Alanine	7	7
Cysteine	2	2
Valine	4	4
Methionine	2	1
Isoleucine	6	5
Leucine	6	7
Tyrosine	4	4
Phenylalanine	4	4
Tryptophan	1	1

aligned on the basis of the absence of an amino acid residue susceptible to trypsin or chymotrypsin hydrolysis as well as on the basis of the composition of fragment T 98-99.

Now with reference to the composition in Table 8, it can be seen that all the amino acid residues have been accounted for in all of the segments and all of the reassembled segments contain the complete amino acid composition as given. Under these circumstances, chymotryptic fragment C 49-55 suffices to establish the overlap between the reconstructed segment spanning residues 1-55 and the segment including residues 55 to 104. The only residue in common to the two segments is the lysine in position 55. As stated earlier, an overlap established on the basis of a single amino acid residue must be viewed as being an equivocal overlap unless that residue is unique to the protein. In the case of cytochrome *c* where lysyl residues account for

15 to 20% of the total amino acid composition, other information must be included in order to consider the overlap assignment to be valid. Having accounted for all the amino acids and eliminating the possibility of two separate lysyl residues, it is safe to assume that the C-terminal lysine of segment 1-55 is identical with the N-terminal lysine of 55-104.

Dogfish cytochrome *c* with 14 amino acid differences from the rabbit sequence just presented illustrates the value in turning towards marker residues as aids in aligning peptide fragments for sequence reconstruction. In place of the single serine in the rabbit sequence, five residues must be located in the dogfish protein; only two histidine residues but three arginine and one methionine are to be accounted for. Of seven serine-containing peptides obtained through chymotryptic digestion, two, C 27-35 and C 99-104, may be eliminated from consideration because they are completely contained in the larger peptide fragments C 27-36 and C 98-104. Peptide fragment C 98-104 is quickly placed at the C-terminus of the protein because it has as its C-terminal residue an amino acid which is not recognized as a site of hydrolysis by either trypsin or chymotrypsin. Peptide C 47-48 may be treated as in the case of the identical fragment in the rabbit protein described above. Some difficulties might be anticipated in utilizing C 53-59 to establish an overlap with tryptic peptide fragments because of the alternate residue spacing of lysine residues. The tryptic peptide containing serine would contain only two residues and would give only an equivocal overlap without further information. The remaining two serine-containing chymotryptic fragments would afford no particular difficulties in orienting them into their proper location in the sequence. They form part of rather long tryptic fragments which can be readily differentiated on the basis of their amino acid compositions.

VI. Qualities of Computer Programs

It should be evident from the above discussion that the alignment of enzymically or chemically derived peptide fragments into the full protein structure is a complex operation and requires a degree of mechanical skill and logic as well as concrete chemical information regarding amino acid composition, sequence of at least one set of peptides, enzyme or chemical hydrolysis specificity, etc. The complexity of the operation increases rapidly as the number of amino acids in the protein increases. In addition, the number of errors in the fragments as well as other uncertainties increase. Hence, the logical and combinational problems become quite severe and the assistance of a computer becomes desirable or even necessary in the analysis. By exhaustively analyzing the possibilities of protein reconstructions, the computer may assist in determining the best next step to try in the chemical analysis processes eliminating much in the way of needless efforts. A number of programs have been made available for sequence analysis. These differ somewhat in the information required and in their approach to the problem.

A programming system for reconstructing the complete amino acid sequence of a protein consists of a number of independent sub-programs, each of which is designed to perform a specific type of operation which in turn becomes the input for one of the other sub-programs [84a, 651, 992].

The first program which need be mentioned may be called: Program ENLIST. From this program are obtained lists of peptide fragments resulting from various enzymic or chemical methods of hydrolysis. Thus KYLIST might represent the list of peptides obtained through chymotryptic digestion while TRLIST might include the trypsin hydrolysis peptide fragments. Depending upon the nature or flexibility of the programming system, the sequence of amino acids in each peptide or only their amino acid composition need be known.

Program MARKER searches all of the peptides generated by the ENLIST program and orients those peptides along the lines described for the manual operation above.

Table 9. *Composition of theoretical peptides obtained from two different methods of enzymic digestion*

Breakdown C		Breakdown T	
c_1	[R] [A,B]	t_1	[A] [B,C,D]
c_2	[D] [B] [C,A]	t_2	[A] [C,A,C]
c_3	[A] [C] [D] [X,A] [C]	t_3	[X] [A,C,B]
c_4	[B] [D] [B] [D] [A,B,D]	t_4	[B] [A,A,C,D]
c_5	[C] [A] [Z]	t_5	[A] [B,C,D]

Table 10. *Combinations of c fragments associated with each t fragment obtained from Table 9*

t_1	t_2	t_3	t_4	t_5
$\mathbf{c}_1\mathbf{c}_2$	$\mathbf{c}_2\mathbf{c}_3$	$\mathbf{c}_3\mathbf{c}_4$	c_1c_3	$\mathbf{c}_1\mathbf{c}_2$
c_1c_4	c_2c_5		c_4c_3	c_2c_4
c_2c_4	c_3c_5		$\mathbf{c}_4\mathbf{c}_5$	c_3c_2
c_4c_2				$\mathbf{c}_3\mathbf{c}_4$
				c_4c_2
				$\mathbf{c}_4\mathbf{c}_5$

Program MAKLAP is an extension of the program just discussed in systematically examining all other peptide fragments for overlap of the segments already generated from peptides containing marker residues.

Program SEARCH supplements the two programs described above by systematically looking at each pair, or higher combination, of amino acids that occur in each peptide fragment. This information might supplement program MAKLAP and program MARKER in effecting the proper alignment of peptides. A further modification of this program could calculate the probability for the occurrence of each pair of amino acids from the amino acid frequency data as given in Table 8. From these probabilities, the program makes a list of those groups of amino acids known to occur together which are of improbable occurrence. Such information is included in the data utilized in orienting peptide fragments. Program MAKLAP finds the maximum possible overlap between any of the longer segments reconstructed from peptide fragments.

The variety of sub-programs possible is as great as the imagination of the programmer and can be as simple as or complex as the information available. Discussion of the programming methods will be clarified if a simple illustration is considered. Suppose that two lists of peptides obtained from different methods of enzymic digestion is made as in Table 9. Each peptide fragment c_i of breakdown method C must either overlap several fragments of Breakdown T or else be included within some fragment of T. One can prepare a list for each c_t of all possible such associated t fragments; Table 10 shows such a list for the present example. Entries in the list are made according to whether or not some arrangement of the amino acid residues in

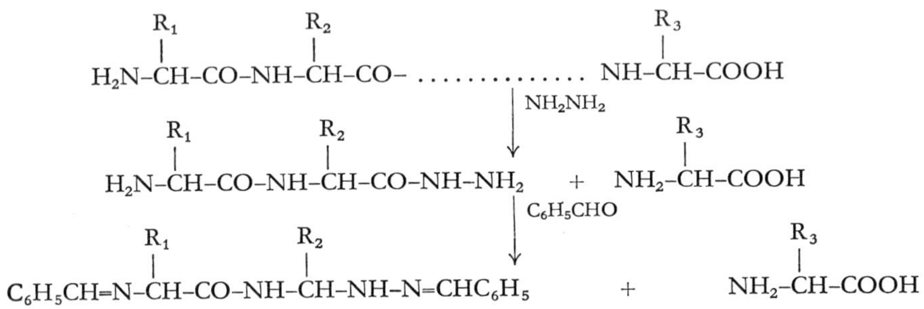

Fig. 9-2. Hydrazinolysis method of Akabori (Ref. [14])

peptide fragments c_4 and c_5 are contained in fragment t_4. The maximum overlap of amino acids in t_4 and c_4 is (B,A,D). This leaves (A,C) of t_4 to be accounted for in peptide fragment c_5. This is clearly possible, resulting in the overlap:

```
                    c4                          c5
|--------------------------------------|  |-----------------|
[B]  [D]  [B]  [D]  [B]  [A,D]  [C]  [A]  [Z]
                    [B]  [A,D]  [C]  [A]
                    |-----------------------|
                               t4
```

In order to determine all the entries in the list, such empirical trials must be made by the computer for every pair of peptide fragments c_ic_j for each t_k.

The lists of all such possibilities can be reduced, however, if one leaves only those combinations which are consistent. For example, t_3 can arise only from c_3c_4. Accordingly c_3 must precede c_4 and c_4 must follow c_3. All other possibilities involving c_3 and c_4 must therefore be eliminated. This would include examples such as c_1c_4 and c_2c_4 as in the t_1 list of Table 10 or c_3c_5 in the t_2 list, etc. This reduces the lists to those pairs of peptide fragments which are underlined in Table 10. This determines the structure as:

$$t_1t_2t_3t_4t_5$$

On proceeding similarly through the t_2,t_3,t_4 and t_5 lists the final structure can be deduced as:

(R) (A) (B) (D) (B) (A) (C) (A) (C) (D) (X) (A) (C) (B) (D) (B) (D) (B) (A,D) (C) (A) (Z).

The solutions of most problems, both scientific and otherwise, are accomplished by first breaking down the more complex complete entity into smaller, more fundamental segments, which can be more readily analyzed, and then by finding suitable interrelationships between the segments recombining them into a whole, which now becomes understandable. As has been described above, the method used in determining the amino acid sequence of proteins have followed this same pattern. The complex protein is first cleaved into smaller segments (peptides) either by chemical means such as cyanogen bromide hydrolysis or by means of proteolytic digestion. After the amino acid sequence of each smaller segment has been determined, the segments can be uniquely aligned properly to restore their orientation in the parent polypeptide chain by the method of homology or by peptide overlap. In this way the once formidable task of determining the complete amino acid sequence of a protein can be achieved.

Chapter 10

Peptide Synthesis

Duane Gish

I. Introduction

The modern era of peptide synthesis began in 1932 with the introduction by Bergmann and Zervas of the carbobenzoxy derivative for the protection of the α-amino group of amino acids and peptides [76]. Before that time some peptides had been synthesized by methods that were laborious and very limited in their application. The introduction of the carbobenzoxy group provided protection for the amino function that could be removed under very mild conditions (catalytic hydrogenolysis), conditions which left the peptide bond unaffected. Today we have numerous other very useful protecting agents, and new and widely applicable coupling methods have been developed, but peptide synthesis is essentially the same procedure it was in 1932. The carbobenzoxy group is still one of the most widely used groups for protection of the amino function.

Peptide synthesis in principle is a simple, straightforward process. Reactive groups of the amino acids or peptides which are not to participate in the coupling reaction are protected. The protected components are coupled, either by prior activation of the carboxyl group or by the use of various coupling reagents:

$$\mathrm{R^1NH\overset{R^2}{\overset{|}{C}}HCOX + H_2N\overset{R^3}{\overset{|}{C}}HCOOR^4 \xrightarrow{-HX} R^1NH\overset{R^2}{\overset{|}{C}}HCONH\overset{R^3}{\overset{|}{C}}HCOOR^4} \tag{1}$$

In Eq. 1, R^1 represents the protecting group on the α-amino function which is not to participate in the reaction, R^4 constitutes the protecting group on the non-participating α-carboxyl function, while R^2 and R^3 represent amino acid side chains. The carboxyl-activated amino acid derivative

$$\mathrm{R^1NH\overset{R^2}{\overset{|}{C}}HCOX\ in\ (1)}$$

may be isolated in the form of a stable intermediate, as is the case with an activated ester, or it may be formed *in situ*, as with the azide procedure and with numerous coupling reagents. If a dipeptide is desired, all protecting groups are removed to complete the synthesis. If the peptide chain is to be lengthened further, the appropriate protecting group is selectively removed and another partially protected amino acid or peptide is coupled:

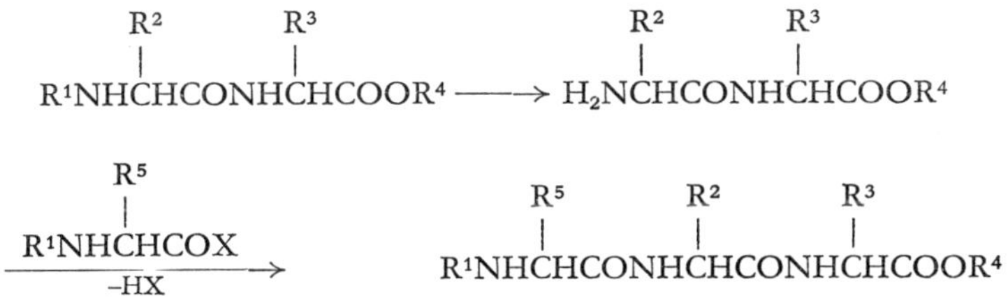

(2)

When the chain has reached the desired length, all protecting groups are removed.

While the basic scheme of peptide synthesis is simple, the actual synthesis of peptides is far from straightforward, as will be apparent from the ensuing discussion. The synthesis of a peptide is usually a research project. The wide choice of protecting groups and coupling methods tests the chemist's ingenuity in devising a synthetic scheme that will offer the best chance for success with reasonable effort.

In the above equations, R^2, R^3 and R^5 might represent reactive side-chains such as those bearing a mercapto group (cysteine), an amino group (lysine), or a carboxyl group (aspartic and glutamic acids). Protecting groups can be so chosen that reactive side chains of amino acids may remain covered while protecting groups on other amino and carboxyl groups may be selectively removed. Proper choice of the method for activating the carboxyl group in the coupling reaction may be the difference between a good yield and a poor yield, or perhaps no yield at all. The choice of solvent and other reaction conditions often determine the yield obtained and the nature and extent of undesirable side reactions.

Although some important innovations have been introduced, such as the Merrifield solid phase method [677], no procedure is universally applicable. Difficulties are commonly encountered. SCHROEDER reported that of the 29 amino acid residues in glucagon, 20 were found to cause difficulties in the synthesis of this polypeptide [900]. Often the coupling reaction does not go to completion, resulting in reduced yield and an impure product which may be difficult to purify.

Side reactions are frequently encountered during the coupling step or during removal of the protecting groups. Intramolecular cyclization may take place during activation of the carboxyl group, as has been encountered with proline [359] and arginine [127]. Coupling reactions may take an undesired course, such as the formation of acylureas during activation of the carboxyl group with dicyclohexylcarbodiimide [944], or the cleavage of mixed anhydrides in an unwanted manner during use of the mixed anhydride procedure [285, 1082]. Asparagine may undergo some dehydration to the nitrile compound when its carboxyl group is activated [347]. The use of the Merrifield solid phase technique [677], while usually shortening the time required for synthesis, introduces difficulties unique to this method. These will be discussed under the section devoted to that technique.

Special difficulties may be encountered during removal of protecting groups when certain amino acid residues are present. The catalytic hydrogenolysis of carbobenzoxy groups usually cannot be accomplished with sulfur-containing compounds except under special conditions [1127], and even in peptides which contain no sulfur, catalytic hydrogenolysis of carbobenzoxy groups becomes more difficult as the peptide chain lengthens. The use of sodium in liquid ammonia to remove protecting groups may result in certain side reactions, such as the cleavage of the proline ring in the peptide [810]. The use of HBr in acetic acid or other solvents may cause difficulties with such amino acids as methionine [16] and tryptophan [133], unless special measures are taken.

Racemization may occur during peptide synthesis. This may always be expected to some extent when the carboxyl group of a peptide is activated, unless its C-terminal amino acid is glycine or proline. Since the azide method avoids racemization, it is almost always the method of choice when peptides terminating in other than glycine or proline are coupled. Racemization is avoided by use of the stepwise method in which appropriately protected amino acids are coupled one at a time to the growing peptide chain.

In spite of these many difficulties, such factors as the availability of a wide variety of protecting groups and coupling methods, and the innovation of techniques as the solid phase method have combined to make syntheses practical which would not have been attempted two decades ago. The most ambitious projects to date have been the syntheses of ACTH (39 amino acid residues) by Schwyzer and co-workers [930] and of insulin (two polypeptide chains, one of 21 residues and one of 30 residues, linked by disulfide bridges) by Katsoyannis and co-workers [525] and by a group of Chinese chemists [585]. While we may exceed these accomplishments with presently available techniques, it seems highly unlikely that practical syntheses for polypeptide chains approaching 100 residues will be devised using these techniques. To reach such a goal, a new technique, which would constitute a real breakthrough, must be devised. It is doubtful that true proteins of definite desired sequences will ever be synthesized without the use of the living cell, or apparatus derived therefrom.

Brief mention should be made here of two recently introduced procedures that are departures from the classical approach to peptide synthesis. In neither method are the usual protected derivatives required. Hirschmann and co-workers have developed a controlled synthesis of peptides in aqueous medium using N-carboxy-α-amino acid anhydride [446a]. In this synthesis, reaction between the anhydrides to form a peptide bond and subsequent release of the α-amino group from the carbamate formed during the reaction is accomplished by control of pH. This is a very convenient method and may prove to be of great value if it can be extended to the synthesis of long-chain polypeptides. A stepwise synthesis of peptides *via* an enzymatic method has been announced by Snellman [978]. Again, its extension to long-chain polypeptides remains to be demonstrated. These two procedures will be more fully discussed in later sections.

This review is necessarily limited in scope and discussion will generally be restricted to the more commonly used protecting groups and coupling methods. In the past few years, several fine books on peptide synthesis have been published [123, 570, 898], and these should be consulted by those desiring a comprehensive review of peptide synthesis and application to specific sequences. Recent reviews include those of Schwyzer [912], Law [599], Anderson [28], Zahn [1168], and Bodanszky and Bodanszky [116].

II. Protecting Groups

A. Amino-Protecting Groups

1. Urethane-Type Protecting Groups

a) Benzyloxycarbonyl (Carbobenzoxy) and Substituted Benzyloxycarbonyl Groups

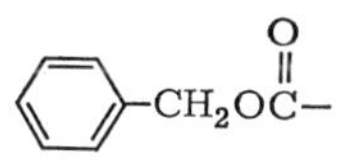

Carbobenzoxy Group

Carbobenzoxy-amino acids are easily prepared in aqueous alkaline solution from carbobenzoxy chloride and the amino acid (Eq. 3):

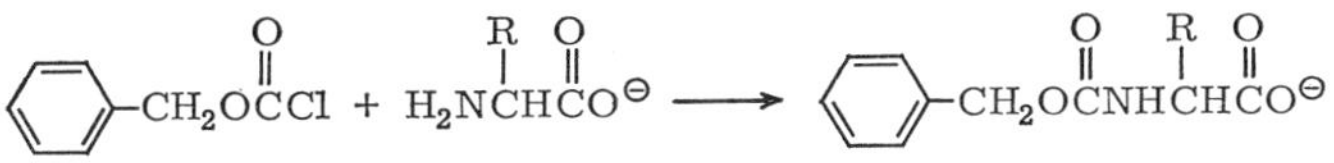

(3)

With certain amino acids, modification of the usual procedure may be necessary, such as careful control of pH. The monograph of GREENSTEIN and WINITZ [372] describes the preparation of carbobenzoxy derivatives and lists the properties of a large number of the carbobenzoxy-amino acids. However, the commercial availability of all of the carbobenzoxy-amino acids at reasonable prices renders their preparation in the laboratory unnecessary.

The carbobenzoxy group is usually stable towards alkaline hydrolysis and ester groups in amino acid or peptide derivatives can ordinarily be readily hydrolyzed with an equivalent of alkali without disturbance of the carbobenzoxy group. Exceptions are noted when glycine is adjacent in the peptide chain to the carbobenzoxy-amid acid. Elimination of benzyl alcohol with formation of a hydantoin derivative may take place, and this can be followed by further hydrolysis of the hydantoin to a urea derivative [362] (Eq. 4):

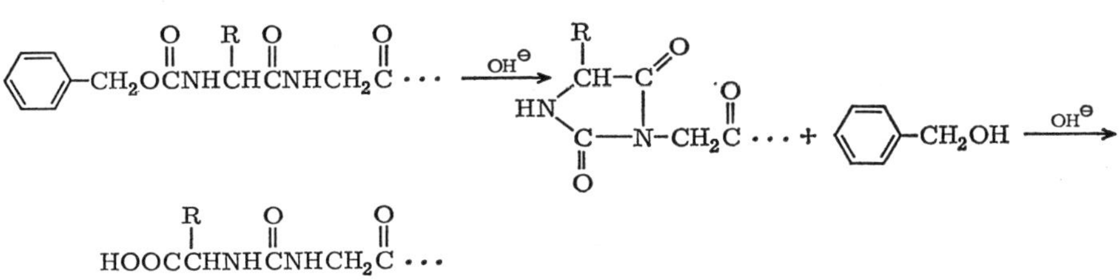

(4)

For the hydrolysis of this type of ester, excess alkali must be avoided. The same type of side reaction has been observed during amidation with ammonia of carbobenzoxy peptide esters containing glycine next to the carbobenzoxy-amino acid [100, 239, 767, 1029].

The carbobenzoxy group can be removed from non-sulfur containing compounds by catalytic hydrogenation. The hydrogenation can be performed at atmospheric pressure or at elevated pressures. As the peptide chain lengthens, the hydrogenation becomes more and more difficult until it may fail altogether. Catalytic hydrogenolysis usually fails with sulfur-containing amino acids and peptides. MEDZIHRADSZKY and SCHWERGER have recently reported that addition of cyclohexylamine makes possible decarbobenzoxylations of compounds containing methionine in other than the N-terminal position [668]. OKAMOTO and co-workers have now discovered that the use of boron trifluoride etherate along with a palladium catalyst is effective for the hydrogenolysis of methionine-containing peptides [752].

Fortunately, other methods for removing the carbobenzoxy group have been devised. The most commonly employed alternative to hydrogenation is treatment with hydrogen bromide in glacial acetic acid or other organic solvents [17, 65, 95, 485]. The concentration of hydrogen bromide may vary from 1 *N* to a saturated solution. The time required varies from a few minutes to up to about 1 h at room temperature. Side reactions may occur during this treatment. When acetic acid is used, the hydroxy amino acids may be acetylated. To avoid this side reaction, trifluoroacetic acid or other organic solvents are used. In the case of methionine, treatment with hydrogen bromide and acetic acid may lead to the benzyl sulfonium bromide addition product [389], while treatment with hydrogen bromide in nitromethane may lead to the displacement of the S-methyl group to form S-benzylhomocysteine [17]. The addition of ethylmethylsulfide prevents this side reaction [389]. Tryptophan-containing peptides may give rise to side products due to alkylation and to the presence of bromine in the reaction mixture. Addition of diethyl phosphite leads to satisfactory results [389].

Groups which are stable under conditions which permit complete decarbobenzoxylation with hydrogen bromide include the formyl, trityl, tosyl and phthalyl groups. The *tert*-butyloxycarbonyl group is readily cleaved by this treatment. The methyl and ethyl esters are stable to hydrogen bromide, while the benzyl ester is stable to limited treatment. The *p*-nitrobenzyl ester is considerably resistant to this treatment.

Other methods of cleaving the carbobenzoxy group include treatment with concentrated hydrochloric acid at 37 °C, esters are simultaneously hydrolyzed [678], ethanolic hydrogen chloride [54], sodium in liquid ammonia [951], hot trifluoroacetic acid [1121], liquid hydrogen fluoride [851] and triethylsilane [97].

Modifications of the carbobenzoxy group that have found significant application in peptide synthesis include the *p*-nitrocarbobenzoxy group [179, 256, 347, 839], the *p*-(phenylazo)-carbobenzoxy and *p*-(*p*'-methoxyphenylazo)-carbobenzoxy groups [926, 929, 932], and the *p*-chlorocarbobenzoxy group [132, 551 to 553]. The derivatives of all of these groups show better crystallizability than the corresponding carbo-

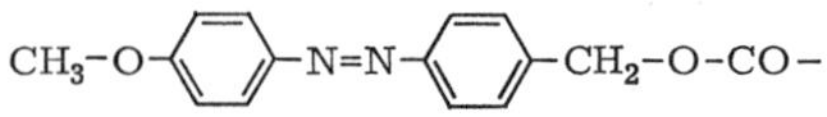

p-(p'-Methoxyphenylazo)-carbobenzoxy Group

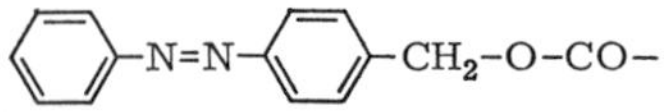

p-(Phenylazo)-carbobenzoxy Group

benzoxy derivatives. Another advantage of the *p*-(phenylazo)-carbobenzoxy and *p*-(*p*′-methoxyphenylazo)-carbobenzoxy groups is the formation of colored derivatives, while derivatives bearing the *p*-nitrocarbobenzoxy group can be detected by UV absorption near 265 mμ. Their common disadvantages, compared to the corresponding carbobenzoxy derivatives, are their greater cost and less ready availability and the somewhat greater difficulty in removing these protecting groups by catalytic hydrogenation or by treatment with HBr-acetic acid.

b) *tert-Butyloxycarbonyl Group*

The *tert*-butyloxycarbonyl (*t*-BOC) group has found wide use since its introduction into peptide synthesis by McKay and Albertson [665] and by Anderson and McGregor [32] in 1957. Today it rivals the carbobenzoxy group in application.

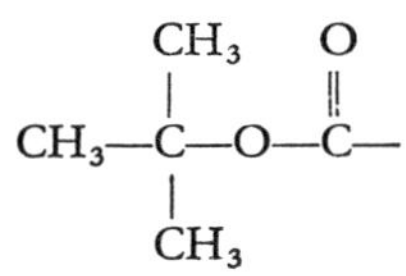

tert-Butyloxycarbonyl Group

The *t*-BOC amino acid derivatives can be conveniently prepared using *tert*-butyloxycarbonyl azide [180, 181, 928]. Schnabel has published an improved method for the synthesis of *t*-BOC amino acids using the azide [855]. Explosions have been reported during distillation of *t*-BOC azide, so this purification step should be omitted. The azide and the *t*-BOC amino acids and various active esters of the *t*-BOC amino acids are available from several commercial sources.

The great versatility of the *t*-BOC group is due to its ease of removal under midly acidic conditions which leave the carbobenzoxy group intact, while, on the other hand, its stability to hydrogenation, sodium in liquid ammonia, and aqueous alkali leaves it intact under conditions that remove the carbobenzoxy and other groups. This have proven especially valuable when multifunctional amino acids, such as lysine and arginine, are present in the peptide being synthesized. The *t*-BOC group can be removed with anhydrous trifluoroacetic acid [573], formic acid [400], and HCl in various solvents such as benzene [32], diethylphosphite [32], acetic acid [665], nitromethane [665], and ethyl acetate [929]. Mild conditions can be chosen so that the *t*-BOC group is readily split off while negligible splitting of the carbobenzoxy group occurs. When it is desired to selectively remove the carbobenzoxy group in the presence of the *t*-BOC group, catalytic hydrogenolysis is most often used, while sodium in liquid ammonia is used when catalytic hydrogenolysis cannot be used. Examples of this combination of protecting groups may be found in papers by Schwyzer and co-workers [924, 925, 929]. The use of HBr in acetic acid or other solvents permits the simultaneous removal of both groups. In spite of the lability of the *t*-BOC group to acid, *t*-BOC amino acid and peptide hydrazides can be converted smoothly to the corresponding azides in the presence of HCl if the temperature is sufficiently lowered [484].

c) Other Aliphatic Urethane Groups

Numerous other aliphatic urethane groups have been tested for utility as masking groups in peptide synthesis. Almost all suffer from some serious disadvantage when compared to the *t*-BOC group. One group which may find important application in peptide synthesis is the 2-*p*-tolylsulfonylethoxycarbonyl group, which was suggested

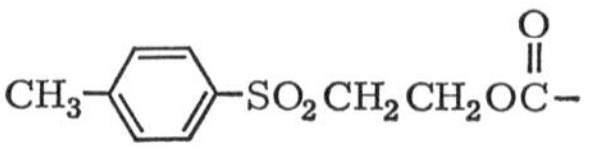

2-p-Tolylsulfonylethoxycarbonyl Group

for the synthesis of peptides by KADER and STIRLING [508]. This group is stable to acid and catalytic hydrogenolysis, but is cleaved by rather mildly basic conditions. It is sufficiently alkali stable to permit peptide synthesis to take place under usual conditions and to permit the formation of hydrazides with hydrazine. It nevertheless can be cleaved under alkaline conditions which leave an ester group intact. Treatment of a *p*-tolylsulfonylethoxycarbonyl dipeptide ethyl ester with 3 moles of *N* aqueous ethanolic sodium hydroxide for 1 h at 20 °C removed both the *p*-tolylsulfonylethoxycarbonyl group and the ester group, while treatment with 1 mole of ethanolic sodium ethoxide for 15 min at 20 °C cleaved only the *p*-tolylsulfonyl-ethoxycarbonyl group [508]. Treatment of the same protected dipeptide ester with concentrated HCl in acetic acid at reflux for 10 min yielded the *p*-tolylsulfonylethoxycarbonyl dipeptide acid in 92% yield.

2. Alkyl-Type Protecting Groups

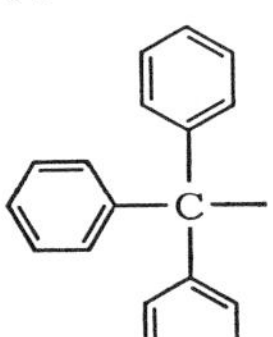

Triphenylmethyl Group

The triphenylmethyl (trityl) group is the only one of several alkyl type protecting groups investigated that has proven to be of fairly wide application in peptide synthesis [25, 438, 1005, 1179]. Ordinarily, monoalkylated amino groups can be acylated, but the bulky trityl group offers sufficient steric hindrance so that the amino group is completely protected. In fact, the steric hindrance offered by this group is great enough that some coupling methods do not succeed with certain of the trityl amino acids. Hydrolysis of esters of triphenylmethylated amino acids requires drastic conditions and the formation of the hydrazide from the ester fails with most amino acids. The exception is glycine, and in some cases, alanine. However, no difficulty is encountered when the carbodiimide method is employed with trityl amino acids, and all coupling methods applicable to peptides can be used when trityl peptides are to be coupled.

The trityl group can be removed by hydrolysis with aqueous acetic acid under conditions that do not affect the carbobenzoxy, *t*-butyloxycarbonyl, and most other groups [133, 827, 924]. An example of its use in conjunction with the *t*-BOC group is found in the paper by RITTEL [827]. It is removed by other acidic conditions, such as hydrogen chloride in various solvents [922, 1179] and trifluoroacetic acid [927]. Due to greater stability of the trityl group when it is on other functional groups, the trityl group on the α-amino group can be selectively removed from α, ε-ditrityllysine [95], N,O-ditrityltyrosine [1005], N,S-ditritylcysteine [26], and *α,imid*-ditritylhistidine [1005].

Catalytic hydrogenolysis may also be used to remove the trityl group [1179]. Since the hydrogenation of this group proceeds slowly, such groups as the carbobenzoxy and benzyl ester groups can be removed selectively by stopping the hydrogenation after the appropriate amount of hydrogen has been taken up [1005].

3. Acyl-Type Protecting Groups

The use of acyl-type protecting groups in peptide synthesis has been fairly extensive. Certain of the acyl amino acids, such as the acetyl and benzoyl amino acids, tend to undergo racemization and have found little use. Until the development of more convenient groups, such groups as the formyl and tosyl groups were used extensively for protection of the ε-amino group of lysine. Other acyl-type groups, including the trifluoroacetyl and *o*-nitrophenylsulfenyl groups, because of their ease of removal under specialized conditions, have found important applications.

a) p-Toluenesulfonyl (Tosyl) Group

The tosyl group is stable to reagents commonly used to cleave other protecting

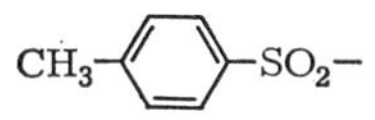

p-Toluenesulfonyl Group

groups. When it was discovered by DU VIGNEAUD and BEHRENS [254] that the tosyl group could be cleaved from amino acids and peptides by sodium in liquid ammonia, the applicability of this group to peptide synthesis became apparent. While the phthalyl, formyl, and *t*-BOC groups are stable to sodium in liquid ammonia, the carbobenzoxy and benzyl groups are cleaved. Some difficulties may be encountered during the treatment with sodium in liquid ammonia. The proline ring may be cleaved with resultant conversion of proline to δ-aminovaleric acid [810]. When multiple tosyl groups are present, the compound may be difficultly soluble in liquid ammonia, or a product may precipitate before all tosyl groups have been removed. The product must be desalted if an ammonium salt or acetic acid has been used to neutralize the excess sodium. Desalting may be accomplished by direct addition of a resin [667].

Tosyl amino acids are prepared without difficulty [320, 520, 1049] and may be coupled by a variety of methods [56, 889]. Before its displacement by the more convenient *t*-BOC group, the tosyl group was widely used as a protecting group for the ε-amino group of lysine. For an example, one is referred to its use in the ACTH series by LI and co-workers [616].

b) Phthalyl Group

The phthalyl group, introduced into peptide chemistry by Sheehan and Frank [943] and King and Kidd [548], has found limited use in peptide synthesis. It has been used for syntheses in the glucagon series [92, 895] and has found useful application as protection for the ε-amino group of lysine [916b]. The phthalyl group can be

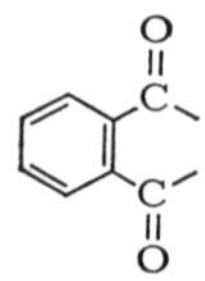

Phthalyl Group

removed only by hydrazinolysis. The use by Schwyzer and co-workers of hydrazine hydrate in neutral solution for cleavage of the phthalyl group was an important modification, for it permits cleavage under very mild conditions [916b].

c) Formyl Group

Because of the tendency of formyl amino acids towards racemization and because of difficulties encountered with some coupling methods, the formyl group has not obtained a prominent place in peptide synthesis. Formyl amino acids are prepared

```
  O
  ‖
HC—
```

Formyl Group

by the treatment of amino acids with a mixture of acetic anhydride and formic acid [255, 948] and can be coupled via the carbodiimide method to form optically active peptides [948]. The formyl group can be removed under mild acid conditions [255, 948], or with acetyl chloride in benzyl alcohol [1097]. It is stable to alkali. Yajima and co-workers have found that the formyl group can be cleaved from the ε-amino group of lysine with hydrazine hydrate [1159]. This group was used by Hofmann and co-workers as protection for the ε-amino group of lysine for the synthesis of peptides related to ACTH [452], and, in like manner, by Yajima and co-workers for the synthesis of monkey β-MSH [1159].

d) Trifluoroacetyl Group

The trifluoroacetyl (TFA) group, first applied to peptide synthesis by Weygand

```
  F  O
  |  ‖
F—C—C—
  |
  F
```

Trifluoroacetyl Group

and CSENDES [1112], shares in common with the *p*-tolylsulfonylethoxycarbonyl group the property of being readily cleaved with mild alkali while being stable to mild acid and catalytic hydrogenolysis, conditions which cleave other conveniently used protecting groups. The review by WEYGAND [1113] should be consulted for methods applicable to this group. Most peptide coupling methods can be employed. The stability to mild acid conditions does not apply in the case of the hydroxy amino acids, for TFA-serine and TFA-threonine are rapidly cleaved under very mildly acidic conditions [1114]. In contrast to the selective removal of the *p*-tolylsulfonylethoxycarbonyl group with alkali in the presence of an ester group [508], this is not possible except in special cases with the TFA group. Acid labile esters, such as the *t*-butyl ester, can be resorted to in order to circumvent this difficulty. This was demonstrated in the synthesis employed by MOORE and co-workers [697].

e) o-Nitrophenylsulfenyl (NPS) and the 2,4-Dinitrophenylsulfenyl (DNPS) Groups

The *o*-Nitrophenylsulfenyl [1174, 1175] and the 2,4-dinitrophenylsulfenyl [314] groups exhibit the same ease of removal with mild acid as the trityl group, but offer significant advantages over the latter group. For example, these groups do not

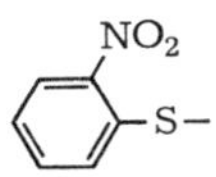

o-Nitrophenylsulfenyl Group

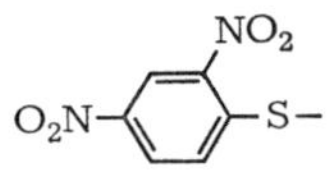

2,4-Dinitrophenylsulfenyl Group

exhibit the steric hindrance shown by the trityl group and the NPS and DNPS amino acids can be coupled in good yield via the diphenylphosphoryl or carbodiimide method, or as the active ester [314, 1174]. The NPS and DNPS groups are rapidly split by 2 moles of HCl in anhydrous solvents and can be split off selectively in the presence of such acid-sensitive groups as the *t*-BOC and *t*-butyl ester groups [1175]. The NPS and DNPS groups can also be removed with Raney nickel [670].

When the NPS and DNPS groups are cleaved with HCl, *o*-nitrophenylsulfenyl chloride and 2,4-dinitrophenylsulfenyl chloride are formed. These react with tryptophan [37] at the 2-position of the indole ring, as shown by FONTANA *et al.* [315]. These alkyl halides would also tend to react with methionine, and the Raney nickel desulfuration method is not applicable to such peptides. To circumvent these difficulties, thiol compounds and other strong nucleophiles can be employed in organic solvents to remove the NPS and DNPS groups (Eq. 5) [148, 315, 538]:

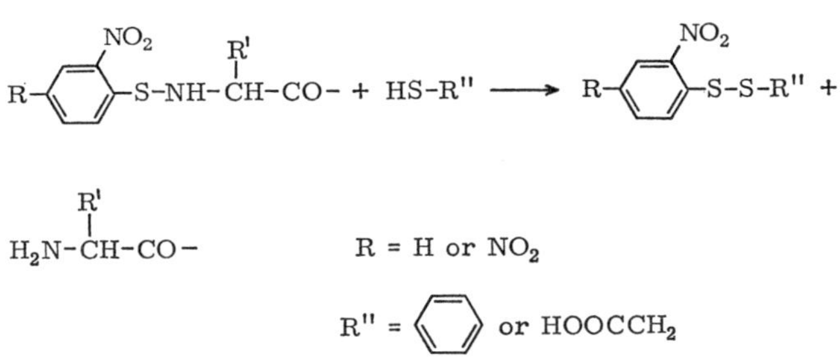

(5)

One hour or less at room temperature is sufficient to complete cleavage of the groups. Yields vary from 70 to 80%. WUENSCH and co-workers, in order to protect tryptophan residues from alkylation, carry out the cleavage in the presence of indoles, especially N-methylindole [1153]. SCOFFONE and co-workers have applied the sulfenyl chlorides as selective reagents for tryptophan in peptides and proteins [318, 934, 935]. The NPS and DNPS groups will find many important applications in peptide synthesis. A recent example is the use of the NPS group in the synthesis of human gastrin [703]. NPS-Amino acid derivatives are commercially available.

f) 5,5-Dimethyl-3-oxo-1-cyclohexen-1-yl Group

The application of the 5,5-dimethyl-3-oxo-1-cyclohexen-1-yl (dimedonyl or Dim)

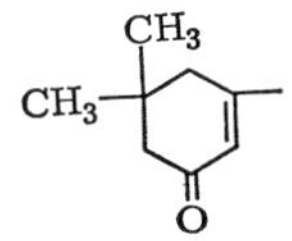

5,5-Dimethyl-3-oxo-1-cyclohexen-1-yl Group

group to peptide synthesis has been described by HALPERN and co-workers [230, 397 to 399]. This protecting group is stable to catalytic hydrogenolysis, sodium in liquid ammonia and to hydrolysis by acids or bases. It can be removed by treatment with bromine or nitrous acid. Nitrous acid appears to be an especially mild and useful reagent [399]. DEER [231] has recommended the dimedonyl group for use with the Merrifield solid phase method. In the final coupling step a dimedonyl amino acid was used. Hydrolysis of the resin ester bond followed by filtration and acidification gave dimedonyl peptides of high purity by direct crystallization. The protecting group can then be removed by treatment with nitrous acid. Because of its unique properties, the dimedonyl group will find application in special cases.

B. Carboxyl-Protecting Groups

As is the case with amino-protecting groups, a variety of carboxyl-protecting groups are available that can be selectively removed by various methods. It is thus possible to remove the carboxyl-protecting group without affecting amino-protecting groups, or *vice versa*, and to choose different protecting groups for α- and β- or γ-carboxyl groups so that one carboxyl-protecting group can be removed without affecting the others.

1. Methyl and Ethyl Esters

Methyl and ethyl esters, equally applicable to peptide synthesis, can be prepared with an excess of the alcohol in the presence of a strong acid, such as HCl, as catalyst [223]. This was the standard method of preparation until the more convenient method of BRENNER and HUBER [153] was introduced. In this method thionyl chloride is first added to the alcohol followed by the amino acid. Acetyl chloride may be used in place of thionyl chloride [405]. Peptides may be similarly esterified [300].

The β-ester of aspartic acid and the γ-ester of glutamic acid may be prepared by acid-catalyzed esterification if careful conditions are followed [77, 686]. The α-esters of aspartic acid and glutamic acid can be obtained from the intramolecular anhydrides of various N-protected derivatives by treatment with the alcohol [1117].

The removal of methyl and ethyl ester groups is usually accomplished by saponification. Ordinarily, from 20 to 60 min at room temperature with a small excess of alkali is sufficient. Various organic solvents in combination with aqueous alkali may be used. These include acetone [121], dioxane-acetone [519], dimethylsulfoxide [1021], pyridine [134], dioxane [134], methanol [134], and ethanol [146]. Caution must be used when peptides containing serine are saponified [389]. Difficulty has also been encountered during alkaline hydrolysis of α-N-*p*-nitrocarbobenzoxy-S-benzyl-*L*-cysteinyl-*L*-proline benzyl and methyl esters [348]. It is probable that β-elimination of the sulfur occurred due to the presence of the electrophilic nitro group, since no difficulty was encountered when the *p*-nitrocarbobenzoxy group was replaced with the carbobenzoxy group.

When difficulties occur with alkaline hydrolysis, as is sometimes the case with γ-esters of glutamic acid [59, 617], β-esters of aspartic acid [617] and with larger peptides, acid hydrolysis may be used. Katsoyannis used dioxane-aqueous HCl at 90 °C for 1.5 h to hydrolyze a carbobenzoxy peptide ethyl ester [519]. The carbobenzoxy peptide was obtained in 94% yield. Sheehan and co-workers used aqueous HCl at 100 °C for 2 h, or an acetone-aqueous HCl mixture at reflux for 2 h to hydrolyze phthalyl peptide esters to the phthalyl peptides [941]. Yields were excellent. Use of concentrated HCl at about 40 °C removes both the ester and the carbobenzoxy group [678].

2. *tert*-Butyl Esters

The introduction of the *t*-butyl ester group for peptide synthesis by Taschner and co-workers [1044], Anderson and Callahan [31] and by Roeske [833] made available a carboxyl-protecting group that could be removed under the same mild acid conditions used for the cleavage of the *t*-BOC group. Its resistance to alkaline hydrolysis, ammonolysis and catalytic hydrogenolysis, and rapid cleavage under conditions that leave the carbobenzoxy and other groups intact, insured its wide use in peptide synthesis. These esters can be prepared by esterification of protected amino acids, followed by removal of the protecting group [31, 899, 1044, 1045], or by direct esterification of free amino acids [833, 1043]. They are commercially available from a number of sources.

The *t*-butyl ester group may be removed by such treatment as HCl in organic solvents [899, 1046], aqueous HCl [827], trifluoroacetic acid [926, 1046] and HBr in acetic acid [31]. The utility of this group in peptide synthesis is illustrated by the synthesis of a decapeptide, described by Schwyzer and Kappeler [923], and the synthesis of β-MSH by Schwyzer and co-workers [921].

Carbobenzoxy-aspartic acid α-benzyl-β-*t*-butyl ester can be hydrolyzed with alkali to the β-*t*-butyl ester in good yield [917]. When the α-carboxyl group of aspartic acid is in peptide linkage, however, the β-*t*-butyl ester becomes labile to alkali [921]. In this instance, the carboxyl group of the C-terminal residue may be protected by the *p*-nitrobenzyl ester. Subsequent removal of the *p*-nitrobenzyl ester group by catalytic hydrogenolysis leaves the β-*t*-butyl ester group unaffected.

3. Benzyl Esters

The benzyl ester group, the stability of which resembles very closely that of the carbobenzoxy group, was first used by BERGMANN and co-workers [81]. This group can be removed by catalytic hydrogenolysis, sodium in liquid ammonia, alkaline hydrolysis and by prolonged treatment with HBr in acetic acid. The carbobenzoxy group can be removed selectively from peptide benzyl esters by HBr in acetic acid or other solvents if treatment is sufficiently brief, although there will be some loss of the benzyl ester group [17, 65].

Benzyl esters of amino acids are most conveniently prepared from benzyl alcohol with organic acids such as *p*-toluene sulfonic acid and benzene sulfonic acids as catalyst and solvents which permit the azeotropic removal of the water formed [73, 95, 632, 1180]. SHIELDS and co-workers prepared benzyl and *p*-nitrobenzyl esters by passing the condensed azeotrope through a drying agent containing an indicator [949]. This permits the most convenient preparation, and completion of the esterification is easily detected.

The α-benzyl esters of aspartic and glutamic acid can be prepared by partial hydrolysis of the diesters or through the intramolecular anhydrides [79, 163, 846]. The papers by SCHROEDER and KLIEGER [896, 897] illustrate the use of the α-esters for preparation of the ω-*t*-butyl esters by selective removal of the benzyl group. The γ-benzyl ester of glutamic acid can be prepared by selective esterification of glutamic acid [205], while the β-benzyl ester of aspartic acid may be obtained by partial acid hydrolysis of N-protected aspartic acid dibenzyl ester [73].

Examples of the application of benzyl esters in peptide synthesis include syntheses in the glucagon series [894], the insulin series [523], and the synthesis of α-MSH [390].

4. Modified Benzyl Esters

a) p-Nitrobenzyl Esters

The introduction of the *p*-nitrobenzyl ester group into peptide synthesis [911, 928] made available a carboxyl-protecting group with the same selectivity as the benzyl ester, with the added property of being much more stable to hydrogen bromide. This added stability permits the selective removal of the carbobenzoxy group with hydrogen bromide without significant effect on the *p*-nitrobenzyl ester group [911, 928]. It was found that even the *p*-nitrocarbobenzoxy group could be selectively cleaved from a *p*-nitrobenzyl ester with hydrogen bromide [949]. Amino acid esters of p-nitrobenzyl alcohol are prepared similarly to esters of benzyl alcohol [897, 949]. KATSOYANNIS has applied this ester in the insulin series [523].

b) 2,4,6-Trimethylbenzyl Esters

STEWART employed the 2,4,6-trimethylbenzyl (TMB) ester group [1008 to 1012, 1015] in order to provide a carboxyl-protecting group which would possess characteristics similar to the benzyl ester, yet could be cleaved by HBr under milder conditions, allowing the carbobenzoxy and TMB groups to be split simultaneously [1008]. Treatment of the TMB ester group with cold trifluoroacetic acid for a few minutes cleaves this group while leaving the carbobenzoxy group intact [1009, 1012].

Further, the TMB ester group is stable to methanolic HCl [1009], a reagent that rapidly splits off the amino-protecting nitrophenylsulfenyl group. The *t*-BOC group is readily split by HCl in organic solvents [32, 665, 929], as is the *t*-butyl ester group. The TMB ester may be used with these and other groups to extend the versatility of peptide synthetic methods by permitting greater selectivity in removing various protecting groups.

Stewart has pointed out [1012] a possible application of methylated polystyrene resins in the Merrifield solid phase peptide synthesis technique. The procedure now used in the Merrifield technique utilizes an ester linkage between the peptide chain and benzyl groups on the resin. At the completion of the synthesis, the completed peptide chain must be cleaved from the resin with HBr. As mentioned in a previous section, the use of HBr can give rise to serious side reactions. The use of methylated polystyrene resins in the Merrifield technique would enable the completed peptide chain to be split off by brief treatment with trifluoroacetic acid, a very mild reagent which would eliminate side reactions encountered with the use of HBr.

Stewart has applied the TMB ester group in conjunction with the *o*-nitrophenylsulfenyl and carbobenzoxy groups in the synthesis of several peptides containing sequences found in ovine insulin [1014]. The TMB group gives promise of important applications in peptide synthesis in standard techniques as well as the solid phase method.

5. Miscellaneous Esters

Other esters which have been proposed as carboxyl-protecting groups include the *p*-methoxybenzyl [1118], the phthalimidomethyl [725] and the benzhydryl ester [446b].

6. Amides

In a number of naturally occurring polypeptides such as α-MSH and the posterior pituitary hormones, vasopressin and oxytocin, the C-terminal amino acid is present as the amide. In many naturally occurring peptides, aspartic acid and glutamic acid occur as the ω-amides. Since the amide bond is about as inert as the peptide bond, the amide group can be introduced at the start of the synthesis of the chain and serve as the carboxyl-protecting group. ω-Amides are not as stable as α-amides, however, and this fact can give rise to certain side reactions. As previously mentioned, the amide group of asparagine may undergo dehydration to the nitrile when the carboxyl group of N-protected asparagine is activated [349]. ω-Amides may undergo partial hydrolysis when peptide esters are hydrolyzed with acid [920]. Sondheimer and Holley have reported the conversion of the α-methyl esters of carbobenzoxy-asparagine and carbobenzoxy-glutamine to the imides when these compounds were treated with alkali [984]. Nevertheless, when these special situations are taken into account, the amide group may serve conveniently as a carboxyl-protecting group when the amide is the final product desired.

Amides are usually prepared by saturating a methanol or ethanol solution of a methyl or ethyl ester with ammonia [79, 967]. Chambers and Carpenter have described many of the amino acid amides [189]. Benzyl esters may also be amidated [821]. As mentioned in an earlier section, hydantoin and urea derivatives tend to form when peptides containing glycine adjacent to an N-carbobenzoxy amino acid is amidated [227, 234, 332, 435a].

7. Substituted Hydrazides

The phenylhydrazide group has been used as a carboxyl-protecting group [531, 690, 1095]. At the completion of synthesis of the peptide, the phenylhydrazide can be converted to the acid by oxidation in aqueous solution with copper acetate [1095], ferric chloride [690], or manganese dioxide in acetic acid [531]. The phenyl group cannot be removed from the hydrazide in order to permit further extension of the peptide chain via the azide method. MILNE and co-workers, however, have discovered that oxidation of phenylhydrazides of protected amino acids and peptides in non-aqueous solvents converts the phenylhydrazide into a phenyldiimide. The phenyl-diimides couple with amino acid and peptide esters or amino acid and peptide phenylhydrazides to form peptide bonds [691, 692]. This will be discussed in more detail in a later section.

HOFMANN and co-workers [454, 455] used the *carbobenzoxyhydrazide* group for carboxyl protection in conjunction with the phthalyl, tosyl and carbobenzoxy groups for protection of the α-amino group. Selective removal of the amino-protecting group permitted extension of the peptide chain, the carbobenzoxyhydrazide group serving as carboxyl protection. Selective removal of the carbobenzoxy group from the carbobenzoxyhydrazide group of phthalyl and tosyl peptides permitted extension of the chains via the azide.

The *tritylhydrazide* group was used by WEYGAND and STEGLICH with the trifluoro-acetyl group as the amino-protecting function [1122]. Since the trityl group could be removed selectively with acid, and the trifluoroacetyl group could be removed selectively with alkali, the peptide chain could be extended through either the amino group or via the azide through the carboxyl group.

The introduction of other protecting groups has increased the versatility of this method. PRESTON and WEINSTEIN have reversed the procedure of WEYGAND and STEGLICH. They have used an acid labile group, the *o*-nitrophenylsulfenyl group, for amino-protection, and the *trifluoroacetylhydrazide* group for carboxyl-protection [800]. Treatment with HCl permitted extension of the peptide chain through the α-amino group, while removal of the trifluoroacetyl group with alkali produced the free hydrazide for conversion to the azide and extension of the chain through the carboxyl group.

BOISSONNAS and co-workers [135] and SCHWYZER and co-workers [933] have employed the *tert-butyloxycarbonylhydrazide group* for carboxyl protection, using it in conjunction with the carbobenzoxy group for amino protection. Further examples of this combination can be found in the papers by GUTTMANN [387] and SCHWYZER and AUNG TUN-KYI [916a].

By careful control of reaction conditions CHEUNG and BLOUT have succeeded in using the unprotected *hydrazide* group itself for carboxyl protection [194]. They took advantage of the difference in basicity of the α-amino group ($pK'_a = 7.4$) and the hydrazide function ($pK'_a = 2.8$) in α-amino acid hydrazides. Using Woodward's reagent [1148a] for activation of the carboxyl group, they coupled carbobenzoxy-glycine with glycine hydrazide to give carbobenzoxy-glycyl-glycine hydrazide in 62% yield. Under proper conditions, the hydrazide could be oxidized with N-bromo-succinimide to the carbobenzoxy peptide free acid. Conversion to the azide would have permitted extension of the chain via the carboxyl group, while catalytic hydro-genation would have permitted extension via the amino group. Where short chain peptide hydrazides are to be prepared, this simpler preparative method may be preferable, but for more complex peptide hydrazides the use of the more elegant and safer protected hydrazides may be advisable.

8. Protection by Salt Formation

A protected amino acid or peptide with an activated carboxyl group can be reacted with an amino acid or peptide when in the form of its salt (Eq. 6):

$$\text{R-NH}\overset{\overset{\text{R}^1}{|}}{\text{C}}\text{H}\overset{\overset{\text{O}}{\|}}{\text{C}}\text{X} + \text{H}_2\text{N}\overset{\overset{\text{R}^2}{|}}{\text{C}}\text{HCOO}^-\text{B}^+ \longrightarrow \text{RNH}\overset{\overset{\text{R}^1}{|}}{\text{C}}\text{H}\overset{\overset{\text{O}}{\|}}{\text{C}}\text{NH}\overset{\overset{\text{R}^2}{|}}{\text{C}}\text{HCOO}^-\text{B}^+ + \text{HX} \quad (6)$$

An aqueous or mixed aqueous-organic solvent mixture must usually be used since the salts of amino acids and lower peptides are generally insoluble in anhydrous solvents. The salts of some peptides, particularly higher peptides, with organic bases such as triethylamine or tributylamine, are soluble in dimethylformamide, and the use of this anhydrous solvent eliminates hydrolysis of the activated carboxyl component.

The advantage of this method is its simplicity and the avoidance of difficulties that sometimes occur during hydrolysis of peptide esters. On the other hand, purification of the product may be difficult due to the fact that the product and the starting carboxyl component (by hydrolysis) are both acid. When the carboxyl component is an activated amino acid derivative and the amino component is a long peptide chain, the product can often be purified without difficulty due to the considerable difference in solubility of the small carboxyl component and the long chain product.

The various methods used for activating the carboxyl components include the acid chloride [80, 520, 627], azide [543, 887], mixed anhydride [287, 1171], and activated esters [391, 392, 435a].

C. Sulfur-Protecting Groups

The sulfhydryl group in cysteine must always be protected during peptide synthesis because of its ease of oxidation to the disulfide, and because, as a nucleophile, it competes with the α-amino group for the acylating agent. The *S-benzyl group* [951, 1146] was one of the earliest groups used and has been the most widely utilized, although today there is an increasing number of more convenient groups available for S-protection. The advantage of this group is that it is stable to most reagents used to remove a wide variety of N-protecting groups, while its disadvantage is that it must be removed by treatment with sodium in liquid ammonia, a procedure less convenient than other methods and which may lead to side reactions.

The S-benzyl group has been used in conjunction with many groups for protection of the α-amino function. These include the tosyl group [86, 257, 258, 461, 505, 635], the carbobenzoxy group [112, 211, 410, 411, 464, 481, 751, 823], the trifluoroacetyl group [1115], the trityl group [202], the *tert*-butyloxycarbonyl group [32], the formyl group [464], the *p*-nitrocarbobenzoxy group [256, 389, 547], the *p*-chlorocarbobenzoxy group [551, 552], and the phthalyl group [52, 306, 547].

Modified benzyl groups used for S-protection include the *p-nitrobenzyl group* [86, 519], the *methoxybenzyl group* [15, 850], and the *p-methoxy-α,α-diethylbenzyl group*

[558]. All of these groups when used for S-protection are stable to HBr in acetic acid under conditions which cleave the N-carbobenzoxy group. Hot trifluoroacetic acid is used to remove S-*p*-methoxybenzyl and S-*p*-methoxy-α,α-diethylbenzyl groups [15, 558], while Berse and co-workers reported that the *p*-nitrobenzyl group could be removed by catalytic hydrogenation [86].

The *trityl group*, first applied by Amiard and co-workers [26], and by Zervas and Theodoropoulos [1179], has found wide use for S-protection. The elegant work of Zervas and co-workers [1176, 1177], and of Hiskey and co-workers [447a, c, d, 448a, c], illustrate the use of the trityl group in combination with the carbobenzoxy, *t*-butyloxycarbonyl, *o*-nitrophenylsulfenyl, trityl, phthalyl, and benzhydryloxycarbonyl groups for N-protection, the methyl, ethyl, benzyl, *t*-butyl and benzhydryl groups for carboxyl protection, and the benzyl, benzhydryl, benzoyl, and diphenylmethyl groups for S-protection. The use of such a variety of protecting groups in conjunction with the S-trityl group, and the use of mild reagents for its selective removal, has resulted in the S-trityl group assuming an important role as an S-protecting group.

Veber and co-workers [1086] have recently introduced the *acetamidomethyl group* for S-protection. This new group is stable to trifluoroacetic acid, HBr in acetic

$$CH_3\overset{\overset{\displaystyle O}{\|}}{C}NHCH_2\text{—}$$

Acetamidomethyl Group

acid, HCl in ethanol, anhydrous HF at 0 °C, and is stable in aqueous solution in the pH range 0 to 13. It is stable to hydrazine and thus applicable when the azide coupling method is used. The acetamidomethyl group is removed in high yield by mercuric ion at pH 4 at room temperature. It may be conveniently used in conjunction with the carbobenzoxy and *t*-butyloxycarbonyl groups.

Other groups used for S-protection include the carbobenzoxy [74, 519, 1176, 1178], diphenylmethyl [447b, 1176], benzoyl [447b, 1178], benzhydryl [447a, d, 448b, c], benzylthiomethyl [162, 448d, 519, 1163], acetyl [1178], ethylcarbamoyl [388, 834], ethylthio [479], and the 1-phenylcyclohexyl, 9-methylfluoren-9-yl, and piperonyl groups [558]. Where multiple disulfide bridges are to be synthesized, different S-protecting groups can be so chosen that particular –SH groups may be selectively uncovered. This assures formation of each disulfide bond in the manner desired.

D. Hydroxyl-Protecting Groups

Often it is possible to prepare peptides of serine, threonine, tyrosine, and hydroxyproline without protection of the hydroxyl groups. At times, however, acylation of the hydroxyl group, as well as other side reactions, occurs during the peptide bond-forming step. Paul reported acylation of the tyrosine hydroxyl group when *t*-BOC-valine was coupled with tyrosine ethyl ester [774]. Stewart and co-workers have reported 0-acylation of threonine when using the carbodiimide reagent in the Merrifield solid phase method [1016, 1017, 1165].

Groups finding use as protecting function for the hydroxyl group of tyrosine include the carbobenzoxy [486, 762], benzyl [1154], tosyl [522], and the *t*-butyl groups [175, 893, 1155, 1156], all in conjunction with the carbobenzoxy group for protection of the α-amino group. Active esters of α-N-NPS-0-*t*-butyl-tyrosine have been reported [1152].

O-Benzyl-serine was first prepared by OKAWA [755] and GRASSMANN and co-workers [367] by rather difficult procedures. The recently announced procedure for the preparation of O-benzyl-threonine [696], if applicable to serine, would make preparation of O-benzyl-serine much more convenient. In spite of the difficulty of preparation, O-benzyl-serine has been used in the preparation of numerous peptides [319, 367, 477, 595, 753, 754].

The *tert*-butyl group has proven to be a very convenient and widely applicable group for the protection of the hydroxyl group of serine and threonine. These derivatives are prepared by acid-catalyzed esterification with isobutylene [90, 91, 175, 893]. WÜNSCH and co-workers have prepared a variety of active esters of the α-N-carbobenzoxy-O-*t*-butyl and α-N-nitrophenylsulfenyl-O-*t*-butyl derivatives of serine and threonine [1152, 1157]. The O-*t*-butyl derivatives of serine and threonine have been applied to syntheses in the glucagon series [92, 893, 1157].

A new hydroxyl-protecting group which may prove to be very useful is the 2,2,2-trifluoro-1-benzyloxycarbonylaminoethyl group, introduced by WEYGAND and co-workers [1123]. These derivatives of serine and threonine are prepared by treating

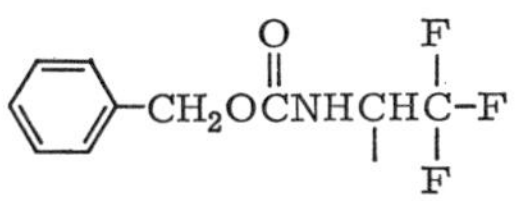

2,2,2-Trifluoro-1-benzyloxycarbonylaminoethyl group

the N-acylamino acid with the chloride. This group can be split off by catalytic hydrogenation, HBr in acetic acid, or with anhydrous HF.

Other groups that have been suggested for hydroxyl protection include the acetyl [1123] and the tetrahydropyranyl groups [483, 852].

III. Peptide Bond Formation

Less than 20 years ago, only two general methods for forming the peptide bond were available, these being the acid chloride and azide procedures. During the past two decades several new general methods and a great variety of specific new reagents have been developed. Just during the past year, numerous new activated derivatives of N-protected amino acids have been suggested. It has become difficult, if not impossible, for each individual chemist to test all suggested methods.

Almost all methods that have assumed importance involve activation of the carboxyl group of an N-protected amino acid or peptide. Some derivatives, such as the acid chlorides, azides, and mixed anhydrides, are unstable and are prepared *in situ* or used immediately after preparation. On the other hand, most of the activated esters are practically indefinitely stable, and are isolated and stored in crystalline form. This is of considerable advantage, not only because they can be prepared in advance in pure form, but also because they are thus obtainable from commercial sources. In other methods, the protected carboxyl and amino components are mixed and coupled by the addition of stable reagents, such as the ester phosphites, carbodiimides, Woodward's reagent, and carbonyldiimidazole.

All coupling methods involve the possibility of side reactions. Careful control of temperature, time, solvent, and pH may be used to minimize these side reactions. For instance, the use of mild [849] and hindered bases [114] has been used to minimize racemization during coupling with active esters. The possibility of racemization is an ever present danger, and intermediates, solvents, and activating reagents must be so chosen to minimize or eliminate this side reaction.

A. Acid Chloride Method

Acid chlorides, first used by Fischer for peptide synthesis [299], are rarely used today. The carbobenzoxy amino acid chlorides are unstable, splitting out benzyl chloride to form the N-carboxyanhydride [80]. Acyl peptide chlorides may give rise to oxazolones [763]. Tosyl amino acid chlorides are stable, crystallizable derivatives, but may undergo fragmentation in aqueous alkali [62]. The acid chloride procedure has been eclipsed by more convenient methods. Earlier, however, this procedure was used to prepare a variety of peptides a few of which may be cited [80, 347, 410, 627].

B. The Azide Procedure

Curtius introduced the azide method [222] at about the same time Fischer began use of the acid chloride procedure. Unlike the acid chloride method, however, the azide procedure is still in wide use. As previously indicated, the azide method is the only method which is free of racemization when the carboxyl group of a protected peptide is activated. In addition, the azide procedure is still the method of choice in the preparation of histidyl [453, 887, 1109] and seryl peptides [7, 240, 303, 304, 416, 887].

Hydrazides are most often prepared by the reaction of hydrazine hydrate with methyl or ethyl esters [232, 346, 389, 521, 1115, 1150], or, less often, with benzyl esters [521, 872]. Alcohol is frequently used as solvent, although dimethylformamide may be used where solubility problems are encountered. The hydrazides are treated with nitrous acid to give the reactive azides.

$$\mathrm{RNH\overset{\overset{\displaystyle R^1}{|}}{C}HCOOCH_3 \xrightarrow{H_2NNH_2} RNH\overset{\overset{\displaystyle R^1}{|}}{C}HCONHNH_2 \xrightarrow{HNO_2} RNH\overset{\overset{\displaystyle R^1}{|}}{C}HCON_3} \qquad (7)$$

The azides react with the amino component with release of gaseous HN_3 (Eq. 8).

$$\mathrm{RNH\overset{\overset{\displaystyle R^1}{|}}{C}HCON_3 + H_2N\overset{\overset{\displaystyle R^2}{|}}{C}HCOOR^3 \longrightarrow RNH\overset{\overset{\displaystyle R^1}{|}}{C}HCONH\overset{\overset{\displaystyle R^2}{|}}{C}HCOOR^3 + NH_3} \qquad (8)$$

The reaction may be carried out in anhydrous solvents or under Schotten-Bauman conditions.

Side reactions are sometimes encountered during preparation of hydrazides. It has been reported that benzyl mercaptan may be split off during formation of hydrazides of S-benzyl-cysteine derivatives [635], although hydrazides of such derivatives have been formed without difficulty, even at reflux temperature [521]. Threonine- and serine-containing peptides suffer slight racemization [886]. Side reactions may be caused by the guanido group of arginine [331], although such side reactions will not occur when the guanido group is simply protonated. Phthalyl and trifluoroacetyl amino acid and peptide esters cannot be converted to hydrazides by reaction with hydrazine since such treatment cleaves these groups. Hydrazides of these derivatives may be prepared by reacting the activated derivatives with carbobenzoxy or *t*-butyloxy-carbonyl hydrazides, as mentioned in an earlier section. Side reactions which occur under alkaline conditions must be considered a possibility during preparation of hydrazides.

Since azide formation takes place in strongly acidic media, mixtures of aqueous acetic acid and hydrochloric acid are often used. Where solubility difficulties are encountered, mixtures of aqueous hydrochloric acid with dimethylformamide may be used [389]. To suppress side reactions, the nitrite is added to the solution of the hydrazide at − 5 to − 10 °C. The azide, which forms within minutes, is extracted into organic solvents, dried briefly after washing, and used immediately for coupling. Sometimes the azide may be filtered off, washed, and used in moist form. When dimethylformamide is used as solvent, the solution may be neutralized with a tertiary base and the amino component added directly to this solution [389]. Azides of peptides containing acid-sensitive groups, such as the trityl or *t*-butyloxycarbonyl groups, may be prepared without difficulty if the temperature is sufficiently low and exposure to acid is brief [480, 484, 923].

The substitution of *t*-butyl nitrite [462] or isoamyl nitrite [661] in the place of sodium nitrite permits the formation of the azide in mixtures of tetrahydrofuran and concentrated hydrochloric acid or in organic solvents with anhydrous HCl. Side reactions are minimized.

Side reactions encountered with azides include rearrangement to the isocyanate and hydrolysis to the amide. These side reactions have been reviewed and studied in detail by Schnabel [884] and Honzl and Rudinger [462].

C. Mixed Anhydride Method

Mixed anhydrides between N-protected amino acids and a wide variety of organic and inorganic acids have been used in peptide synthesis. Those that have found the greatest application have been the anhydrides with monoesters of carbonic acid and with organic acids. The discussion here will be limited to these two types. Mixed anhydrides were introduced by Wieland and co-workers [1130, 1131]. A series of papers from several laboratories soon appeared, and the mixed anhydride procedure became permanently established in peptide chemistry.

1. Mixed Anhydrides with Monoesters of Carbonic Acid

This method was introduced independently by Wieland and Bernhard [1132], Boissonnas [130], and Vaughan [1081]. The anhydrides are formed by reaction of N-protected amino acids with various esters of chloroformic acid (Eq. 9).

$$\mathrm{RNH\overset{R^1}{\overset{|}{C}}HCOOH + Cl\overset{O}{\overset{\|}{C}}OR^2 \xrightarrow{\text{base}} RNH\overset{R^1}{\overset{|}{C}}H\overset{O}{\overset{\|}{C}}\text{–}O\text{–}\overset{O}{\overset{\|}{C}}OR^2} \qquad (9)$$

Aminolytic cleavage of the anhydride yields only CO_2 and an alcohol, in addition to the peptide (Eq. 10).

$$\mathrm{RNH\overset{R^1}{\overset{|}{C}}H\overset{O}{\overset{\|}{C}}\text{–}O\text{–}\overset{O}{\overset{\|}{C}}OR^2 + H_2N\overset{R^3}{\overset{|}{C}}HCOOR^4 \longrightarrow RNH\overset{R^1}{\overset{|}{C}}H\overset{O}{\overset{\|}{C}}NH\overset{R^3}{\overset{|}{C}}HCOOR^4 + R^2OH + CO_2} \qquad (10)$$

Wieland and Bernhard [1132] and Vaughan and Osato [1082] have carried out studies on the influence of the alkyl group on the course of the reaction. Other than methyl and phenyl esters which give unsatisfactory results, a wide variety of alkyl groups may be used. Ethyl chloroformate and isobutyl chloroformate have been most frequently used.

Normally the condensation of the acyl amino acid and chloroformate requires only a few minutes at − 5 to − 10 °C [1048, 1083]. Temperatures as high as + 10 °C and reaction times up to 30 min have been used [289], although Wieland and Bernhard have reported that disproportionation of the anhydride occurs above 0 °C [1132]. Tetrahydrofuran is favored as a solvent, although many other solvents have been used, including toluene [163, 838, 1078], ethyl acetate [606], dimethylformamide [485, 523], dioxane [289, 450] and chloroform [25, 95, 163, 635]. Triethylamine is generally used to neutralize the HCl formed, although other tertiary amines have been used, including tri-*n*-butylamine [450, 1081], tri-*n*-propylamine [881], and N-ethylpiperidine [839].

The amino component is often added as a solution of its ester in an organic solvent. The reaction is generally allowed to proceed in the cold for a few hours and then overnight at room temperature. Sometimes brief heating is used to rapidly complete the reaction [1078, 1083]. Alternately, the amino component may be added as an aqueous solution of its salt [1172].

Few side reactions are encountered with this method of bond formation. Aminolytic cleavage of the anhydride at the wrong place may occur on rare occasions [346, 549, 1082, 1129]. No racemization occurs when protected amino acids are coupled, except for such N-acylated derivatives as the acetyl or benzoyl compounds. However, when the carboxyl group of a protected peptide is activated via the anhydride, some racemization must always be expected, although Vaughan and Eichler have reported the preparation of optically pure peptides by the mixed anhydride method [1084].

2. Mixed Anhydrides with Organic Acids

Mixed anhydrides of protected amino acids with carboxylic acids tend to cleave in a manner that does not lead to formation of the peptide bond [285, 1130]. The organic acids that have been used with good success have been very limited. Good results have been obtained with the sterically hindered isovaleric [334, 349, 521, 821] and pivalic acid [604, 1173]. In the coupling of tosyl-glutaminyl-asparagine with S-benzyl-cysteine benzyl ester, where partial dehydration of the asparagine occurred with the use of tetraethyl pyrophosphite and dicyclohexylcarbodiimide, no dehydration product could be isolated when the mixed anhydride method was used [334]. The mixed anhydride with isovaleric acid gave the best yield.

3. Mixed Anhydrides with Esters of Phosphorous Acid

The ester chlorophosphites and pyrophosphites are used to form reactive amides with the amino component as well as reactive anhydrides with the carboxyl components (Eq. 11). For simplicity, however, these compounds will be considered in this section. The procedure involving prior formation of the reactive amide is known as

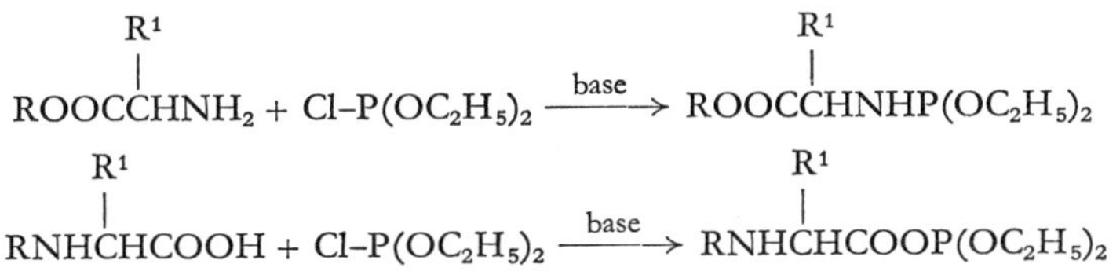

(11)

the "amide" procedure and that involving formation of the anhydride as the "anhydride" procedure. More often, the "standard" procedure is used, involving addition of the chlorophosphite and base to a mixture of the amino and carboxyl components.

ANDERSON and co-workers introduced diethyl chlorophosphite for this purpose [34], and later, the more advantageous *o*-phenylene chlorophosphite [29, 35]. The most widely used of the phosphites has been tetraethyl pyrophosphite [30, 31].

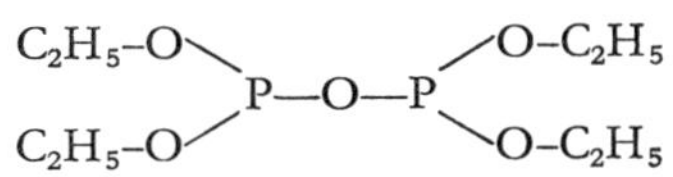

Tetraethyl Pyrophosphite

It has been used to prepare a number of naturally occurring polypeptides [258, 348, 461, 521, 839]. In spite of the convenience of this reagent, and lack of side reactions, tetraethyl pyrophosphite is in little, if any, use today. Perhaps this is because the reagent is not commercially available. The peptide chemist, like any other chemist, prefers to use what he can find on the shelf.

D. The Carbodiimide Method

Since its introduction into peptide synthesis by SHEEHAN and HESS in 1955 [945], the carbodiimide method has become second to none. Its special applicability in the Merrifield solid phase procedure has added to its already wide use. The overall reaction may be represented as a dehydration (Eq. 12). The initial reaction involves

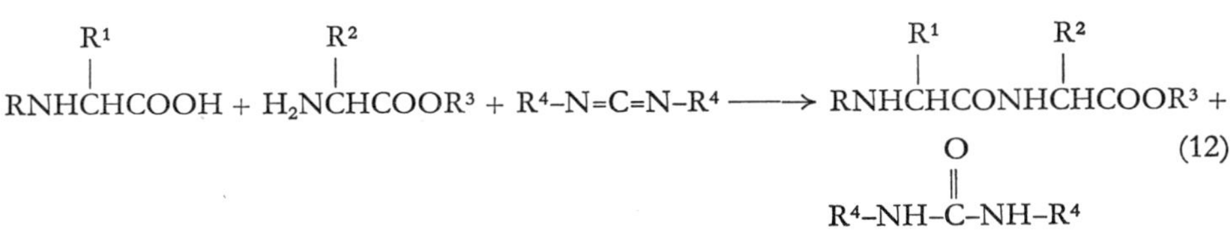

(12)

the formation of an intermediate [540, 883], further reaction leading either to the desired peptide or to an acyl urea (Eq. 13).

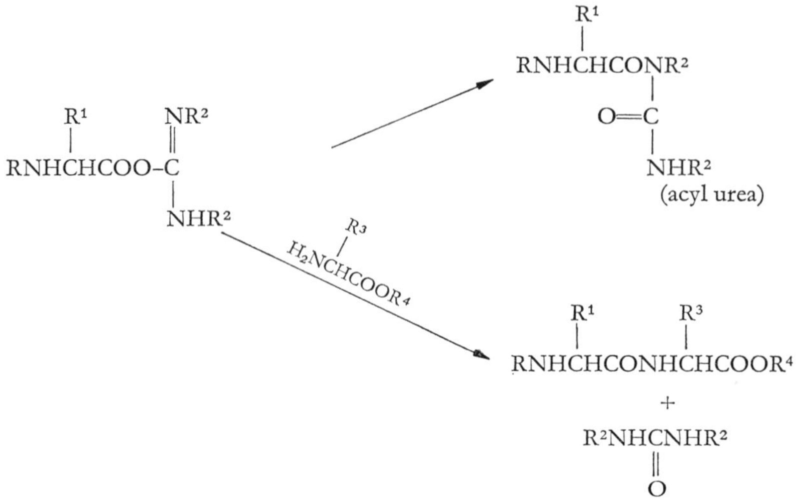

(13)

This tendency to form acyl ureas is the main side reaction encountered with the carbodiimide method [485, 921, 944, 1170]. Because low temperature reduces this tendency, a temperature of 0 °C is generally used. It is possible to use a wide variety of solvents, including dimethylformamide [119, 133, 485, 1156], tetrahydrofuran [95, 678], aqueous tetrahydrofuran [93, 253], acetonitrile [389, 921, 944], chloroform [853], pyridine [133, 430], and methylene chloride [679, 944, 1156]. The formation of the acyl urea occurs more readily in dimethylformamide than in methylene chloride and acetonitrile. The use of strong bases also favors this side reaction [883]. When the carboxyl group of asparagine is coupled via the carbodiimide method, dehydration of the amide group may occur [349].

The first carbodiimide used for peptide synthesis, and still most commonly used, was N,N′-dicyclohexylcarbodiimide. Dicyclohexylurea, the by-product of the reaction with this reagent, is only slightly soluble in most solvents, but difficulty is sometimes encountered in removing the last traces of the urea from the product [946]. Sheehan and co-workers have prepared a number of water-soluble and acid-soluble carbodiimides by incorporating tertiary or quaternary bases in the structure [942, 946]. The acylureas formed when these reagents are used are water-soluble and can be removed from the product by extraction with dilute acid.

Sheehan and associates have used the water-soluble carbodiimide 1-ethyl-3-(3-dimethylaminopropyl)carbodiimide for the synthesis of polypeptides containing up to seven amino acid residues without isolation of intermediates [947]. A reaction time of 1 h was allowed for each coupling reaction, and unreacted starting materials plus the urea and N-acylurea formed were washed out of the product with dilute acid and base. After removal of the aminoprotecting group, the product was used in the next step without further purification. The time required for synthesis of the peptides varied from 2.5 to 3.5 days.

Wolman and co-workers have reported the use of an insoluble carbodiimide, polyhexamethylenecarbodiimide, as a condensing agent in peptide synthesis [1144]. After the condensation was complete, the polymer was removed by filtration, and the product was then worked up in the usual manner. Yields of about 90% were reported for dipeptides.

E. Isoxazolium Salts

In 1961, WOODWARD and co-workers introduced the use of isoxazolium salts for activation of the carboxyl group [1148b]. The N-ethyl-5-phenylisoxazolium-3'-sulfonate compound proved to be the most useful, and has become known as Woodward's reagent K, or simply, Woodward's reagent. A comprehensive report was published by these same workers in 1966 [1149].

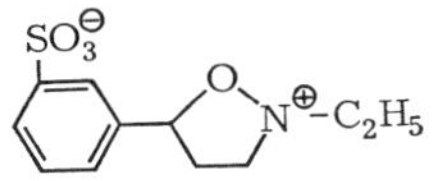

Woodward's Reagent

For activation, the N-protected amino acid or peptide is dissolved in solvent (preferably acetonitrile or nitromethane) containing triethylamine, and this solution is added to a suspension of Woodward's reagent in the same solvent. The suspension is stirred at 0 °C until the reagent has dissolved. The amino acid ester or peptide ester (or the hydrochloride salt plus triethylamine) is added to this solution and the mixture is stirred overnight at room temperature. All reagents are used in equimolar amounts. Yields vary from about 80 to 95% [1149]. Even when the carboxyl group of carbobenzoxy-asparagine is activated, so that some dehydration of the amide group could be expected [349], the yields are 75 to 80% [1149]. Reference to the synthesis of a number of naturally-occurring peptides using this reagent is given in the report by WOODWARD [1149].

Very little racemization is obtained when the carboxyl group of peptides is activated [1149]. The by-products of the reaction are water soluble, and therefore, easily removed from the product. Arginyl, seryl, asparaginyl, and glutaminyl peptides, ordinarily difficult to prepare in good yield, are obtained in excellent yield [1149]. Limitations on the choice of solvents is a disadvantage of the method.

KEMP and CHIEN have recently reported the activation of the carboxyl group by reaction with 2-ethyl-7-hydroxybenzisoxazolium fluoroborate [532]. The reaction product is an active ester, and this reagent will be considered in a later section devoted to active esters.

F. N,N-Carbonyldiimidazole

PAUL and ANDERSON, prompted by previous observations of WIELAND and SCHNEIDER [1134], and STAAB [991], applied the reagent N,N-carbonyldiimidazole for the activation of the carboxyl group [33, 775]. Carbon dioxide and imidazole are

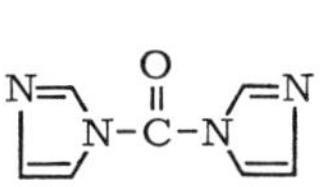

N,N-Carbonyldiimidazole

harmless by-products of the reaction, and carbon dioxide evolution drives the reaction to completion (Eq. 14).

The formation of the activated derivative requires about 30 min, after which the

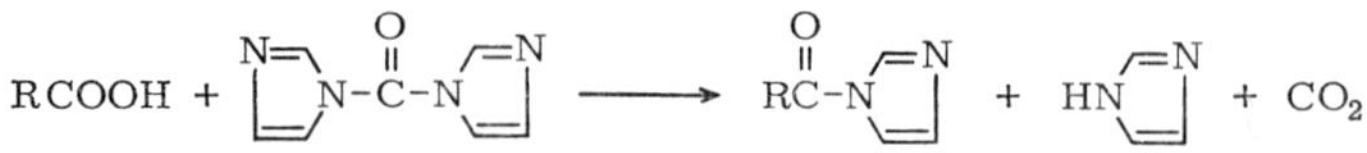

(14)

amino acid or peptide ester is added to the solution. Various solvents may be used, but for the activation step, the solvents must be dry since carbonyldiimidazole is highly sensitive to moisture [775]. The very convenient preparative procedure and the good yields obtained make this method highly useful. A disadvantage is the great sensitivity of the reagent to moisture. A number of naturally occurring peptides have been synthesized using this reagent [456, 512, 513, 776].

G. Active Ester Method

Active, or activated, esters have become, along with the carbodiimides, the most widely used reagents in peptide synthesis. Active esters were introduced by WIELAND and associates [1133].

A wide variety of active esters with great stability, enhanced reactivity and reduced tendency towards racemization make the method convenient and versatile. These active esters can be obtained in pure, crystalline form, and have a long shelf life. As a result, most are commercially available, greatly encouraging their use.

Ordinary alkyl esters are insufficiently activated for use as carboxyl components in peptide synthesis. An electrophilic alcoholic component, however, as, for example, in cyanomethyl esters [918], thiophenyl esters [918], and nitrophenyl esters [112, 113], correspondingly increases the electrophilic nature of the carbonyl carbon atom facilitating reaction with an amino group.

Systematic investigations by several groups have examined the influence of substituent groups on reactivity of esters [112, 113, 913 to 915, 918, 919, 1133]. The thiolesters and the cyanomethyl esters have such strong disadvantages in the synthesis of higher peptides that they will not be considered. The presence of sulfur in thiolesters tends to poison the catalyst in subsequent hydrogenations, while the cyanomethyl esters are more difficult to prepare than other equally active activated esters. JAKUBKE has published a recent review of the use of active esters in peptide synthesis [490].

1. O-Aryl Esters

a) p-Nitrophenyl Esters

Of the O-aryl esters, the *p*-nitrophenyl esters, first used by BODANSKY [112, 113], have assumed a leading position. A variety of methods have been used to prepare these esters [115, 280, 296, 351, 482, 848, 1101], but the most convenient and widely used method is that employing dicyclohexylcarbodiimide [117, 118, 278, 836]. The N-protected amino acid is treated with *p*-nitrophenol and dicyclohexylcarbodiimide. WOLMAN and co-workers have used a novel method for obtaining N-protected amino acid active esters directly from the amino acids [1145]. The amino acid sodium salt is treated with benzyl *p*-nitrophenyl carbonate, producing the carbobenzoxy amino acid and *p*-nitrophenol. Addition of an equivalent of dicyclohexylcarbodiimide then forms the *p*-nitrophenyl ester. Carbobenzoxy amino acid *p*-nitrophenyl esters were obtained in 65 to 80% yield. In a similar manner, using *t*-butyl *p*-nitrophenyl carbonate, the *t*-butyloxycarbonyl amino acid *p*-nitrophenyl esters were prepared in 70 to 75% yield. The paper by MEIENHOFER may be consulted for a list of *p*-nitrophenyl esters [669].

The *p*-nitrophenyl esters react with amino acid and peptide esters at room temperature in a variety of solvents. The reaction is complete within a few hours at the most, and the yields are generally excellent. The reaction is catalyzed by imidazole [660, 1135] and by certain bifunctional catalysts [94]. Imidazole was reported to give some racemization in a few cases, while 1,2,4-triazole gave no racemization and was a more efficient catalyst [94].

GOODMAN and STUEBEN prepared tripeptides by a novel method via the active esters [358]. Carbobenzoxy amino acid *p*-nitrophenyl esters were treated with HBr in acetic acid to yield crystalline amino acid *p*-nitrophenyl ester hydrobromide salts. These were coupled with carbobenzoxy amino acids via the mixed anhydride or dicyclohexylcarbodiimide method to give the carbobenzoxy dipeptide *p*-nitrophenyl esters. Addition of amino acid esters directly to the reaction mixture containing the active esters gave the carbobenzoxy tripeptide esters. KOVACS and RODIN have suggested a variation of this procedure [580]. They have found that carbobenzoxy amino acid and peptide *p*-nitrophenyl esters can be catalytically hydrogenated in the presence of an equivalent of hydrochloric acid to the amino acid and peptide *p*-nitrophenyl ester hydrochlorides, if the hydrogenation is stopped after uptake of one mole of hydrogen.

No side reactions are encountered during the coupling step with *p*-nitrophenyl esters, but intramolecular cyclization of some derivatives may occur. The *p*-nitrophenyl ester of carbobenzoxy-nitro-arginine cyclizes to the lactam [777], and *p*-nitrophenyl esters of carbobenzoxy dipeptides with C-terminal proline cyclize to the carbobenzoxy diketopiperazine [359]. The amino acid *p*-nitrophenyl esters tend to racemize when in solution in the presence of strong bases [126]. Racemization is not a problem in the coupling reaction when no excess base is added, but some racemization may occur during the preparation of carbobenzoxy peptide *p*-nitrophenyl esters.

Preparation of the *p*-nitrophenyl esters of carbobenzoxy threonine and serine by the carbodiimide method yields mixtures due to O-acylation [124]. The 2,4-dinitrophenyl esters of carbobenzoxy threonine [830] and serine [930], however, can be prepared in good yield, due to the enhanced nucleophilicity of the phenolic group in dinitrophenol. BODANSZKY and ONDETTI also discovered that the 2,4-dinitrophenyl ester of carbobenzoxy-nitroarginine could be prepared without lactam formation [124].

A large number of naturally occurring peptides have been prepared via *p*-nitrophenyl esters. These include ACTH [930] and its analogues [747, 1047], insulin [524, 671], secretin [125], oxytocin [120], lysine-vasopressin [122], partial sequences of glucagon [1151], and bradykinin [740, 848].

b) Chlorophenyl Esters

KUPRYSZEWSKI [587 to 589] and PLESS and BOISSONNAS [792] have investigated the application of various chlorophenyl esters to peptide synthesis. PLESS and BOISSONNAS synthesized a great variety of chloro-, bromo-, and iodophenyl esters. Their results favored the use of the 2,4,5-trichlorophenyl ester, while good results were also obtained with the pentachlorophenyl ester. These active esters may be synthesized by the phosphorous oxychloride method [588], with tri-(2,4,5-trichlorophenyl)-phosphite [792], or with dicyclohexylcarbodiimide and the phenol [792]. Kovacs and associates have reported the synthesis of a series of *t*-butyloxycarbonyl amino acid pentachlorophenyl esters [579] and carbobenzoxy amino acid pentachlorophenyl esters [578], including among the latter those of carbobenzoxy-nitro-arginine and carbobenzoxy-serine. In a series of papers, they have reported on the polymerization of amino acid and peptides using pentachlorophenyl esters (see reference [578]).

Sakakibara and Inukai have applied 2,4,5-trichlorophenyl trifluoroacetate and pentachlorophenyltrifluoroacetate to the synthesis of the carbobenzoxy amino acid chlorophenyl esters [848]. They also reported the direct synthesis of peptides by adding the trifluoroacetates to a mixture of the carboxyl and amino components.

Wolman and co-workers have synthesized 2,4,5-trichloro-, 2,4,6-trichloro-, and pentachlorophenyl esters of carbobenzoxy amino acids directly from amino acids via the benzyl chlorophenyl carbonates and dicyclohexylcarbodiimide in the manner described for the *p*-nitrophenyl esters in the preceeding section [1145]. In an analogous manner using *t*-butyl 2,4,5-trichlorophenyl carbonate, Broadbent and associates have reported the preparation of *t*-butyloxycarbonyl amino acid 2,4,5-trichlorophenyl esters [158].

Stewart has prepared the 2,6-dichloro-4-nitrophenyl esters of several carbobenzoxy amino acids but found them unsuitable for peptide synthesis [1013].

Kapoor has claimed the synthesis of carbobenzoxy dipeptide and tripeptide pentachlororphenyl esters by addition of a carbobenzoxy amino acid or dipeptide pentachlorophenyl ester, along with an equivalent of triethylamine and 2-hydroxypyridine, to a solution of the hydrochloride of an amino acid pentachlorophenyl ester [511].

2,4,5-Trichlorophenyl esters of carbobenzoxy amino acids have been used in the preparation of several naturally occurring peptides [38, 215, 406].

c) 8-Hydroxyquinoline and 5-Chloro-8-hydroxyquinoline Esters

Jakubke and Voigt have prepared a number of the 8-hydroxyquinoline esters of amino acids with various protecting groups and have tested their utility in peptide synthesis [491]. They could be prepared by either the mixed anhydride or the carbodiimide method. They reported the preparation of peptides in good yield and free of racemization, even when coupling carbobenzoxy peptides. The 8-hydroxyquinoline and by-products could be removed by extraction with dilute acid.

Jakubke and Voigt have also reported the preparation of the 5-chloro-8-hydroxyquinoline esters of carbobenzoxy amino acids and the preparation of a number of peptides using these esters [492]. These esters are stable, crystalline derivatives and exhibited the same favorable characteristics as the 8-hydroxyquinoline esters. In contrast to the *p*-nitrophenyl esters, the 8-hydroxy- and 5-chloro-8-hydroxyquinoline esters are remarkably resistant to racemization in the presence of strong base.

2. N-Hydroxysuccinimide Esters

The N-hydroxysuccinimide active esters, introduced into peptide synthesis by Anderson and co-workers [36], have proven to be very useful. They are most conveniently synthesized by the carbodiimide method. A series of carbobenzoxy and *t*-butyloxycarbonyl amino acid hydroxysuccinimide esters have been reported [36]. They are stable, crystalline compounds. Reaction with amino acid and peptide esters is usually complete within 1 h. Hydroxysuccinimide is water-soluble and thus easily removed from the products. The hydroxysuccinimide esters may also be reacted with amino acids and peptides in the presence of bicarbonate in aqueous organic solvent mixtures in very good yield.

WEYGAND and associates have reported that addition of 2 moles of N-hydroxysuccinimide, when using the dicyclohexylcarbodiimide method to couple carbobenzoxy peptides, eliminates practically all racemization and formation of acylurea [1116]. WÜNSCH and DREES increased the yield from 31% to 75% by the addition of N-hydroxysuccinimide when coupling a pentapeptide to a tripeptide via the dicyclohexylcarbodiimide method [1151].

H-Hydroxysuccinimide esters have been used in the synthesis of phyllokinin [82], in the glucagon series [1155, 1157, 1158], in the insulin series [1169], in the ACTH series [760, 671], and for the synthesis of an octapeptide including the "reactive" serine of chymotrypsin [595].

3. 1-Hydroxypiperidine Esters

HANDFORD and co-workers found that the 1-hydroxypiperidine esters of N-protected amino acids and peptides react readily at room temperature in high yield with amino acid and peptide esters [503]. They can be prepared by the carbodiimide method, the mixed anhydride method, the acid chloride method [503], and with WOODWARD's reagent [504]. The presence of acetic acid [503] or 1,2,4-triazole [504] considerably enhances the reaction rate during the peptide bond-forming step. These esters are optically stable in the presence of base and little or no racemization occurs when coupling carbobenzoxy peptide 1-hydroxypiperidine esters [503, 504]. The synthesis of a series of dipeptides and tetrapeptides using 1-hydroxypiperidine esters has been described [504].

WEYGAND and co-workers have described a novel method for the preparation of dipeptides and tripeptides using 1-hydroxypiperidine esters [1119]. Amino acid N-carboxyanhydrides are condensed with the hydrochloride salt of 1-hydroxypiperidine to yield amino acid 1-hydroxypiperidine ester hydrochlorides. These are coupled as the amino components with carbobenzoxy amino acids by a variety of methods to produce carbobenzoxy dipeptide 1-hydroxypiperidine esters. These are further condensed with amino acid esters to yield carbobenzoxy tripeptide esters. Yields were generally good with no detectable racemization.

4. Derivatives of N-Ethyl Benzisoxazolium Salts

Based on earlier studies of KEMP and WOODWARD [533], KEMP and CHIEN have prepared N-ethyl-7-hydroxybenzisoxazolium fluoroborate and have used it for peptide synthesis [532].

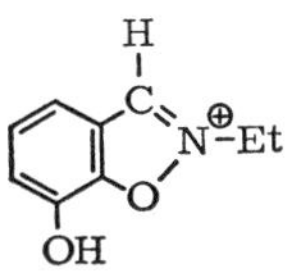

N-Ethyl-7-Hydroxyisoxazolium Cation

N-Protected amino acids in the presence of base react to form stable 3-acyloxy-2-hydroxy-N-ethylbenzamides. These compounds are optically stable in the presence of base and react with amino acid esters or salts in good yield.

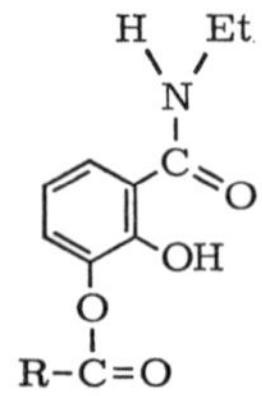

3-Acyloxy-2-hydroxy-N-ethylbenzamide

RAJAPPA and AKERKAR have synthesized 5,7-dichloro-N-ethyl benzisoxazolium fluoroborate and have prepared active esters with this compound, including a crystalline ester of carbobenzoxy-serine [809]. They postulate that these active esters are 2-acyloxy-3,5-dichloro-N-ethylbenzamides:

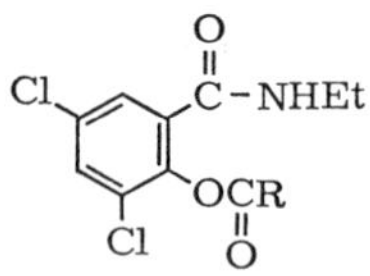

Carbobenzoxy-seryl-glycine *p*-nitrobenzyl ester was produced in about 80% yield using this active ester of carbobenzoxy-glycine.

5. Miscellaneous Active Esters

Other active esters that have been reported include those of N-hydroxyphthalimide [726, 727, 1047, 1101], N-hydroxyglutarimide [495], 3-nitroacetophenonoxime [626], and 1-phenyl-3-methylpyrazolone-(5) [626].

6. Resin Active Esters

Several research groups have prepared derivatives of resins which yield active esters when coupled with N-protected amino acids. WIELAND and BIRR [1128] used a gel crosslinked with *p*-*p*'-dihydroxydiphenyl sulfone:

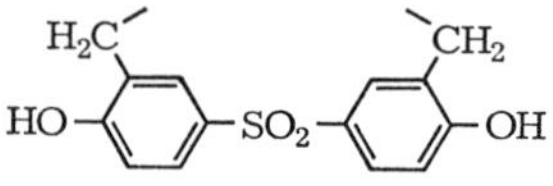

Active esters with carbobenzoxy and *t*-butyloxycarbonyl amino acids were formed via the mixed anhydride or carbodiimide method. The N-protected amino acid resin ester column was percolated with a solution of amino acid or peptide ester in dimethylformamide to yield di- and tripeptide esters in 56 to 71% yield.

Fridkin and co-workers [329] prepared active esters with carbobenzoxy amino acids and a poly-4-hydroxy-3-nitrostyrene crosslinked with divinylbenzene:

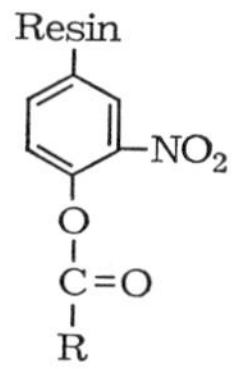

The resin active esters were prepared using the carbodiimide method. The carbobenzoxy amino acid resin active esters were stirred with a dimethylformamide solution of amino acid and peptide esters to yield polypeptides in good yield.

Laufer and associates [596] prepared copoly-(ethylene-N-hydroxymaleimide):

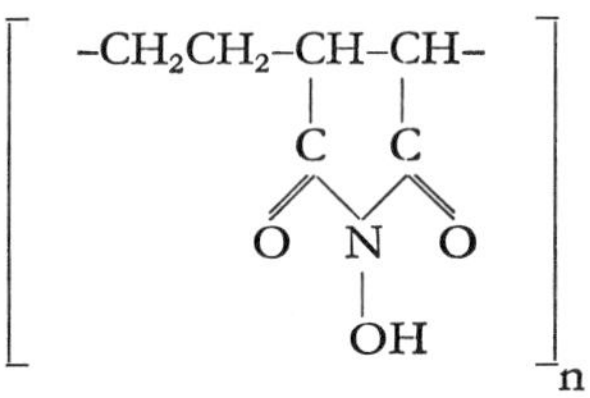

t-Butyloxycarbonyl amino acids were condensed with this resin to give active esters of the N-hydroxysuccinimide type. They were prepared via the mixed anhydride or carbodiimide method. Peptides were prepared by stirring the active ester resin with amino acid and peptide esters in dimethoxyethane or dimethylformamide. Insoluble by-products were filtered off and evaporation of the solvent gave chromatographically pure peptides in nearly quantitative yields. The preparation of a series of peptides by the method was described [596].

H. Coupling via Oxidation

1. Oxidation of N-Acyl Amino Acid Phenylhydrazides

Milne and Most have discovered that phenylhydrazides of carbobenzoxy amino acids can be oxidized in nonaqueous solutions with N-bromosuccinimide or lead tetraacetate to the crystalline carbobenzoxy amino acid phenyldiimides [691, 692]. These react with amino acid or peptide phenylhydrazides to yield carbobenzoxy peptide phenylhydrazides. They formulate these reactions as in Eq. (15):

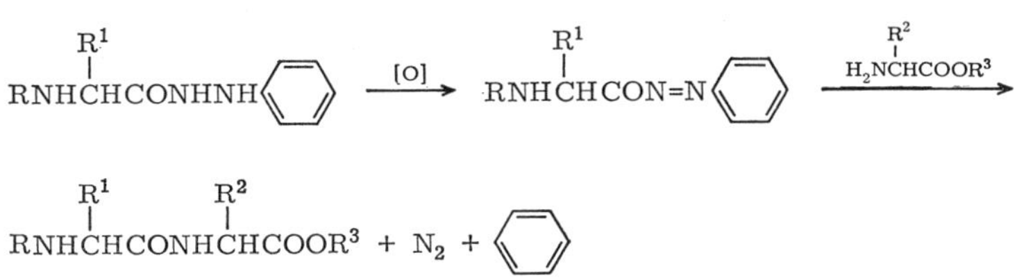

(15)

They also employed *o*-chlorophenylhydrazides for the same purpose [692]. Yields were good. The use of a carboxyl-protecting group that could participate directly in a coupling reaction would be very useful. However, diadvantages of these derivatives are the instability of the phenyldiimides and the significant amount of racemization encountered in some couplings. Furthermore, amino acids, which are sensitive to oxidation, such as tryptophan, could not be used.

2. Oxidation-Reduction Condensation

MUKAIYAMA and co-workers [708] have found that when a solution of the copper (II) or mercury (II) salt of a carboxylic acid and a sulfenamide is treated with triphenylphosphine, a carboxamide is formed (Eq. 16):

$$[(R^1CO_2)_2Cu] + 2R^2SNR^3R^4 + 2Ph_3P \longrightarrow [(R^2S)_2Cu] + 2R^1CONR^3R^4 + 2Ph_3P{=}O \quad (16)$$

When equimolar amounts of carbobenzoxy-phenylalanine, mercury (II) chloride, glycine ethyl ester, di-*o*-nitrophenyl disulfide, and triphenylphosphine in methylene chloride were stirred at room temperature for 3 h, the carbobenzoxy dipeptide ester was obtained in 89% yield.

o-Nitrophenylsulfenyl amino acid esters, being sulfenamides, could be used as the amino component. Thus, carbobenzoxy-phenylalanine was condensed with NPS-glycine ethyl ester to give carbobenzoxy-phenylalanylglycine ethyl ester in 68% yield.

I. Leuch's Anhydride Method

HIRSCHMANN and co-workers have reported the controlled synthesis of peptides in aqueous medium using amino acid N-carboxyanhydrides (Leuch's anhydrides) and amino acid N-thiocarboxyanhydrides [238, 241, 446b, 1081]. Previously, N-carboxyanhydrides had been used successfully only for the preparation of polyamino acids [517] (Eq. 17).

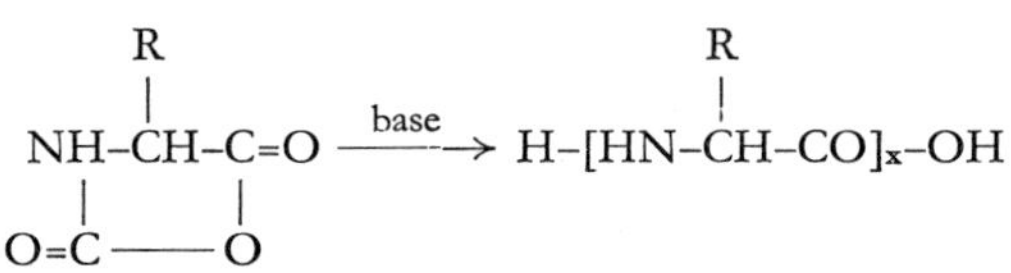

(17)

Polymers of various chain lengths are obtained. Now it has been found possible to use N-carboxyanhydrides in a stepwise synthesis by careful control of pH. A crystalline N-carboxyanhydride is added to a cold aqueous solution of an amino acid at pH 10.2 with high speed mixing, and after 2 min, the solution is acidified to yield the dipeptide (Eq. 18).

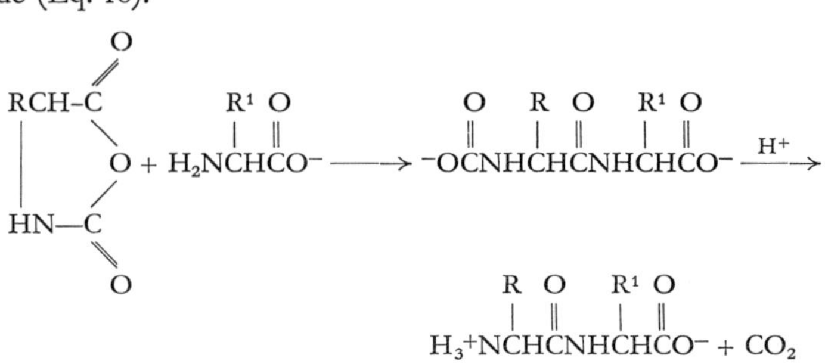

(18)

In the condensation step, the amino group of the resulting dipeptide is in the form of a carbamate ion, which protects the amino group from further reaction during the brief time required for completion of the coupling. Acidification decomposes the carbamate and releases the amino group for further coupling. After readjustment of the pH to 10.2 and addition of a second N-carboxyanhydride, the condensation may be repeated.

The advantages of this method is the lack of a requirement for blocking groups, and the speed of the reaction. Thus, free peptides are produced directly, and the time required for each step is but a few minutes. This would indeed be an ideal method for peptide synthesis, if it could be extended to the synthesis of higher peptides. The synthesis of many dipeptides, in yields of 90% or better, have been reported [446b], as well as the synthesis of a tripeptide [1087].

The method is free of racemization when N-carboxyanhydrides are used [1087], but racemization occurs when N-thiocarboxyanhydrides are employed [241]. The use of the N-thiocarboxyanhydrides of glycine and histidine is advantageous. In a synthesis using the N-thiocarboxyanhydride of glycine, the yield obtained was increased from 37% to 95% as compared to the use of the N-carboxyanhydride. The use of the N-thiocarboxyanhydride of histidine was possible without protection for the imidazole group, which is not possible with the N-carboxyanhydride. Of the amino acids providing the amino component for the reaction, only lysine and cysteine requires protecting groups on the side chains. It is reported that the N-carboxyanhydrides of amino acids, except those of proline and glycine, are stable for several months when stored at low temperature.

The successful use of a multistep synthesis without isolation of intermediates requires nearly 100% reaction at each step without the occurrence of side reactions. Two side reactions occur during synthesis using the N-carboxyanhydride method. These are the formation of hydantoic acids and overreaction [446b]. These side reactions may impose serious limitations on the length of the peptide chain obtainable by this method.

J. Enzymatic Synthesis

Snellman has reported the isolation of an enzyme from beef liver which can be used in a stepwise synthesis of polypeptides [978]. In the presence of ascorbic acid and a nucleoside triphosphate, especially guanosine triphosphate (GTP), the enzyme couples the amino component to the carboxyl group of amino acid esters, alcohol being split out. The series of reactions which take place are formulated as follows (Eqs. 19 to 21):

$$H_2NR + GTP \longrightarrow GMP + (RNH\text{-}OP_2O_6^{--}) \tag{19}$$

$$H_2N\overset{R^1}{\overset{|}{C}}HCOOR^2 + (RNH\text{-}OP_2O_6^{--}) \longrightarrow (H_2N\overset{R^1}{\overset{|}{C}}HCO\overset{R}{\overset{|}{N}}\text{-}P_2O_6^{--}) + R^2OH \tag{20}$$

$$(H_2N\overset{R^1}{\overset{|}{C}}HCO\overset{R}{\overset{|}{N}}\text{-}P_2O_6)^{--} \longrightarrow H_2N\overset{R^1}{\overset{|}{C}}HCONHR + P_2O_6^{--} \tag{21}$$

An amino component with an amino group pK of 7.8 to 8 is required. Since the pK's of the α-amino group of amino acids range from 9 to 10, and those of dipeptides are about 8.5, amino acids and dipeptides cannot be used as the amino component. Tripeptides, with pK's for the α-amino group of 7.8 to 8 are suitable.

The reaction is run as follows. To the enzyme solution is added an equivalent of tripeptide, followed by an equivalent of GTP. One minute later, one equivalent of the methyl ester of the first amino acid to be coupled is added. After 2 or 3 min, a second equivalent of GTP is added, followed in 1 min by the ester of the second amino acid. This is repeated until the synthesis of the peptide chain is complete. SNELLMAN has claimed by this method to have synthesized histidyl-phenylalanyl-arginyl-tryptophyl-glycyl-glycyl-glycine in 90% yield, and to have synthesized α-MSH and four of its analogues and the first 22 amino acid sequence of ACTH. This method apparently has great potential, if scale-up is possible. Further evaluation must await future publications.

IV. Merrifield Solid Phase Method

The introduction of the Merrifield solid phase method of peptide synthesis [674] has had a greater impact on this field than any other development since the introduction of the carbobenzoxy group. The inexperienced and the naive may feel that the Merrifield method has removed all problems, and that the automated synthesis of pure proteins is now assured. Others assume a more cautious attitude, realizing that every method involving the formation of chemical bonds is subject to limitations and side reactions. In a recent review [83], several workers in the field gave varying opinions. BODANSZKY characterized the solid phase method as a promising approach to the synthesis of long-chain polypeptides, while mentioning that several problems remain to be solved. ONDETTI stated that many people have experienced difficulty in applying the method. SCHWYZER characterized the Merrifield method as "the coming strategy".

The solid phase method has, in principle, greatly simplified the course required to synthesize a polypeptide, and has reduced the time to complete a synthesis to a fraction of that formerly required. Overall yields have been substantially increased. These results have been brought about by the fact that no intermediates are isolated, necessitating no time-consuming and yield-reducing purification procedures. As in other methods in which no intermediates are isolated, success depends upon nearly quantitative yield at each coupling step and the absence of side reactions. The occurrence of side reactions during coupling steps and during cleavage of protecting groups and of the completed peptide chain from the resin undoubtedly accounts for many of the problems that have been encountered. Nevertheless, outstanding success has been reported in the synthesis of a number of naturally occurring peptides by this method, including bradykinin and analogues [675, 676], angiotensin and analogues [542, 650], oxytocin [89], insulin [448f, 639], and a decapeptide from tobacco mosaic virus protein [1016], to mention just a few. The method has now been automated [680], and the A chain of insulin was synthesized in 8 days using this apparatus [639].

In the solid phase method, an N-protected amino acid is bound to a resin by esterification (Eq. 22). The N-protecting group is removed (Eq. 23), and a second N-protected amino acid is coupled to the amino group of the resin-bound amino acid (Eq. 24). All side products and unreacted soluble materials are washed out of the resin, the N-protecting group is cleaved, and the coupling step is repeated with a third N-protected amino acid (Eq. 25). The process is continued until the synthesis is complete. The peptide is now cleaved from the resin (Eq. 26), and purified by conventional techniques (in Eqs. 22 to 26, BOC = *t*-butyloxycarbonyl, and DCCI = dicyclohexylcarbodiimide).

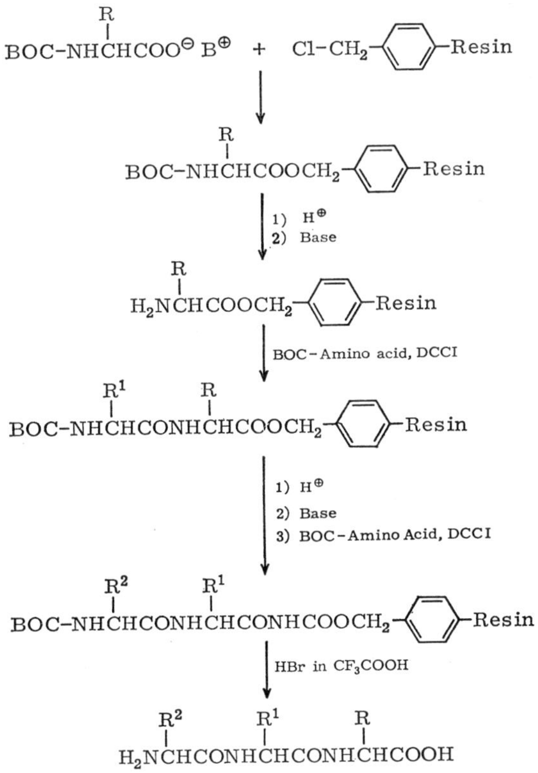

(22)

(23)

(24)

(25)

(26)

The resin used by MERRIFIELD [677] is a copolymer made from 98% styrene and 2% divinylbenzene. To provide a reactive group for attachment of the first amino acid, the benzene rings of the resin are partially chloromethylated. When this chloromethylated resin is treated with the triethylamine salt of an N-protected amino acid, a benzyl ester-type bond is formed. Such a bond is stable during the synthetic steps, but can be cleaved with HBr in acetic acid or trifluoroacetic acid, which simultaneously removes the N-protecting group and cleaves the peptide from the resin (Eq. 26).

In earlier work [674], the carbobenzoxy group was used for N-protection, but the necessity to use HBr in acetic acid to remove this protecting group caused a number of problems. Use of the *t*-butyloxycarbonyl group, which is cleaved with HCl in acetic acid, a much milder reagent, alleviated these difficulties. The *t*-butyloxycarbonyl group is used almost exclusively, although the *o*-nitrophenylsulfenyl group can also be used with advantage [448f].

The coupling reagent most widely used has been dicyclohexylcarbodiimide. Using methylene chloride as the solvent, and about a 50% excess of the *t*-butyloxycarbonyl amino acid and dicyclohexylcarbodiimide, the reaction is complete and quantitative in a few minutes [677]. Dimethylformamide is also used as solvent. Active esters can also be used in the coupling steps. The exclusive use of *p*-nitrophenyl esters have been reported by HOERNLE for the synthesis of the A-chain of bovine insulin [448f], and by BODANSZKY and SHEEHAN for the synthesis of several polypeptides [128, 129]. BEYERMAN and co-workers used both *p*-nitrophenyl and N-hydroxysuccinimide esters, with the addition of 1,2,4-triazole to catalyze the condensation [89]. BODANSZKY and SHEEHAN [129] used a hydroxymethylated resin rather than a chloromethylated resin. This permitted the esterification to be accomplished using dicyclohexylcarbodiimide, a much milder method than is necessary for esterification to a chloromethylated resin.

Difficulties encountered in this method may be due to various factors. These include incomplete coupling, creating chains of various lengths and amino acid sequence, side reactions during coupling, such as might occur between side chains, and side reactions induced by the treatment with anhydrous acid during the removal of protective groups and cleaving the peptide from the resin. These problems which no doubt increase with chain length, may finally determine the ultimate limit to which the chain may be lengthened. The synthesis of true proteins is not yet in sight.

References

1. Abderhalden, E., Brockmann, H.: Biochem. Z. **225**, 386 (1930).
2. — Komm, E.: Z. physiol. Chem., Hoppe-Seyler's **139**, 147 (1924).
3. Acher, R., Crocker, C.: Biochim. Biophys. Acta **9**, 704 (1952).
4. — Laurila, U. R.: Bull. soc. chim. biol. **35**, 413 (1953).
5a. Ackers, G. K.: Biochemistry (U.S.A.) **3**, 723 (1964).
5b. Adams, J. M.: J. Mol. Biol. **33**, 571 (1968).
6. — Capecchi, M. R.: Proc. Natl. Acad. Sci. (US) **55**, 147 (1966).
7. Adkins, H., Roos, R. M., Schroeder, D. C., Maloney, C. L., Gilbert, W. W.: J. Am. Chem. Soc. **76**, 147 (1954).
8. Africa, B., Carpenter, F. H.: Biochem. Biophys. Res. Commun. **24**, 113 (1966).
9. Akabori, S., Ikenaka, T.: J. Biochem. (Tokyo) **42**, 603 (1956).
10. — — Okada, Y., Kohno, K.: Proc. Japan Acad. **29**, 509 (1953).
11. — Narita, K.: In: Protein chemistry (Akabori, S., Mizushima, S., Eds.), Vol. 2, p. 170. Tokyo: Kyoritsu Shuppan 1954 (in Japanese).
12. — Ohno, K., Ikenaka, T., Nagata, A., Haruna, I.: Proc. Japan Acad. **29**, 561 (1953).
13. — — — Okada, Y., Hanafusa, H., Haruna, H., Tsugita, A., Sugae, K., Matsushima, T.: Bull. Chem. Soc. Japan **29**, 507 (1956).
14. — — Narita, K.: Bull. Chem. Soc. Japan **25**, 214 (1952).
15. — Sakakibara, S., Shimonishi, Y., Nobuhara, Y.: Bull. Chem. Soc. Japan **37**, 433 (1964).
16. Albertson, N. F., McKay, F. C.: J. Am. Chem. Soc. **75**, 5323 (1953).
17. — — J. Am. Chem. Soc. **75**, 5325 (1953).
18. Alberty, R. A.: In: The proteins (Neurath, N., Bailey, K., Eds.), Vol. I, Part A. (1st ed.). New York: Academic Press 1956.
19. Alexander, N. M.: Anal. Chem. **30**, 1292 (1958).
20. Alving, R. E., Laki, K.: Biochemistry (U.S.A.) **5**, 2597 (1966).
21. — Moczar, E., Laki, K.: Biochem. Biophys. Res. Commun. **23**, 540 (1966).
22. Ambler, R. P.: Biochem. J. **96**, 32p (1965).
23. — In: Methods in enzymology (Hirs, C. H. W., Ed.), Vol.XI., p. 155, 436. New York: Academic press 1967.
24. — Brown, L. H.: Biochem. J. **104**, 784 (1967).
25. Amiard, G., Heymes, R., Velluz, L.: Bull. soc. chim. France **1955**, 191.
26. — — — Bull. soc. chim. France **1956**, 698.
27. Anderer, F. A., Uhlig, H., Weber, E., Schramm, G.: Nature **186**, 922 (1960).
28. Anderson, G. W.: Ann. Rep. Med. Chem. **1966**, 296.
29. — Blodinger, J., Young, R. W., Welcher, A. D.: J. Am. Chem. Soc. **74**, 5304 (1952).
30. — — Welcher, A. D.: J. Am. Chem. Soc. **74**, 5309 (1952).
31. — Callahan, F. M.: J. Am. Chem. Soc. **82**, 3359 (1960).
32. — McGregor, A. C.: J. Am. Chem. Soc. **79**, 6180 (1957).
33. — Paul, R.: J. Am. Chem. Soc. **80**, 4423 (1958).
34. — Welcher, A. D., Young, R. W.: J. Am. Chem. Soc. **73**, 501 (1951).
35. — Young, R. W.: J. Am. Chem. Soc. **74**, 5307 (1952).
36. — Zimmerman, J. E., Callahan, , F. M.: J. Am. Chem. Soc. **86**, 1839 (1964).
37. Anderson, J. C., Barton, M. A., Hardy, P. M., Kenner, G. W., MacLeod, J. K., Preston, J., Sheppard, R. C.: Acta Chim. Acad. Sci. Hung. **44**, 187 (1965).
38. — Kenner, G. W., MacLeod, J. K., Sheppard, R. C.: Tetrahedron Letrs., Suppl. **8**, (**39**) (1966).
39. Ando, T., Tobita, T., Yamasaki, M.: J. Biochem. (Tokyo) **45**, 285 (1958).

40. ANFINSEN, C. B., HABER, E.: J. Biol. Chem. **236**, 1361 (1961).
41. — SELA, M., COOKE, J. P.: J. Biol. Chem. **237**, 1825 (1962).
42. — — TRITSCH, H.: Arch. Biochem. Biophys. **65**, 156 (1956).
43. ANSON, M. L.: J. Gen. Physiol. **20**, 663 (1937).
44. ATFIELD, G. N., MORRIS, C. J. D. R.: Biochem. J. **81**, 606 (1961).
45. AZARI, P.: Arch. Biochem. Biophys. **115**, 230 (1966).
46. BAHL, O. P., SMITH, E. L.: J. Biol. Chem. **240**, 3585 (1965).
47. BAILEY, J. L.: In: Techniques in protein chemistry, p. 190. Amsterdam, London, New York: Elsevier Publ. Co. 1967.
48. — J. Biochem. (Tokyo) **60**, 173 (1955).
49. — COLE, R. D.: J. Biol. Chem. **234**, 1733 (1959).
50.— MOORE, S., STEIN, W. H.: J. Biol. Chem. **221**, 143 (1956).
51. BAKER, L. E.: J. Biol. Chem. **193**, 809 (1951).
52. BALENOVIC, K., GAŠPERT, B.: Chem. & Ind. (London) **1957**, 115.
53. BAPTIST, W. H., BULL, H. B.: J. Am. Chem. Soc. **75**, 1727 (1953).
54. BARKDOLL, A. E., ROSS, W. F.: J. Am. Chem. Soc. **66**, 951 (1944).
55. BARMAN, T. E., KOSHLAND, JR., D. E.: J. Biol. Chem. **242**, 5771 (1967).
56. BARRAS, B. C., ELMORE, D. T.: J. Chem. Soc. **1957**, 3134.
57. BARRON, E. S. G.: Advances in Enzymol. **11**, 201 (1951).
58. BARRETT, G. C.: Chem. Commun. **1967**, 487.
59. BATTERSBY, A. R., ROBINSON, J. C.: J. Chem. Soc. **1955**, 259.
60. BAYER, E., REUTHER, K. H., BORN, R.: Angew. Chem. **69**, 640 (1957).
61. BEAVEN, G. H., HOLIDAY, R. E.: Advances in Protein Chem. **7**, 319 (1952).
62. BEECHAM, A. F.: J. Am. Chem. Soc. **79**, 3257 (1957).
63. BENASSI, C. A., SCOFFONE, E., GALIAZZO, G., JORI, G.: Photochem. Photobiol. **6**, 857 (1967).
64. BENCZE, W. L., SCHMID, K.: Anal. Chem. **25**, 1193 (1957).
65. BEN-ISHAI, D., BERGER, A.: J. Org. Chem. **17**, 1564 (1952).
66. BENISEK, W. F., COLE, R. D.: Biochem. Biophys. Res. Commun. **20**, 655 (1965).
67. — RAFERTY, M. A., COLE, R. D.: Biochemistry (U.S.A.) **6**, 3780 (1967).
68. BENESCH, R., BENESCH, R. E., GUTCHO, M., LAUFER, L.: Science **123**, 981 (1956).
69. — — Biochim. Biophys. Acta **23**, 658 (1957).
70. BENESCH, R. E., LARDY, H. A., BENESCH, R.: J. Biol. Chem. **216**, 663 (1955).
71. BENSON, J. V., GORDON, M. J., PATTERSON, J. A.: Anal. Biochem. **18**, 228 (1967).
72. — JONES, R. T., CORMICK, J., PATTERSON, J. A.: Anal. Biochem. **16**, 91 (1966).
73. BERGER, A., KATCHALSKI, E.: J. Am. Chem. Soc. **73**, 4084 (1951).
74. — NOGUCHI, J., KATCHALSKI, E.: J. Am. Chem. Soc. **78**, 4483 (1956).
75. BERGMANN, M., FRUTON, J. S.: J. Biol. Chem. **145**, 247 (1942).
76. — ZERVAS, L.: Chem. Ber. **65**, 1192 (1932).
77. — — Z. physiol. Chem., Hoppe-Seyler's **221**, 51 (1933).
78. — — J. Biol. Chem. **113**, 341 (1936).
79. — — ROSS, W. F.: J. Biol. Chem. **111**, 245 (1935).
80. — — SCHLEICH, H., LEINERT, F.: Z. physiol. Chem., Hoppe-Seyler's **212**, 72 (1932).
81. — — SALZMANN, L.: Ber. deut. chem. Ges. **66**, 1288 (1933).
82. BERNARDI, L., BOSISIO, G., DE CASTIGLIONE, R., GOFFREDO, O.: Experientia **22**, 425 (1966).
83. BERNHARD, R.: Scientific Research, June 24, 1968.
84a. BERNHARD, S. A., DUDE, W. L., BRADLEY, D. F.: In: Data acquisition and processing in biology and medicine, Vol. 2 (ENSLEIN, K., Ed.). New York: The Macmillan Co. 1964.
84b. BERNIER, I., JOLLES, P.: Compt. rend. **253**, 745 (1961).
85. BERSE, C., BOUCHER, R., PICHE, L.: J. Org. Chem. **22**, 805 (1957).
86. — MASSIAH, T., PICHE, L.: J. Org. Chem. **26**, 4514 (1961).
87. BETHELL, D., METCALFE, G. E., SHEPPARD, R. C.: Chem. Commun. **1965**, 189.
88. BETTELHEIM, F. R.: J. Biol. Chem. **212**, 235 (1955).
89. BEYERMAN, H. C., BOERS-BOONEKAMP, C. A. M., VAN DEN BRINK-ZIMMERMANNOVA, H. M.: Rec. trav. chim. **87**, 257 (1968).
90. — BONTEKOE, J. S.: Proc. Chem. Soc. 249 (1961).

91. — — Rec. trav. chim. **81**, 691 (1962).
92. — — Rec. trav. chim. **81**, 699 (1962).
93. — — Koch, A. C.: Rec. trav. chim. **78**, 935 (1959).
94. — van den Brink, W. M., Weygand, F., Prox, A., König, W., Schmidhammer, L., Nintz, E.: Rec. trav. chim. **84**, 213 (1965).
95. Bezas, B., Zervas, L.: J. Am. Chem. Soc. **83**, 719 (1961).
96. Bier, M., Teitelbaum, P.: Ann. N.Y. Acad. Sci. **72**, 641 (1959).
97. Birkhofer, L., Bierwirth, E., Ritter, A.: Chem. Ber. **94**, 821 (1961).
98. — Dolike, M.: J. Chromatog. **26**, 270 (1967).
99. Biserte, G., Osteux, R.: Bull. soc. chim. biol. **33**, 50 (1951).
100. Blackburn, S., Lee, G. R.: Biochem. J. **58**, 227 (1954).
101. — Lowther, A. G.: Biochem. J. **48**, 126 (1951).
102. Blau, K., Darbre, A.: Biochem. J. **88**, 8 (1963).
103. Block, R. J.: In: A laboratory manual of analytical methods of protein chemistry (Alexander, P., Block, R. J., Eds.), Vol. 2, p. 1. New York: Pergamon Press 1960.
104. — Arch. Biochem. Biophys. **31**, 266 (1951).
105. — Bolling, D.: In: The amino acid composition of proteins and foods. Illinois: Charles C. Thomas 1951.
106. — Durrum, E. L., Zweig, G.: In: Paper chromatography and paper electrophoresis. New York: Academic Press 1958.
107. Blömback, B., Blömback, M., Edman, P.: Acta Chem. Scand. **17**, 1184 (1963).
108. — — — Hessel, B.: Biochim. Biophys. Acta **115**, 371 (1966).
109. — — Hessel, B., Iwanaga, S.: Nature **215**, 1445 (1967).
110. — Doolittle, R. F.: Acta Chem. Scand. **17**, 1816 (1963).
111. Bock, R. M., Ling, N.-S.: Anal. Chem. **26**, 1543 (1954).
112. Bodanszky, M.: Acta Chim. Acad. Sci. Hung. **10**, 335 (1957).
113. — Nature **175**, 685 (1955).
114. — Bath, R. J.: Chem. Commun. **1968**, 766.
115. — Birkhimer, C. A.: Chem. & Ind. (London) **1962**, 1620.
116. — Bodanszky, A.: Am. Scientist **55**, 185 (1967).
117. — Denning, G. S., du Vigneaud, V.: In: Biochemical preparations (Brown, G. B., Ed.), Vol. 10, p. 122. New York: Wiley 1963.
118. — du Vigneaud, V.: In, Biochemical preparations (Coon, M. J., Ed.), Vol. 9, p. 110. New York: Wiley 1962.
119. — — J. Am. Chem. Soc. **81**, 2504 (1959).
120. — — J. Am. Chem. Soc. **81**, 5688 (1959).
121. — — J. Am. Chem. Soc. **81**, 6072 (1959).
122. — Meienhofer, J., du Vigneaud, V.: J. Am. Chem. Soc. **82**, 3195 (1960).
123. — Ondetti, M. A.: In: Peptide synthesis. New York: Interscience Publishers 1966.
124. — — Chem. & Ind. (London) **1966**, 26.
125. — — Levine, S. D., Williams, N. J.: J. Am. Chem. Soc. **89**, 6753 (1967).
126. — Sheehan, J. T.: Ann. N.Y. Acad. Sci. **88**, 655 (1960).
127. — — Chem. & Ind. (London) **1960**, 1268.
128. — — Chem. & Ind. (London) **1964**, 1423.
129. — — Chem. & Ind. (London) **1966**, 1597.
130. Boissonnas, R. A.: Helv. Chim. Acta **34**, 874 (1951).
131. — Haselbach, C. H.: Helv. Chim. Acta **36**, 576 (1953).
132. — Preitner, G.: Helv. Chim. Acta **36**, 875 (1953).
133. — Guttmann, St., Huguenin, R. L., Jaquenod, P.-A., Sandrin, E.: Helv. Chim. Acta **41**, 1867 (1958).
134. — — Jacquenod, P.-A., Waller, J. P.: Helv. Chim. Acta **38**, 1491 (1955).
135. — — — Helv. Chim. Acta **43**, 1349 (1960).
136. Bolling, D., Sober, H. A., Block, R. J.: Federation Proc. **8**, 185 (1949).
137. Boulton, A. A., Bush, I. E.: Biochem. J. **92**, 11p (1964).
138. Bovarnick, M., Clarke, H. T.: J. Am. Chem. Soc. **60**, 2426 (1938).
139. Boyd, W. J.: Biochem. J. **23**, 78 (1929).
140. Boyden, A., Gemeroy, D.: Zoologica **35**, 145 (1950).

141. Boyer, P. D.: J. Am. Chem. Soc. **76**, 4331 (1954).
142. — In: The enzymes (Boyer, P. D., Lardy, H., Myrback, K., Eds.), Vol. **1**, p. **511**. New York: Academic Press 1959.
143. Boyer, S. H., Hathaway, P., Pascasio, F., Orton, C., Bordley, J.: Science **153**, 1539 (1966).
144. Bradbury, J. H.: Biochem. J. **68**, 482 (1958).
145. Brand, E., Edsall, J. T.: Ann. Rev. Biochem. **16**, 724 (1947).
146. — Erlanger, B. F., Sachs, H., Polatnick, J.: J. Am. Chem. Soc. **73**, 3510 (1951).
147. — Kassel, B.: J. Biol. Chem. **145**, 365 (1942).
148. Brandenberg, D.: Tetrahedron (**49**), 6201 (1966).
149. Braun, V., Schroeder, W. H.: Arch. Biochem. Biophys. **118**, 241 (1967).
150. Braunitzer, G.: Chem. Ber. **88**, 2025 (1955).
151. — Biochim. Biophys. Acta **19**, 574 (1956).
152. Brautlecht, C. A.: J. Biol. Chem. **10**, 139 (1911).
153. Brenner, M., Huber, W.: Helv. Chim. Acta **36**, 1109 (1953).
154. — Niederwieser, A., Pataki, G.: Experientia **17**, 145 (1961).
155. — — — In: Thin layer chromatography, p. 427. (Stahl, E., Ed.). New York: Academic Press Inc. 1965.
156. Brewer, C. F., Riehm, J. P.: Anal. Biochem. **18**, 248 (1967).
157. Britten, R. J., Roberts, R. B.: Science **131**, 32 (1960).
158. Broadbent, W., Morley, J. S., Stone, B. E.: J. Chem. Soc. (C) **1967**, 2632.
159. Brown, F., Hall, L. P.: Nature **166**, 66 (1950).
160. Brown, J. R., Hartley, B. S.: Biochem. J. **101**, 214 (1966).
161. — Kauffman, D. L., Hartley, B. S.: Biochem. J. **103**, 497 (1967).
162. Brownlee, P. J. E., Cox, M. E., Handford, B. O., Marsden, J. C., Young, G. T., J. Chem. Soc. **1964**, 3832.
163. Bryant, P. M., Moore, R. H., Pimlott, P. J., Young, G. T.: J. Chem. Soc. **1959**, 3868.
164. Buckingham, D. A., Collman, J. P.: Inorg. Chem. **6**, 1803 (1967).
165. — — Happer, D. A. R., Marzilli, L. G.: J. Am. Chem. Soc. **89**, 1082 (1967).
166. Buckley, III, C. E., Whitney, P. L., Tanford, C.: Proc. Natl. Acad. Sci. (US) **50**, 827 (1963).
167. Bull, H. B., Hahn, J. W.: J. Am. Chem. Soc. **70**, 2132 (1948).
168. — — Baptist, V. H.: J. Am. Chem. Soc. **71**, 550 (1949).
169. Burawoy, A., Liversedge, F., Vellins, C. E.: J. Chem. Soc. **1954**, 4481.
170. Burkhardt, III, W. A., Wilcox, P. E.: Biochem. Biophys. Res. Commun. **28**, 803 (1967).
171. Burns, J. A., Curtis, C. F., Kacser, H.: J. Chromatog. **20**, 310 (1965).
172. Burton, H.: Biochim. Biophys. Acta **29**, 193 (1958).
173. Butler, P. J. G., Harris, J. I., Hartley, B. S., Leberman, R.: Biochem. J. **103**, 78 (1967).
174. Cahnmann, H. J., Arnon, R., Sela, M.: J. Biol. Chem. **241**, 3247 (1966).
175. Callahan, F. M., Anderson, G. W., Paul, R., Zimmerman, J. E.: J. Am. Chem. Soc. **85**, 201 (1963).
176a. Canfield, R. E.: J. Biol. Chem. **238**, 2691 (1963).
176b. — J. Biol. Chem. **238**, 2698 (1963).
177. — Anfinsen, C. B.: J. Biol. Chem. **238**, 2684 (1963).
178. — — In: The proteins (Neurath, J., Ed.), 2nd Ed., Vol. 1, p. 311. New York: Academic Press 1963.
179. Carpenter, F. H., Gish, D. T.: J. Am. Chem. Soc. **74**, 3818 (1952).
180. Carpino, L. A.: J. Am. Chem. Soc. **79**, 4427 (1957).
181. — Giza, C. A., Carpino, B. A.: J. Am. Chem. Soc. **81**, 955 (1959).
182. Carter, J. R.: J. Biol. Chem. **234**, 1705 (1959).
183. Cecil, R.: Biochim. Biophys. Acta **18**, 154 (1955).
184. — In: The proteins (Neurath, H., Ed.), 2nd Ed., Vol. 1, p. 380. New York: Academic Press 1963.
185. — McPhee, J. R.: Biochem. J. **59**, 234 (1955).
186. — — Biochem. J. **60**, 496 (1955).

187. — — Advances in Protein Chem. **14**, 255 (1959).
188. — Wake, R. G.: Biochem. J. **82**, 401 (1962).
189a. Chambers, R. W., Carpenter, F. H.: J. Am. Chem. Soc. **77**, 1522 (1955).
189b. Chauvet, J. P., Acher, R.: Febs Letters **1**, 305 (1968).
190. Chan, S. K., Margoliash, E.: J. Biol. Chem. **241**, 507 (1966).
191. Cherbuliez, E., Baehler, B., Rabinowitz, J.: Helv. Chim. Acta **43**, 1871 (1960).
192. — Sussmann, A. R., Rabinowitz, J.: Helv. Chim. Acta **44**, 319 (1961).
193. Chervenka, C. H., Wilcox, P. E.: J. Biol. Chem. **222**, 612 (1956).
194. Cheung, H. T., Blout, E. R.: J. Org. Chem. **30**, 315 (1965).
195. Chibnall, A. C.: J. Intern. Soc. Leather Trades' Chemists **30**, 1 (1946).
196. — Mangan, J. L., Rees, M. W.: Biochem. J. **68**, 111 (1958).
197. — — — Biochem. J. **68**, 114 (1958).
198. — Rees, M. W.: Biochem. J. **68**, 105 (1958).
199. — — Biochem. J. **68**, 111 (1958).
200. — — Williams, E. F.: Biochem. J. **37**, 354 (1943).
201. — Spahr, P. F.: Biochem. J. **68**, 135 (1958).
202. Chillemi, D., Scarso, L., Scoffone, E.: Gazz. chim. ital. **87**, 1356 (1957).
203. Chu, R. C. L., Yasunobu, K. T.: Biochim. Biophys. Acta **89**, 148 (1964).
204. Clarke, H. T.: J. Biol. Chem. **97**, 235 (1932).
205. Clayton, D. W., Kenner, G. W., Sheppard, R. C.: J. Chem. Soc. **1956**, 371.
206. Cleland, W. W.: Biochemistry (U.S.A.) **3**, 480 (1964).
207. Cole, R. D., Stein, W. H., Moore, S.: J. Biol. Chem. **233**, 1359 (1958).
208. Collman, J. P., Buckingham, D. A.: J. Am. Chem. Soc. **85**, 3039 (1963).
209. — Kimura, R.: J. Am. Chem. Soc. **89**, 6096 (1967).
210. Consden, R.: Nature **162**, 1007 (1948).
211. — Gordon, A. H.: Biochem. J. **46**, 8 (1950).
212. — — Martin, A. J. P.: Biochem. J. **38**, 224 (1944).
213. — — — Synge, R. L. M.: Biochem. J. **41**, 596 (1947).
214. Cope, A. C., Towe, P. H.: J. Am. Chem. Soc. **71**, 3423 (1949).
215. Costopanagiotis, A. A., Hanford, B. O., Weinstein, B.: J. Org. Chem. **33**, 1261 (1968).
216. Crammer, J. L., Neuberger, A.: Biochem. J. **37**, 302 (1943).
217. Crestfield, A. M.: Anal. Chem. **28**, 117 (1956).
218. — Moore, S., Stein, W. H.: J. Biol. Chem. **238**, 622 (1963).
219. — Stein, W. H., Moore, S.: Arch. Biochem. Biophys. Suppl. **1**, 217 (1962).
220. Criddle, R. S., Edwards, D. L., Petersen, T. G.: Biochemistry (U.S.A.) **5**, 578 (1966).
221. Cruickshank, P. A., Sheehan, J. C.: Anal. Chem. **36**, 1191 (1964).
222. Curtius, T.: Ber. deut. chem. Ges. **35**, 3226 (1902).
223. — Goebel, F.: J. prakt. Chem. **37**, 150 (1888).
224. Dakin, H. D.: J. Chem. Soc. **107**, 434 (1915).
225. Dautrevau, M., Biserte, G.: Compt. rend. **240**, 1153 (1955).
226. Davidson, B. E., Sajgo, M., Noller, H. F., Harris, J. I.: Nature **216**, 1181 (1967).
227. Davis, N. C., Smith, E. L.: J. Biol. Chem. **223**, 935 (1956).
228. — — J. Biol. Chem. **224**, 261 (1957).
229. Deal, W. C., Rutter, W. J., van Holde, K. E.: Biochemistry (U.S.A.) **2**, 246 (1963).
230. Deer, A., Fried, J. H., Halpern, B.: Australian J. Chem. **20**, 797 (1967).
231. — Angew. Chem. Int. Ed. Engl. **5**, 1041 (1966).
232. Dekker, C. A., Fruton, J. S.: In: Methods in medical research (Gerard, R. W., Ed.), Vol. 3, p. 272. Chicago: Year Book Pub., Inc. 1950.
233. — Stone, D., Fruton, J. S.: J. Biol. Chem. **181**, 719 (1949).
234. — Taylor, S. P., Fruton, J. S.: J. Biol. Chem. **180**, 155 (1949).
235. deLange, R. J., Smith, E. L., Fambrough, D. M., Bonner, J.: Proc. Natl. Acad. Sci. (US) **61**, 1145 (1968).
236. Delpierre, G. R., Fruton, J. S.: Proc. Natl. Acad. Sci. (US) **56**, 1817 (1966).
237. deMarco, C., Graziani, M. T., Mosti, R.: Anal. Biochem. **15**, 40 (1966).
238. Denkewalter, R. G., Schwam, H., Strachan, R. G., Beesley, T. E., Veber, D. F., Schoenewaldt, E. F., Barkemeyer, H., Paleveda, Jr., W. J., Jacob, T. A., Hirschmann, R.: J. Am. Chem. Soc. **88**, 3163 (1966).

239. Desnuelle, P., Casal, A.: Biochim. Biophys. Acta **2**, 64 (1948).
240. Determann, H., Torff, H.-J., Zipp, O.: Ann. Chem., Justus Liebigs **670**, 141 (1963).
241. Dewey, R. S., Schoenewaldt, E. F., Joshua, H., Paleveda, Jr., W. J., Swam, H., Barkemeyer, H., Arison, B. H., Veber, D., F. Denkewalter, R. G., Hirschmann, R.: J. Am. Chem. Soc. **90**, 3254 (1968).
242. Deyl, Z., Rosmus, J.: J. Chromatog. **20**, 514 (1965).
243. Dickman, S. R., Crockett, A. L.: J. Biol. Chem. **220**, 957 (1956).
244. — Westcott, W. L.: J. Biol. Chem. **210**, 481 (1954).
245. Dijkstra, A., Billiet, H. A., van Doninck, A. H., van Velthuysen, H., Maat, L., Beyerman, H. C.: Rec. trav. chim. **86**, 65 (1967).
246. Dixon, G. H., Wardlaw, A. C.: Nature **188**, 721 (1960).
247. Djerassi, C., Undheim, K., Sheppard, R. C., Terry, W. G., Sjoberg, B.: Acta Chem. Scand. **15**, 903 (1961).
248. Dohan, J. S., Woodward, G. E.: J. Biol. Chem. **129**, 393 (1939).
249. Doolittle, R. F.: Biochem. J. **94**, 742 (1965).
250. — Armentrout, R. W.: Biochemistry (U.S.A.) **7**, 516 (1968).
251. Dopheide, T. A. A., Trikojus, V. M.: Nature **201**, 1128 (1964).
252. Dus, K., Lindroth, S., Pabst, R., Smith, R. M.: Anal. Biochem. **18**, 532 (1967).
253. du Vigneaud, V., Bartlett, M. F., Jöhl, A.: J. Am. Chem. Soc. **79**, 5572 (1957).
254. — Behrens, O. K.: J. Biol. Chem. **117**, 27 (1937).
255. — Dorfmann, R., Loring, H. S.: J. Biol. Chem. **98**, 577 (1932).
256. — Gish, D. T., Katsoyannis, P. G.: J. Am. Chem. Soc. **76**, 4751 (1954).
257. — Hunt, M.: J. Biol. Chem. **125**, 269 (1938).
258. — Ressler, C., Swan, J. M., Roberts, C. W., Katsoyannis, P. G.: J. Am. Chem. Soc. **76**, 3115 (1954).
259. — — Trippitt, S.: J. Biol. Chem. **205**, 949 (1953).
260. Easley, C. W.: Biochim. Biophys. Acta **107**, 386 (1965).
261. Edelhoch, H.: Biochemistry (U.S.A.) **6**, 1948 (1967).
262. Edman, P.: Arch. Biochem. Biophys. **22**, 475 (1949).
263. — Acta Chem. Scand. **4**, 277 (1950).
264. — Acta Chem. Scand. **4**, 283 (1950).
265. — Nature **177**, 667 (1956).
266. — Acta Chem. Scand. **10**, 761 (1956).
267. — Proc. Roy. Austral. Chem. Inst. **1957**, 434.
268. — Acta Chem. Scand. **7**, 700 (1953).
269. — Ann. N.Y. Acad. Sci. **88**, 602 (1960).
270. — Begg, G.: Europ. J. Biochemistry **1**, 80 (1967).
271. — Lauber, K.: Acta Chem. Scand. **10**, 466 (1956).
272. — Sjöquist, J.: Acta Chem. Scand. **10**, 1507 (1956).
273. Edmundson, A. B.: Nature **198**, 354 (1963).
274. — Nature **205**, 883 (1965).
275. Edward, J. T., Nielsen, S.: Chem. & Ind. (London) **1953**, 197.
276. Edwards, B. G.: Arch. Biochem. Biophys. **21**, 103 (1949).
277. Eisen, H. M., Belman, S., Carsten, M. E.: J. Am. Chem. Soc. **75**, 4583 (1953).
278. Elliot, D. F., Russel, D. W.: Biochem. J. **66**, 49p (1957).
279. Ellman, G. L.: Arch. Biochem. Biophys. **82**, 70 (1959).
280. Elmore, D. T., Smyth, J.: Proc. Chem. Soc. **1963**, 18.
281. — Toseland, P. A.: J. Chem. Soc. **1954**, 4533.
282. — — J. Chem. Soc. **1957**, 2460.
283. — — J. Chem. Soc. **1965**, 188.
284. Elving, P. J., Rosenthal, I., Cramer, M. K.: J. Am. Chem. Soc. **73**, 1717 (1951).
285. Emery, A. R., Gold, V.: J. Chem. Soc. **1950**, 1443.
286. — — J. Chem. Soc. **1950**, 1447.
287. — — J. Chem. Soc. **1950**, 1455.
288. Erlanger, B. F., Cohen, W.: J. Am. Chem. Soc. **85**, 348 (1963).
289. — Sachs, H., Brand, E. J.: J. Am. Chem. Soc. **76**, 1806 (1954).
290. Erspamer, V., Anastasi, A.: Experientia **18**, 58 (1962).

291. — — Bertaccini, G., Cei, J. N.: Experientia **20**, 489 (1966).
292. Ertingshausen, G., Adler, H. J., Reichler, A. S.: J. Chromatog. **42**, 355 (1969).
293. Eveleigh, J. W., Thomson, A. R.: Biochem. J. **99**, 49p (1966).
294. Fahrney, D., Gold, A. M.: J. Am. Chem. Soc. **85**, 997 (1963).
295. Farnworth, A. J.: Biochem. J. **60**, 626 (1955).
296. Farrington, J. A., Kenner, G. W., Turner, J. M.: Chem. & Ind. (London) **1955**, 601.
297. Feigl, F.: In: Spot tests in organic analysis, p. 228. Amsterdam, London, New York: Elsevier Publ. Co. 1956.
298. Fischer, E.: Z. physiol. Chem., Hoppe-Seylers' **33**, 151 (1901).
299. — Ber. deut. chem. Ges. **36**, 2982 (1903).
300. — Ber. deut. chem. Ges. **39**, 2893 (1906).
301. — Ber. **39**, 530 (1906).
302. — Lipschitz, W.: Chem. Ber. **48**, 360 (1915).
303. Fischer, R. F., Whetstone, R. R.: J. Am. Chem. Soc. **76**, 5076 (1954).
304. — — J. Am. Chem. Soc. **77**, 750 (1955).
305. Flecher, J. C.: Biochem. J. **102**, 815 (1967).
306. Fleš, D., Markovac-Prpić, A., Tomašić, V.: J. Am. Chem. Soc. **80**, 4654 (1958).
307. Fletcher, A. P., Marshall, R. D., Neuberger, A.: Biochem. J. **88**, 37p (1963).
308. Folin, O., Ciocalteu, V.: J. Biol. Chem. **73**, 627 (1927).
309. Folk, J. E., Gladner, J. A.: J. Biol. Chem. **231**, 379 (1958).
310. — Piez, K. A., Carroll, W. R., Gladner, J. A.: J. Biol. Chem. **235**, 2272 (1960).
311. — Schirmer, E. W.: J. Biol. Chem. **238**, 3884 (1963).
312. Foltman, B., Hartley, B. S.: Biochem. J. **104**, 1064 (1967).
313. Fontana, A.: Unpuplished results.
314. — Marchiori, F., Moroder, L.: Ricerca sci. **36**, 261 (1966).
315. — — — Scoffone, E.: Tetrahedron **26**, 2985 (1966).
316. — Scoffone, E., Benassi, C. A.: Biochemistry (U.S.A.) **7**, 980 (1968).
317. — Veronese, F. M., Scoffone, E.: Biochemistry (U.S.A.) **7**, 3901 (1968).
318. Fowden, L.: Nature **167**, 1030 (1951).
319. Fox, S. W., Hayakawa, T., Harada, K.: Bull. Chem. Soc. Japan **36**, 1050 (1963).
320. — Hurst, T. L., Griffith, J. F., Underwood, O.: J. Biol. Chem. **77**, 3118 (1955).
321. Fraenkel-Conrat, H., Harris, J. I., Levy, A. L.: In: Methods of biochemical analysis Glick, D., Ed.), Vol. 2, p. 359. New York: Interscience 1955.
322. — Mahammad, A., Ducay, E. D., Mecham, D. K.: J. Am. Chem. Soc. **73**, 625 (1951).
323. — Simpson, M. E., Evans, H. M.: J. Biol. Chem. **142**, 107 (1942).
324. — Singer, B.: J. Am. Chem. Soc. **76**, 180 (1954).
325. — Tsung, C. M.: In: Methods in enzymology (Hirs, C. H. W., Ed.), Vol. XI, p. 151. New York: Academic Press 1967.
326. Frater, R., Light, A., Smith, E. L.: J. Biol. Chem. **240**, 253 (1965).
327. Friedmann, E.: Biochim. Biophys. Acta **9**, 65 (1952).
328. — Marrian, D. H., Simon-Reuss, I.: Brit. J. Pharmacol. **4**, 105 (1949).
329. Frikin, M., Patchornik, A., Katchalski, E.: J. Am. Chem. Soc. **88**, 3164 (1966).
330. Fromageot, C., Jutisz, M., Meyer, D., Penasse, L.: Biochim. et Biophys. Acta **6**, 283 (1950).
331. Fruton, J. S.: Advances in Protein Chem. **5**, 1 (1949).
332. — Bergmann, M.: J. Biol. Chem. **145**, 253 (1942).
333. — — Anslow, Jr., W. P.: J. Biol. Chem. **127**, 627 (1939).
334. — Hearn, W. R., Ingram, V. M., Wiggans, D. S., Winitz, M.: J. Biol. Chem. **204**, 891 (1953).
335. Fujita, H.: In: Mathematical theory of sedimentation analysis. NewYork: Academic Press 1962.
336. — Williams, J. W.: J. Phys. Chem. **70**, 309 (1966).
337. Gaetjens, E., Barany, M.: Biochim. Biophys. Acta **117**, 176 (1966).
338. Galiazzo, G. Jori, G., Scoffone, E.: Biochem. Biophys. Res. Commun. **31**, 158 (1968).
339. Gehrke, C. W.: Personal communications.
340. — Lamkin, W. M., Stalling, D. L., Shahrokhi, F.: Biochem. Biophys. Res. Commun. **19**, 328 (1965).

341. — Sharokhi, F.: Anal. Biochem. **15**, 97 (1966).
342. — Stalling, D. L.: Separation Science **2**, 101 (1967).
343. Gergely, E., Iredale, T.: J. Chem. Soc. **1951**, 3502.
344. Gerwin, B. I., Stein, W. H., Moore, S.: J. Biol. Chem. **241**, 3331 (1966).
345. Gilbert, G. A., Jenkins, R. C.: Proc. Roy. Soc. **A 253**, 420 (1959).
346. Gillessen, D., Schnabel, E., Meienhofer, J.: Ann. Chem., Justus Liebigs **667**, 164 (1963).
347. Gish, D. T., Carpenter, F. H.: J. Am. Chem. Soc. **75**, 5872 (1953).
348. — du Vigneaud, V.: J. Am. Chem. Soc. **79**, 3579 (1957).
349. — Katsoyannis, P. G., Hess, G. P., Stedman, R. J.: J. Am. Chem. Soc. **78**, 5954 (1956).
350. Gladner, J. A., Folk, J. E.: J. Biol. Chem. **231**, 393 (1958).
351. Glatthard, T., Matter, M.: Helv. Chim. Acta **46**, 795 (1963).
352. Goddard, D. R., Michaelis, L.: J. Biol. Chem. **112**, 361 (1935).
353. Goldberger, R. F., Anfinsen, C. B.: Biochemistry (U.S.A.) **1**, 401 (1962).
354. Goldstein, J., Konigsbreg, M., Hill, R. J.: J. Biol. Chem. **238**, 2016 (1963).
355. Goldstone, A. D., Smith, E. L.: J. Biol. Chem. **241**, 4480 (1966).
356. — — J. Biol.Chem. **242**, 4702 (1967).
357. Goodman, M., Levine, L.: J. Am. Chem. Soc. **86**, 2918 (1964).
358. — Stueben, K. C.: J. Am. Chem. Soc. **81**, 3980 (1959).
359. — — J. Am. Chem. Soc. **84**, 1279 (1962).
360. Goodwin, T. W., Morton, R. A.: Biochem. J. **40**, 628 (1946).
361. Gordon, A. H., Keil, B., Sebesta, K., Knesse, O., Šorm, F.: Collection Czechoslov. Chem. Commun. **15**, 1 (1950).
362. — Martin, A. J. P., Synge, R. L. M.: Biochem. J. **35**, 1369 (1941).
363. Gornall, A. G., Bordowill, C. J., David, M. M.: J. Biol. Chem. **177**, 751 (1949).
364. Graff, J., Wein, P. J., Winitz, M.: Federation Proc. **22**, 244 (1963).
365. Grannis, G. F.: Arch. Biochem. Biophys. **91**, 255 (1960).
366. Grassmann, W., Hörmann, H.: Z. physiol. Chem., Hoppe-Seyler's **292**, 24 (1953).
367. — Wünsch, E., Deufel, P., Zwick, A.: Chem. Ber. **91**, 538 (1958).
368. Gray, W. R.: In: Methods in enzymology (Hirs, C. H. W., Ed.), Vol. XI, p. 139. New York: Academic Press 1967.
369. — Hartley, B. S.: Biochem. J. **89**, 379 (1963).
370. Green, F. C., Kay, L. M.: Anal. Chem. **24**, 726 (1952).
371. Green, M., Stahmann, M. A.: J. Biol. Chem. **197**, 771 (1952).
372. Greenstein, J. P., Winitz, M.: In: Chemistry of amino acids, Vol. 2, p. 887. New York: Wiley 1961.
373. Gregory, J. D.: J. Am. Chem. Soc. **77**, 3922 (1955).
374. Gregory, H., Hardy, P. M., Jones, D. S., Kenner, G. W., Sheppard, R. C.: Nature **204**, 931 (1964).
375. Groskopf W. R., Holleman, J. W., Margoliash, E., Klotz, I. M.: Biochemistry (U.S.A.) **5**, 3783 (1966).
376. Gross, D.: J. Chromatog. **5**, 194 (1961).
377. Gross, E., Morell, J. L.: J. Biol. Chem. **241**, 3638 (1966).
378. — Witkop, B.: J. Am. Chem. Soc. **83**, 1510 (1961).
379. — — J. Biol. Chem. **237**, 1856 (1962).
380. Guest, J. R., Yanofsky, C.: J. Biol. Chem. **241**, 1 (1966).
381. Guidotti, G.: Biochim. Biophys. Acta **42**, 177 (1960).
382. — Hill, R. J., Konigsberg, W.: J. Biol. Chem. **237**, 2184 (1962).
383. — Konigsberg, W.: J. Biol. Chem. **239**, 1474 (1964).
384. Gurd, F. R. N., Wilcox, P. E.: Advances in Protein Chem. **11**, 311 (1956).
385. Gurin, S., Clarke, H. T.: J. Biol. Chem. **107**, 395 (1934).
386. Gussin, G. N., Capecchi, M. R., Adams, J. M., Argetsinger, J. E., Weber, K., Watson, J. D.: Cold Spring Harbor Symposia Quant. Biol. **36**, 257 (1966).
387. Guttmann, St.: Helv. Chim. Acta **44**, 721 (1961).
388. — Helv. Chim. Acta **49**, 83 (1966).
389. — Boissonnas, R. A.: Helv. Chim. Acta **41**, 1852 (1958).

390. — — Helv. Chim. Acta **42**, 1257 (1959).
391. — — Helv. Chim.Acta **44**, 1713 (1961).
392. — PLESS, J., BOISSONNAS, R. A.: Helv. Chim. Acta **45**, 170 (1962).
393. HAARMAN, W.: Biochem. Z.: **314**, 1 (1943).
394. HABEEB, A. F. S. A., CASSIDY, A. H., SINGER, S. J.: Biochim. Biophys. Acta **29**, 587 (1958).
395. HAGEN, P., BLACK, W.: Federation Proc. **23**, 371 (1960).
396. HALEY, E., CORCORAN, B. J.: Biochemistry (U.S.A.) **6**, 2668 (1967).
397. HALPERN, B., JAMES, L. B.: Nature **202**, 592 (1964).
398. — Australian J. Chem. **18**, 417 (1965).
399. — CROSS, A. D.: Chem. & Ind. (London) **1965**, 1183.
400. — NITECKI, D. E.: Tetrahedron Letrs. **1967**, 3031.
401. HALSEY, Y. D., NEURATH, H.: J. Biol. Chem. **217**, 247 (1955).
402. HAMILTON, L. D. G.: In: Analytical methods of protein chemistry (ALEXANDER, P., BLOCK, R. J., Eds.), Vol. 2, p. 61. New York: Pergamon Press 1960.
403. HAMILTON, P. B.: Anal. Chem. **30**, 914 (1958).
404. — Anal. Chem. **35**, 2055 (1963).
405. HANBY, W. E., WALEY, S. G., WATSON, J.: J. Chem. Soc. **1950**, 3239.
406. HANFORD, B. O., HYLTON, T. A., PRESTON, J., WEINSTEIN, B.: J. Org. Chem. **32**, 1243 (1967).
407. — JONES, J. H., YOUNG, G. T., JOHNSON, T. F. N.: J. Chem. Soc. **1965**, 6814.
408. HANGAARD, N.: J. Biol. Chem. **164**, 265 (1946).
409. HARMAN, R. E., PATTERSON, J. L., VAN DEN HEUVEL, W. J. A.: Anal. Biochem. **25**, 452 (1968).
410. HARRINGTON, C. R., MEAD, T. H.: Biochem. J. **30**, 1598 (1936).
411. — PITT-RIVERS, R. V.: Biochem. J. **38**, 417 (1944).
412. HARRINGTON, W. F., SCHELLMAN, J. A.: Compt. rend. trav. lab. Carlsberg. Sér. chim. **30**, 21 (1956).
413. HARRIS, C. K., TIGANE, E., HANES, C. S.: Can. J. Biochem. Physiol. **39**, 439 (1961.)
414. HARRIS, J. I.: Nature **177**, 471 (1956).
415. — Biochem. J. **71**, 451 (1959).
416. — FRUTON, J. S.: J. Biol. Chem. **191**, 143 (1951).
417. — HINDLEY, J.: J. Mol. Biol. **3**, 117 (1961).
418. — KNIGHT, C. A.: Nature **170**, 613 (1952).
419. — LERNER, A. B.: Nature **179**, 1346 (1957).
420. — LI, C. H.: J. Am. Chem. Soc. **76**, 2607 (1954).
421. — — J. Biol. Chem. **213**, 499 (1955).
422. — PERHAM, R. N.: 7th Int. Congr. Biochem., Tokyo, Aug. 1967, Symposium I.
423. HARRISON, P. M., HOFFMAN, T.: Biochem. J. **80**, 389 (1961).
424. HARSHMAN, S., NAJJAR, V. A.: Biochemistry (U.S.A.) **4**, 2526 (1965).
425. HARTLEY, B. S. KAUFFMAN, D. L.: Biochem. J. **101**, 229 (1966).
426. — MASSEY, V.: Biochim. Biophys. Acta **21**, 58 (1956).
427. HAUPTSCHEIN, M., STOKES, C. S., NODIFF, E. A.: J. Am. Chem. Soc. **74**, 4005 (1952).
428. HEFFTER, B.: Med. Naturw. Arch. **1**, 81 (1907).
429. — Chem. Zentr. **1907**, 822.
430. HELFERICH, B., BÖSHAGEN, H.: Chem. Ber. **92**, 2813 (1959).
431. HELLER, J., SMITH, E. L.: J. Biol. Chem. **241**, 3165 (1966).
432. HELLERMAN, L.: Physiol. Revs. **17**, 454 (1937).
433. — CHINARD, F. P., DEITZ, V. R.: J. Biol. Chem. **147**, 443 (1943).
434. HENNING, U., HELINSKI, D. R., CHAO, F. C., YANOFSKY, C.: J. Biol. Chem. **237**, 1523 (1962).
435a. HEYNES, K., LEGLER, G.: Z. physiol. Chem., Hoppe-Seyler's **321**, 161 (1960).
435b. HILL, R. L.: Advances in Protein Chem. **20**, 37 (1965).
436a. — KONIGSBERG, W.: J. Biol. Chem. **237**, 3151 (1962).
436b. — SCHMIDT, W. R.: J. Biol. Chem. **237**, 389 (1962).
437a. — SMITH, E. L.: J. Biol. Chem. **228**, 577 (1957).
437b. — SPACKMAN, D. H., BROWN, D. M., SMITH, E. L.: Biochem. Preparations **6**, 35 (1958).

438. HILLMANN-ELIES, HILLMANN, G., JATZKEWITZ, H.: Z. Naturforsch. **8**, 445 (1953).
439. HILSCHMANN, N., CRAIG, L. C.: Proc. Natl. Acad. Sci. (US) **53**, 1403 (1965).
440. HILSE, K., SORGER, V., BRAUNITZER, G.: Z. physiol. Chem., Hoppe-Seyler's **344**, 466 (1966).
441a. HIRS, C. H. W.: J. Biol. Chem. **219**, 611 (1956).
441b. — Ann. Rev. Biochem. **33**, 597 (1964).
442. — In: Methods in enzymology, Vol. XI. New York: Academic Press 1967.
443. — In: Automation in analytical chemistry, Technicon Symposia (1966) and ALONZO, N., HIRS, C. H. W.: Anal. Biochem. **23**, 272 (1968).
444. — MOORE, S., STEIN, W. H.: J. Biol. Chem. **195**, 669 (1952).
445a. — — — J. Biol. Chem. **219**, 623 (1956).
445b. — STEIN, W. H., MOORE, S.: J. Biol. Chem. **221**, 151 (1956).
445c. — MOORE, S., STEIN, W. H.: J. Biol. Chem. **235**, 633 (1960).
446a. HIRSCHMANN, R., STRACHAN, R. G., SCHWAN, H., SCHOENEWALDT, E. F., JOSHUA, H., BARKEMEYER, B., VEBER, D. F., PALEVEDA, W. J., JACOB, T. A., BEESLEY, T. E., DENKEWALTER, R. G.: J. Org. Chem. **32**, 3415 (1967).
446b. HISKEY, R. G., ADAMS, JR., J. B.: J. Am. Chem. Soc. **87**, 3969 (1965).
447a. — — J. Org. Chem. **31**, 2178 (1966).
447b. — MIZOGUCHI, T., INUI, T.: J. Org. Chem. **31**, 1192 (1966).
447c. — — SMITHWICK, JR., E. L.: J. Org. Chem. **32**, 97 (1967).
447d. — STAPLES, J. T., SMITH, R. L.: J. Org. Chem. **32**, 2772 (1967).
448a. — HARPOLD, M. A.: J. Org. Chem. **33**, 559 (1968).
448b. — HARPP, D. N.: J. Am. Chem. Soc. **87**, 3965 (1965).
448c. — SMITH, R. L.: J. Am. Chem. Soc. **90**, 2677 (1968).
448d. — TUCKER, W. P.: J. Am. Chem. Soc. **84**, 4789 (1962).
448e. HITCH, S. F.: Ph. D., Thesis, London 1956.
448f. HOERNLE, S.: Z. physiol. Chem., Hoppe-Seyler's **348**, 1355 (1967).
448g. HOFMAN, K., FIMM, F. M., LIMETTI, M., MONTIBELLER, J., ZANETTI, G.: J. Am. Chem. Soc. **88**, 3633 (1966).
449. — HAAS, W., SMITHERS, M. J., WELLS, R. D., WOLMAN, Y., YANAIHARA, N., ZANETTI, G.: J. Am. Chem. Soc. **87**, 620 (1965).
450. HOFMAN, K., PECHHAM, W. D., RHEINER, A.: J. Am. Chem. Soc. **78**, 238 (1956).
451. — WOOLNER, M. E., SPÜHLER, G., SCHWARTZ, E. T.: J. Amer. Chem. Soc. **80**, 1486 (1958).
452. — YAJIMA, H., YANAIHARA, N., LUI, T., LANDE, S.: J. Am. Chem. Soc. **83**, 487 (1961).
453. — KAPPELER, H., FURLENMEIER, A. E., WOOLNER, M. E., SCHWARTZ, E. T., THOMPSON, T. A.: J. Am. Chem. Soc. **79**, 1641 (1057).
454. — LINDENMANN, A., MAGEE, M. Z., KAHN, N. H.: J. Am. Chem. Soc. **74**, 470 (1952).
455. — MAGEE, M. Z., LINDENMANN, A.: J. Am. Chem. Soc. **72**, 7814 (1960).
456. — YANAIHARA, N., LANDE, S., YAJIMA, H.: J. Am. Chem. Soc. **84**, 4470 (1962).
457. HOFMANN, T.: Biochemistry (U.S.A.) **3**, 356 (1964).
458. HOLCOMB, G. N., JAMES, S. A., WARD, D. N.: Biochemistry (U.S.A.) **7**, 1291 (1968).
459. HOLLEY, R. W., HOLLEY, A. D.: J. Am. Chem. Soc. **74**, 5445 (1952).
460. HOLMQUIST, W. R., SCHROEDER, W. A.: Biochemistry (U.S.A.) **5**, 2489 (1966).
461. HONZL, J., RUDINGER, I.: Collection Czechoslov. Chem. Commun. **20**, 1190 (1955).
462. — — Collection Czechoslov. Chem. Commun. **26**, 2333 (1961).
463. HOOD, L., GREY, W. R., DREYER, W. J.: J. Mol. Biol. **22**, 179 (1966).
464. HOOPER, K. C., RYDON, H. N., SCHOFIELD, J. A., HEATON, G. S.: J. Chem. Soc. **1956**, 3148
465. HORINISHI, H., HACHIMORI, Y., KURIHARA, K., SHIBATA, K.: Biochim. et Biophys. Acta **86**, 477 (1964).
466. HORMANN, H., JOSEPH, K. TH.: Biochim. Biophys. Acta **100**, 598 (1965).
467. HORTON, H. R., KOSHLAND, JR., D. E.: J. Am. Chem. Soc. **87**, 1126 (1965).
468. HUNT, J. A., INGRAM, V. M.: Biochim. Biophys. Acta **49**, 520 (1961).
469. HUNTER, I. R., DIMICK, K. R., CORSE, J. W.: Chem. & Ind. (London) **1956**, 294.
470. IKENAKA, T., BAMMERLIN, H., KAUFMANN, H., SCHMID, K.: J. Biol. Chem. **241**, 5560 (1966).
471. — SCHMID, K.: Nature **215**, 66 (1967).

472. ILSE, D., EDMAN, P.: Australian J. Chem. **16**, 411 (1963).
473. INGRAM, V. M.: J. Chem. Soc. **1953**, 3717.
474a. — Nature **166**, 1038 (1950); — J. Biol. Chem. **202**, 193 (1953).
474b. — Biochim. Biophys. Acta **20**, 577 (1956).
475. — Biochim. Biophys. Acta **28**, 539 (1958).
476. INOUYE, K., FRUTON, J. S.: Biochemistry (U.S.A.) **6**, 1765 (1967).
477. — OTSUKA, H.: Bull. Chem. Soc. Japan **34**, 1 (1961).
478. — VOYNICK, I. M., DELPIERRE, G. R., FRUTON, J. S.: Biochemistry (U.S.A.) **5**, 2473 (1966).
479. INUKAI, N., NAKANO, K., MURAKAMI, M.: Bull Chem. Soc. Japan **40**, 2913 (1967).
480. ISELIN, B.: Arch. Biochem. Biophys. **78**, 532 (1958).
481. — FEURER, M., SCHWYZER, R.: Helv. Chim. Acta **38**, 1508 (1955).
482. — RITTEL, W., SIEBER, P., SCHWYZER, R.: Helv. Chim. Acta **40**, 373 (1957).
483. — SCHWYZER, R.: Helv. Chim. Acta **39**, 57 (1956).
484. — — Helv. Chim. Acta **44**, 169 (1961).
485. — — Helv. Chim. Acta **45**, 1499 (1962).
486. ISHIDA, Y., ONISHI, M.: Tokushima Daigaku Yakugokubu Kenkyû Nempo **14**, 5 (1965) (Engl.); — Chem. Abs. **68**, 13371a, (1968).
487. ITANO, H. A., GOTTLIEB, A. J.: Biochem. Biophys. Res. Commun, **12**, 405 (1963).
488. IZUMI, Y.: Anal. Biochem. **12**, 1 (1965).
489. JACOBSEN, C. J., LEONIS, J., LINDERSTROM-LANG, K., OTTESSEN, M.: In, Methods of biochemical analysis (GLICK, D., Ed.), Vol. IV, p. 171. New York: Wiley-Interscience 1955.
490. JAKUBKE, H. D.: Z. Chem. **6**, 52 (1966).
491. — VOIGT, A.: Chem. Ber. **99**, 2419 (1966).
492. — — Chem. Ber. **99**, 2944 (1966).
493. JEPPSSON, J., SJÖQUIST, J.: Anal. Biochem. **18**, 264 (1967).
494. JEPSON, J. B., SMITH, I.: Nature **172**, 1100 (1953).
495. JESCHKEIT, H.: Z. Chem. **8**, 20 (1968).
496. JOHN, W. D., YOUNG, G. T.: J. Chem. Soc. **1954**, 2870.
497. JOHNSON, D. E., SCOTT, S. S., MEISTER, A.: Abstracts of papers, American Chemical Society, Chicago, 1961, 48C.
498. — — — Anal. Chem. **33**, 669 (1961).
499. JOHNSON, T. B., NICOLET, B. H.: J. Am. Chem. Soc. **33**, 1973 (1911).
500. JOLLES, P., FROMAGEOT, C.: Biochim. Biophys. Acta 9, 287 (1952).
501. — — Biochim. Biophys. Acta **9**, 416 (1952).
502. — THAUREAUX, J., FROMAGEOT, C.: Arch. Biochem. Biophys. **69**, 290 (1957).
503. JONES, J. H., YOUNG, G. T.: J. Chem. Soc. **1968**, 53.
504. — LIBEREK, B., YOUNG, G. T.: J. Chem. Soc. **1967**, 2371.
505. JOST, K., RUDINGER, J.: Collection Czechoslov. Chem. Commun. **26**, 2345 (1961).
506. JULIAN, P. L., MEYER, E. W., PRINTY, H. C.: In: Heterocyclic compounds (ELDERFIELD, R. C., Ed.), Vol. 3, p. 84. New York: John Wiley and Sons, Inc. 1952.
507. JUTISZ, M., RITSCHARD, W.: Biochim. Biophys. Acta **17**, 548 (1955).
508. KADER, A. T., STIRLING, C. J. M.: J. Chem. Soc. **1964**, 258.
509. KAGAMIYAMA, H., WATANABE, T., WADA, H.: Biochem. Biophys. Res. Commun. **32**, 678 (1968).
510. KAISER, E., MAXWELL, L. C., LANDMANN, W. A., HUBATA, R.: Arch. Biochem. Biophys. **42**, 94 (1953).
511. KAPOOR, A.: J. Pharm. Sci. **56**, 1532 (1967).
512. KAPPELER, H.: Helv. Chim. Acta **44**, 476 (1961).
513. — SCHWYZER, R.: Helv. Chim. Acta. **43**, 1453 (1960).
514. — — Helv. Chim. Acta **44**, 1136 (1961).
515. KASPER, C. B., MATSUBARA, H., SMITH, E. L.: J. Biol. Chem. **240**, 1131 (1965).
516. — SMITH, E. L.: J. Biol. Chem. **241**, 3754 (1966).
517. KATCHALSKI, E.: Advances in Protein Chem. **6**, 123 (1951).
518. KATO, H., IWANAGA, S., SUZUKI, T.: Experientia **22**, 49 (1966).
519. KATSOYANNIS, P. G.: J. Am. Chem. Soc. **83**, 4053 (1961).

520. — du Vigneaud, V.: J. Am. Chem. Soc. **76**, 3113 (1954).
521. — Gish, D. T., Hess, G. P., du Vigneaud, V.: J. Am. Chem. Soc. **80**, 2558 (1958).
522. — — du Vigneaud, V.: J. Am. Chem. Soc. **79**, 4516 (1957).
523. — Suzuki, K.: J. Am. Chem. Soc. **83**, 4057 (1961).
524. — Tometsko, A. M., Ginos, J. Z., Tilak, M. A.: J. Am. Chem. Soc. **88**, 164 (1966).
525. — — Zalut, C.: J. Am. Chem. Soc. **88**, 166 (1966).
526. Katsuki, S., Scott, J. E., Yamashina, J.: Biochem. J. **97**, 25c (1965).
527. Katz, A. M., Dreyer, W. J., Anfinsen, C. B.: J. Biol. Chem. **234**, 2897 (1959).
528. Katz, J. J.: Nature **174**, 509 (1954).
529. Kawahara, K., Tanford, C.: Biochemistry (U.S.A.) **5**, 1578 (1966).
530. Kawanishi, Y., Iwai, K., Ando, T.: J. Biochem. (Tokyo) **56**, 314 (1964).
531. Kelley, R. B.: J. Org. Chem. **28**, 453 (1963).
532. Kemp, D. S., Chien, S. W.: J. Am. Chem. Soc. **89**, 2743 (1967).
533. — Woodward, R. B.: Tetrahedron **21**, 3019 (1965).
534. Kenner, G. W., Khorana, H. G., Stedman, R. J.: J. Chem. Soc. **1953**, 673.
535. — — J. Chem. Soc. **1952**, 2076.
536. — Laird, A. H.: Chem. Commun. **1965**, 305.
537. — — Stedman, R. J.: J. Chem. Soc. **1953**, 673.
538. Kessler, W., Iselin, B.: Helv. Chim. Acta **49**, 1330 (1966).
539. Keyser, J. W.: Biochem. J. **43**, 488 (1948).
540. Khorana, H. G.: Chem. Rev. **53**, 145 (1953).
541. — J. Chem. Soc. **1952**, 2081.
542. Khosla, M. C., Smeby, R. R., Bumpus, F. M.: Biochemistry (U.S.A.) **6**, 754 (1967).
543. Kienhuis, H., van de Linde, A., van der Holst, J. P. J., Verweij, A.: Rec. trav. chim. **80**, 1278 (1961).
544. Kimmel, J. R., Kato, G. K., Paiva, A. C. M., Smith, E. L.: J. Biol. Chem. **237**, 2525 (1962).
545. — Smith, E. L.: J. Biol. Chem. **215**, 55 (1955).
546. — Thompson, E. O. P., Smith, E. L.: J. Biol. Chem. **217**, 151 (1955).
547. King, F. E., Clark-Lewis, J. W., Wade, R.: J. Chem. Soc. **1957**, 886.
548. — Kidd, D. A. A.: J. Chem. Soc. **1949**, 3315.
549. — Clark-Lewis, J. W., Kidd, D. A. A., Smith, G. R.: J. Chem. Soc. **1954**, 1039.
550. Kirchner, J. C.: In: Thin layer chromatography (Dekker, M., Ed.), p. 307. New York: Interscience Publ. 1967.
551. Kisfaludy, L., Dualszky, S.: Acta Chim. Acad. Sci. Hung. **24**, 301 (1960).
552. — Acta. Chim. Acad. Sci. Hung. **24**, 309 (1960).
553. — Dualszky, S., Medzihradszky, K., Bajusz, S., Bruckner, V.: Acta Chim. Acad. Sci. Hung. **30**. 473 (1962).
554. Kjaer, A., Ericksen, P.: Acta Chem. Scand. **6**, 448 (1952).
555. Klebe, J. F., Finkbeiner, H., White, D. M.: J. Am. Chem. Soc. **88**, 3390 (1966).
556. Klotz, I. M., Keresztes-Nagy, S.: Biochemistry (U.S.A.) **2**, 445 (1963).
557. Koch, G., Weidel, W.: Z. physiol. Chem., Hoppe-Seyler's **303**, 213 (1956).
558. Koenig, W., Geiger, R., Siedel, W.: Chem. Ber. **101**, 681 (1968).
559. Kohn, J.: Clin. Chim. Acta **2**, 297 (1957).
560. Kolthoff, I. M., Lingane, J. J.: In: Polarography, 2nd ed. New York: Interscience 1952.
561. — Stricks, W.:Anal. Chem. **23**, 763 (1951).
562. — — J. Am. Chem. Soc. **72**, 1952 (1950).
563. — Harris, W. E.: Ind. Eng. Chem., Anal. Ed. **18**, 161 (1946).
564. Komatsu, S.: Chem. Zentralblatt **II**, 537 (1911).
565. Konigsberg, W.: In: Methods in enzymology (Colowick, S.P., Kaplan, N. O., Eds.), Vol. XI, p. 461. New York and London: Academic Press 1967.
566. — Goldstein, J., Hill, R. J.: J. Biol. Chem. **238**, 2028 (1963).
567. — Hill, R. J.: J. Biol. Chem. **237**, 2547 (1962).
568. — — J. Biol. Chem. **237**, 3157 (1962).
569. — Weber, K., Notani, G., Zinder, N.: J. Biol. Chem. **241**, 2579 (1966).
570. Kopple, K. D.: In: Peptides and amino acids. New York: W. A. Benjamin, Inc. 1966.

571. — Bächli, E.: J. Org. Chem. **24**, 2053 (1959).
572. Korman, S., Clarke, H. T.: J. Biol. Chem. **221**, 113, 133 (1956).
573. Koshland, Jr., D. E., Karkhanis, Y. D., Latham, H. G.: J. Am. Chem. Soc. **86**, 1448 (1964).
574. — Strumeyer, D. H., Ray, Jr., W. J.: Brookhaven Symposia in Biol. **15**, 101 (1962).
575. Kossel, A., Kutscher, F.: Z. physiol. Chem., Hoppe-Seyler's **31**, 165 (1901).
576. Kostka, V., Carpenter, F. H.: J. Biol. Chem. **239**, 1799 (1964).
577. Kotaki, A., Haroda, M., Yagi, K.: J. Biochem. (Tokyo) **55**, 553 (1964).
578. Kovacs, J., Ceprini, M. Q., Dupraz, C. A., Schmit, G. N.: J. Org. Chem. **32**, 3696 (1967).
579. — Gianotti, R., Kapoor, A.: J. Am. Chem. Soc. **88**, 2282 (1966).
580. — Rodin, R. L.: J. Org. Chem. **33**, 2418 (1968).
581. Kreil, G., Kreil-Kiss, G.: Biochem. Biophys. Res. Commun. **27**, 275 (1967).
582. — Tuppy, H.: Nature **192**, 1121 (1961).
583. Krichevsky, M. I., Schwarz, J., Mage, M.: Anal. Biochem. **12**, 94 (1965).
584. Kubowitz, F., Ott, P.: Biochem. Z. **314**, 94 (1943).
585. Kung Yueh-Ting: Sci. Sinica (Peking) **15**, 544 (1966).
586. Kunkel, H. G., Tiselius, A.: J. Gen. Physiol. **35**, 89 (1951).
587. Kupryszewski, G.: Roczniki Chem. **35**, 595 (1961).
588. — Formela, M.: Roczniki Chem. **35**, 1533 (1961).
589. — Roczniki Chem. **37**, 593 (1963).
590. Kurihara, K., Horinishi, H., Shibata, K.: Biochim. Biophys. Acta **74**, 678 (1963).
591. Kusama, K.: J. Biochem. (Tokyo) **44**, 375 (1957).
592. Lamkin, W. M., Gehrke, C. W.: Anal. Chem. **37**, 383 (1965).
593. Landowne, R. A., Lipsky, C. W.: Nature **199**, 141 (1963).
594. Landmann, W. A., Drake, M. P., Dillaha, J.: J. Amer. Chem. Soc. **75**, 3638 (1953).
595. Laufer, D. A., Blout, E. R.: J. Am. Chem. Soc. **89**, 1246 (1967).
596. — Chapman, T. H., Marlborough, D. I., Vaidya, V. M., Blout, E. R.: J. Am. Chem. Soc. **90**, 2696 (1968).
597. Laursen, R. A.: J. Am. Chem. Soc. **88**, 5344 (1966).
598. Laver, W. G.: Biochim. Biophys. Acta **53**, 469 (1961).
599. Law, H. D.: Ann. Repts Progr. Chem. (Chem. Soc. London) **63**, 517 (1966).
600. Lederer, E., Lederer, M.: In: Chromatography, 2nd ed. Amsterdam: Elsevier 1957.
601. Leonis, J., Levy, A. L.: Compt. rend. trav. lab. Carlsberg **29**, 57 (1954).
602. — — Compt. rend. trav. lab. Carlsberg **29**, 87 (1954).
603. — Li, C. H., Chung, D.: J. Am. Chem. Soc. **81**, 419 (1959).
604. Leplawy, M. T., Jones, D. S., Kenner, G. W., Sheppard, R. C.: Tetrahedron Letrs. (**11**), 39 (1960).
605. Leslie, J., Williams, D. L., Gorin, G.: Anal. Biochem. **3**, 257 (1962).
606. Levin, Y., Berger, A., Katchalski, R.: Biochem. J. **63**, 308 (1956).
607. Levy, A. L.: J. Chem. Soc. **1950**, 404.
608. — Biochim. Biophys. Acta **15**, 589 (1954).
609. — Nature **174**, 126 (1954).
610. — Chung, D.: Anal. Chem. **25**, 396 (1953).
611. — — Biochim. Biophys. Acta **17**, 454 (1955).
612. Levy, M.: In: Methods in enzymology (Colowick, S. P., Kaplan, N. O., Eds.), Vol. IV, p.238. New York: Academic Press 1957.
613. Li, C. H., Bertsch, L.: J. Biol. Chem. **235**, 2638 (1960).
614. — Cummings, J. T.: J. Biol. Chem. **233**, 73 (1958).
615. — Geschwind, I. I., Cole, R. D., Raacke, I. D., Harris, J. I., Dixon, J. S.: Nature **176**, 687 (1955).
616. — Ramachandran, J., Chung, D., Gorup, B.: J. Am. Chem. Soc. **86**, 2703 (1964).
617. Liefländer, M.: Z. physiol. Chem., Hoppe-Seyler's **320**, 35 (1960).
618. Liener, L. E.: J. Biol. Chem. **225**, 1061 (1957).
619. Light, A. L.: In: Methods in enzymology (Hirs, C. H., Ed.), Vol. XI, p. 417. New York: Academic Press 1967.
620. — Greenberg, J.: J. Biol. Chem. **240**, 258 (1965).

621. — Smith, E. L.: In: The proteins, 2nd ed. (Neurath, H., Ed.), Vol. 1, p. 1. New York: Academic Press 1963.
622. — — J. Biol.Chem. **237**, 2537 (1962).
623. Lindley, H.: J. Am. Chem. Soc. **77**, 4927 (1955).
624. — Nature **178**, 647 (1956).
625. Liu, T-Y., Stein, W. H., Moore, S., Elliot, S. D.: J. Biol. Chem. 240, 1143 (1965).
626. Losse, G., Hoffmann, K.-H., Hetzer, G.: Ann. Chem., Justus Liebigs **684**, 236 (1965).
627. — Müller, G.: Chem. Ber. **94**, 2768 (1961).
628. Lowry, D. H., Rosebrough, N. A., Farr, A. L., Randall, R. J.: J. Biol. Chem. **193**, 265 (1951).
629. Lowther, A. G.: Nature **167**, 767 (1951).
630. Ludowieg, J., Dorfman, A.: Biochim. Biophys. Acta **38**, 212 (1960).
631. Lugg, J. W. H.: Biochem. J. **26**, 2144 (1932).
632. Lutz, W. B., Ressler, C., Nettleton, Jr., D. E., du Vigneaud, V.: J. Am. Chem. Soc. **81**, 167 (1959).
633. Mackworth, J. F.: Biochem. J. **42**, 82 (1948).
634. MacLaren, J. A.: Australian J. Chem. **11**, 360 (1958).
635. — Savige, W. E., Swan, J. M.: Australian J. Chem. **11**, 345 (1958).
636. Mahowald, T.: Biochemistry (U.S.A.) **4**, 732 (1965).
637. Malkin, R., Rabinowitz, J. C.: Biochemistry (U.S.A.) **6**, 3880 (1967).
638. Mann, T., Leone, E.: Biochem. J. **53**, 140 (1953).
639. Marglin, A., Merrifield, R. B.: J. Am. Chem. Soc. **88**, 5051 (1966).
640. Margoliash, E.: J. Biol. Chem. **237**, 2161 (1962).
641. — Smith, E. L.: J. Biol. Chem. **237**, 2151 (1962).
642. — — Kreil, G., Tuppy, H.: Nature **192**, 1125 (1961).
643. Markland, F. S., Kreil, G., Ribadeau-Dumas, B., Smith, E. L.: J. Biol. Chem. **241**, 4642 (1966).
644. — Ribadeau-Dumas, B., Smith, E. L.: J. Biol. Chem. **242**, 5174 (1967).
645. — Smith, E. L.: J. Biol. Chem. **242**, 5198 (1967).
646. Markus, G., Grossberg, A. L., Pressman, D.: Arch. Biochem. Biophys. **96**, 63 (1962).
647. — Karush, F.: J. Am. Chem. Soc. **79**, 134 (1956).
648. Marrian, D. H.: J. Chem. Soc. **1949**, 1515.
649. Marrig, C., Luccioni, F., Laurent, B.: Biochim. Biophys. Acta **105**, 606 (1965).
650. Marshall, G. R., Merrifield, R. B.: Biochemistry (U.S.A.) **4**, 2394 (1965).
651. Mason, E. E.: In: Computer applications in Medicine. Springfield (Ill.): C. C. Thomas Co. 1966.
652. Matsubara, H., Sasaki, R. M., Chain, R. K.: Proc. Natl. Acad. Sci. (US) **57**, 439 (1967).
653. — Smith, E. L.: J. Biol. Chem. **241**, 2732 (1963).
654. Matsuo, H.: Protein Nucleic Acid Enzyme **14**, 523, 589 (1969).
655. — Fujimoto, Y., Tatsuno, T.: Tetrahedron Letrs. (**39**), 3465 (1965).
656. — — — Biochem. Biophys. Res. Commun. **22**, 69 (1966).
657. — — — Chem. Pharm. Bull. (Tokyo) **15**, 716 (1967).
658. — Matsubara, H.: Proc. Soc. Exptl Biol. Med. **129**, 564 (1968).
659. Mayo, F. R., Walling, C.: Chem. Revs **27**, 351 (1940).
660. Mazur, R. H.: J. Org. Chem. **28**, 2498 (1963).
661. — Schlatter, J. M.: U.S. Patent 3,280,099 (1966).
662. McCaldin, D. J.: Chem. Revs **60**, 39 (1960).
663. McDowall, M. A., Smith, E. L.: J. Biol. Chem. **240**, 281 (1965).
664. McGregor, W. H., Carpenter, F. H.: J. Org. Chem. **26**, 1849 (1961).
665. McKay, F. C., Albertson, N. F.: J. Am. Chem. Soc. **79**, 4686 (1957).
666. McPhee, J. R.: Biochem. J. **64**, 22 (1956).
667. Mechanic, G. L., Levy, M.: J. Am. Chem. Soc. **81**, 1889 (1959).
668. Medzihradszky, K., Schweiger, H. M.: Acta Chim. Acad. Sci. Hung. **30**, 105 (1962).
669. Meienhofer, J.: Chimia (Switz) **16**, 385 (1962).
670. — Nature **205**, 73 (1965).
671. — Schnabel, E., Bremer, H., Brinkhoff, O., Zabel, R., Sroka, W., Klostermeyer, H., Brandenburg, D., Okuda, T., Zahn, H.: Z. Naturforscher **18**, 1120 (1963).

672. Melville, J.: Biochem. J. **29**, 179 (1935).
673. Merigan, T. C., Dreyer, W. J., Berger, A.: Biochim. Biophys. Acta **62**, 122 (1962).
674. Merrifield, R. B.: J. Am. Chem. Soc. **85**, 2194 (1963).
675. — J. Org. Chem. **29**, 3100 (1964).
676. — Biochemistry (U.S.A.) **3**, 1385 (1964).
677. — Science **150**, 178 (1965).
678. — Wooley, D. W.: J. Am. Chem. Soc. **78**, 4646 (1956).
679. — — J. Am. Chem. Soc. **80**, 6635 (1958).
680. — Stewart, J. M., Jernberg, N.: Anal. Chem. **38**, 1905 (1966).
681. Meites, L.: In: Polarographic techniques. New York: Interscience 1955.
682. Meselson, M., Stahl, F. W., Vinograd, J.: Proc. Natl. Acad. Sci. (US) **1959**, 581.
683. Michl, H.: Monatsh. Chem. **82**, 489 (1951).
684. — J. Chromatog. **1**, 93 (1958).
685. Mikeš, O., Holeyšovsky, V., Tomašek, V., Šorm, F.: Biochem. Biophys. Res. Commun. **24**, 346 (1966).
686. Miller, H. K., Waelsch, H.: Arch. Biochem. Biophys. **35**, 176 (1952).
687. Mills, G. L.: In: Chromatographic and electrophoretic techniques (Smith, I., Ed.), Vol. I., p. 143. London: William Heinemann 1960.
688. — Nature **165**, 403 (1950).
689. — Biochem. J. **50**, 707 (1952).
690. Milne, H. B., Halver, J. E., Ho, D. S., Mason, M. S.: J. Am. Chem. Soc. **79**, 637 (1957).
691. — Kilday, W. D.: J. Org. Chem. **30**, 64 (1965).
692. — Most, Jr., C. F.: J. Org. Chem. **33**, 169 (1968).
693. Milstein, C.: Biochem. J. **100**, 40c (1966).
694. — Sanger, F.: Biochem. J. **79**, 456 (1961).
695. Mirsky, A. E., Anson, M. L.: J. Gen. Physiol. **18**, 307 (1935).
696. Mizoguchi, T., Levin, G., Woolley, D. W., Stewart, J. M.: J. Org. Chem. **33**, 903 (1968).
697. Moore, A. T., Rydon, H. N., Smithers, M. J.: J. Chem. Soc. **1966**, 2349.
698. — Cole, R. D., Gundlach, H. G., Stein, W. H.: Proc. 4th Int. Congr. Biochem., Vienna 1958, Symposium **8**, 51 (1959).
699. — Stein, W. H.: J. Biol. Chem. **176**, 367 (1948).
700. — — J. Biol. Chem. **178**, 53, 79 (1949).
701. — — J. Biol. Chem. **192**, 663 (1951).
702. — — J. Biol. Chem. **211**, 907 (1954).
703. Morley, J. S.: J. Chem. Soc. **1967**, 2410.
704. Morris, A. J.: Biochem. J. **81**, 606 (1961).
705. Morse, D., Horecker, B. L.: Anal. Biochem. **14**, 429 (1966).
706. Mross, G. A., Doolittle, R. F.: Arch. Biochem. Biophys. **122**, 674 (1967).
707. Mueller, J. M., Pierce, J. G., Davoll, H., du Vigneaud, V: J. Biol. Chem. **191**, 309 (1951).
708. Mukaiyama, T., Ueki, M., Maruyama, H., Matsueda, R.: J. Am. Chem. Soc. **90**, 4490 (1968).
709. Munier, R. L., Sarrazin, G.: J. Chromatog. **22**, 347 (1966).
710. Muraoka, S., Takata, J.: Yakugaku Zasshi **77**, 1155 (1957) (in Japanese).
711. Murdock, A. L., Grist, K. L., Hirs, C. W. H.: Arch. Biochem. Biophys. **114**, 375 (1966).
712. Mutt, V., Magnusson, S., Jorpes, J. S., Dahl, E.: Biochemistry (U.S.A.) **4**, 2358 (1965).
713. Nakayima, S., Okuyama, G.: J. Pharm. Soc. Japan **76**, 620 (1956).
714. Narita, K.: Biochim. Biophys. Acta **28**, 184 (1958).
715. — Biochim. Biophys. Acta **30**, 352 (1958).
716. — Biochim. Biophys. Acta **31**, 372 (1959).
717. — Biochem. Biophys. Res. Commun. **5**, 160 (1961).
718. — Ishii, J.: J. Biochem. (Tokyo) **52**, 367 (1962).
719. — Murakami, H., Ikenaka, T.: J. Biochem. (Tokyo) **59**, 170 (1966).

720. — Ohta, Y.: Bull. Chem. Soc. Japan **32**, 1023 (1959).
721. Naughton, M. A., Sanger, F., Hartley, B. S., Shaw, D. C.: Biochem. J. **77**, 149 (1960).
722. Neadle, D. J., Pollitt, R. J.: Biochem. J. **97**, 607 (1965).
723. Needleman, S. B.: Atlas of protein sequences and structure. Nat'l. Biomed. Res. Fdn. **1968**, 86.
724. — Margoliash, E.: J. Biol. Chem. **241**, 853 (1966).
725. Nefkens, G. H. L.: Nature **193**, 974 (1962).
726. — Tesser, G. I.: J. Am. Chem. Soc. **83**, 1263 (1961).
727. — — Nivard, R. J. F.: Rec. trav. chim. **81**, 683 (1962).
728. Neuberger, A.: Advances in Protein Chem. **4**, 339 (1948).
729. — Gottschalk, A., Marshall, A. D.: In: Glycoproteins (Gottschalk, A., Ed.), p. 273. Amsterdam, London, New York: Elsevier Publ. Co. 1966.
730. — Sanger, F.: Biochem. J. **36**, 662 (1942).
731. Neuman, H., Goldberger, R. F., Sela, M.: J. Biol. Chem. **239**, 1536 (1964).
732. — Smith, R. A.: Arch. Biochem. Biophys. **122**, 354 (1967).
733. — Steinberg, J. Z.: Brown, J. R., Goldberger, R. F., Sela, M.: European J. Biochem. **3**, 171 (1967).
734. — — Katchalski, E.: J. Am. Chem. Soc. **87**, 3841 (1965).
735. Neuman, N. P., Lampen, J. O.: Biochemistry (U.S.A.) **6**, 468 (1968).
736. Neurath, H.: In: Methods in enzymology (Colowick, S. P., Kaplan, N. O., Eds.,) Vol. II, p. 77. New York: Acadenic press 1955.
737. — In: The enzymes (Boyer, P. D., Lardy, H., Myrback, K., Eds.), 2nd Ed., Vol. IV., p. 11. New York: Academic Press 1960.
738. Niall, H. D.: Personal communication.
739. Niall, H., Edman, P.: J. Gen. Physiol. **45** (Suppl.), 185 (1962).
740. Nicolaides, E. D., de Wald, H. A.: J. Org. Chem. **26**, 2872 (1961).
741. Niu, C. I., Fraenkel-Conrat, H.: J. Am. Chem. Soc. **77**, 5882 (1955).
742. Nolan, C., Margoliash, E.: J. Biol. Chem. **241**, 1049 (1966).
743. Noltman, E. A., Mahowold, T. S., Kuby, S. A.: J. Biol. Chem. **237**, 1146 (1962).
744. Northrop, J. H.: J. Gen. Physiol. **9**, 1217 (1927).
745. Nygaard, A. P.: Acta Chem. Scand. **10**, 397 (1956).
746. O'Donnell, I. J., Thompson, E. O. P.: Australian J. Biol. Sci. **17**, 271 (1964).
747. Oelofsen, W., Li, C. H.: J. Am. Chem. Soc. **88**, 4254 (1966).
748. Offer, G. W.: Biochim. Biophys. Acta **90**, 193 (1964).
749. Ohira, T.: Nippon Nôgei-kagaku Kaishi **15**, 370 (1939).
750. Ohno, M.: J. Biochem. (Tokyo) **41**, 345 (1954).
751. — Kato, T., Mitsuyasu, N., Waki, M., Makisumi, S., Izumiya, N.: Bull. Chem. Soc. Japan **40**, 204 (1967).
752. Okamoto, M., Kimoto, S., Oskima, T., Kinomura, Y., Kawasaki, K., Yajima, H.: Chem. Pharm. Bull. **15**, 1618 (1967).
753. Okawa, K.: Bull. Chem. Soc. Japan **29**, 486 (1956).
754. — Bull. Chem. Soc. Japan **30**, 976 (1957).
755. — Bull. Chem. Soc. Japan **30**, 110 (1957).
756. Okunuki, K., Hagihara, B., Matsubara, H., Nakayama, T.: J. Biochem. (Tokyo) **43**, 453 (1956).
757. Okuyama, T., Satake, K.: J. Biochem. (Tokyo) **47**, 454 (1960).
758. Olcott, H. S., Fraenkel-Conrat, A.: J. Biol. Chem. **171**, 583 (1947).
759. Olson, O. E., Kuby, S. A.: J. Biol. Chem. **239**, 460 (1964).
760. Otsuka, H., Inouye, K., Shinozaki, F., Kanayama, M.: J. Biochem. (Tokyo) **58**, 512 (1965).
761. — — Kanayama, M., Shinozaki, F.: Bull. Chem. Soc. Japan **39**, 882 (1966).
762. Overell, B. G., Petrow, J.: J. Chem. Soc. **1955**, 232.
763. Pacsu, E., Wilson, E. J.: J. Org. Chem. **7**, 117 (1942).
764. Palmer, G., Massey, V.: Biochim. Biophys. Acta **58**, 349 (1962).
765. Papastamatis, S. C., Williams, J. F.: Nature **167**, 724 (1951).
766. Parker, A. J., Kharasch, N.: J. Am. Chem. Soc. **82**, 3071 (1960).

767. Partridge, S. M., Davis, H. F.: Nature **165**, 62 (1950).
768. Pasternak, R., von Halban, H.: Helv. Chim. Acta **29**, 190 (1946).
769. Pataki, G.: Chimia (Switz) **18**, 24 (1964).
770. Patton, A. R., Chism, P.: Anal. Chem. **23**, 1683 (1951).
771. Patchornik, A., Lawson, W. B., Gross, E., Witkop, B.: J. Am. Chem. Soc. **82**, 5923 (1960).
772. — — Witkop, B.: J. Am. Chem. Soc. **80**, 4747 (1958).
773. — Wilchek, M., Sarid, S.: J. Am. Chem. Soc. **86**, 1457 (1964).
774. Paul, R.: J. Org. Chem. **28**, 236 (1963).
775. — Anderson, G. W.: J. Am. Chem. Soc. **82**, 4596 (1960).
776. — — J. Org. Chem. **27**, 2094 (1962).
777. — — Callahan, F. M.: J. Org. Chem. **26**, 3347 (1961).
778. Pechere, J. F., Dixon, G. H., Maybury, R. H., Neurath, H.: J. Biol. Chem. **233**, 1364 (1958).
779. Penhoet, E., Kochman, M., Valentine, R., Rutter, W. J.: Biochemistry (U.S.A.) **6**, 2940 (1967).
780. Perham, R., Appella, E., Potter, M.: Science **154**, 391 (1966).
781. — Jones, G. M. T.: European J. Biochem. **2**, 84 (1967).
782. Perrin, C. H.: Anal. Chem. **25**, 968 (1953).
783. Peterson, E. A., Sober, H. A.: Anal. Chem. **31**, 857 (1959).
784. — — J. Am. Chem. Soc. **78**, 751 (1956).
785. Phillips, D. M. F.: Biochem. J. **87**, 258 (1963).
786. — Biochem. J. **86**, 397 (1963).
787. — Stephen, J. M. L.: Nature **162**, 152 (1948).
788. Piez, K. A., Morris, L.: Anal. Biochem. **1**, 187 (1960).
789. Pisano, J. J., Bronzert, T. J.: Federation Proc. **28**, 661 (1969).
790. — van den Heuvel, W. J. A., Horning, E. C.: Biochem. Biophys. Res. Commun, **7**, 82 (1962).
791. Plapp, B. V., Raftery, M. A., Cole, R. D.: J. Biol. Chem. **242**, 265 (1967).
792. Pless, J., Boissonnas, R. A.: Helv. Chim. Acta **46**, 1609 (1963).
793. Popenoe, E. A., Lawler, H. C., du Vigneaud, V.: J. Am. Chem. Soc. **74**, 3713 (1952).
794. Porath, J.: Biochim. Biophys. Acta **39**, 193 (1960).
795. — Bennich, H.: Arch. Biochem. Biophys. Suppl. **1**, 152 (1962).
796. Porn, S. C., Dutcher, J. T.: Anal. Chem. **38**, 836 (1956).
797. Porter, R. R.: Biochem. J. **46**, 304 (1950).
798. — Sanger, F.: Biochem. J. **42**, 287 (1948).
799. Press, E. M., Piggott, P. J., Porter, R. R.: Biochem. J. **99**, 356 (1966).
800. Preston, J., Weinstein, B.: Experientia **23**, 265 (1967).
801. Previero, A., Colletti-Previero, M. A., Bull. soc. chim. biol. **49**, 1059 (1967).
802. — Coletti, M. A., Cavadore, J. C.: Biochim. Biophys. Acta **147**, 453 (1967).
803. — Scoffone, E., Pajetta, P., Benassi, C. A.: Gazz. chim. ital. **93**, 841 (1963).
804. Putnam, F. W.: In: The proteins (Neurath, H., Ed.), Part 1, Vol. 2, p. 893. New York: Academic Press 1953.
805. Raftery, M. A., Cole, R. D.: J. Biol. Chem. **241**, 3457 (1966).
806. — — Biochem. Biophys. Res. Commun. **10**, 467 (1963).
807. Rajagopalan, T. G., Moore, S., Stein, W. H.: J. Biol. Chem. **241**, 4940 (1966).
808. — Stein, W. H., Moore, S.: J. Biol. Chem. **241**, 4295 (1966).
809. Rajappa, S., Akerkar, A. S.: Chem. Commun. **22**, 826 (1966).
810. Ramachandran, J.: Nature **206**, 927 (1965).
811. — Epp, A., McConnell, W. B.: Anal. Chem. **27**, 1734 (1955).
812. — McConnell, W. B.: J. Am. Chem. Soc. **78**, 1255 (1956).
813. — Witkop, B.: J. Am. Chem. Soc. **81**, 4028 (1959).
814. Ray, Jr., W. J., Koshland, Jr., D. E.: Brookhaven Symposia in Biol. **13**, 135 (1960).
815. Raymond, S., Weintraub, L. S.: Science **130**, 711 (1959).
816. Redfield, R. R., Anfinsen, C. B.: J. Biol. Chem. **221**, 385 (1956).
817. — — J. Biol. Chem. **221**, 387 (1956).
818. Rees, M. W.: Biochem. J. **40**, 632 (1946).

819. — Short, M. N.: Virology **26**, 596 (1965).
820. Reith, W. S., Waldron, N. M.: Biochem. J. **56**, 116 (1954).
821. Ressler, C., du Vigneaud, V.: J. Am. Chem. Soc. **76**, 3107 (1954).
822. Richards, F. F., Barnes, W. T., Lovins, R. E., Salomone, R., Waterfield, M. D. Nature **221**, 1241 (1969).
823. Ried, W., Franz, G.: Ann. Chem., Justus Liebigs **644**, 141 (1961).
824. Riordan, J. F., Sokolovsky, M., Vallee, B. L.: J. Am. Chem. Soc. **88**, 4104 (1966).
825. — — — Biochemistry (U.S.A.) **6**, 358 (1967).
826. — Wacker, W. E. C., Vallee, B. L.: Biochemistry (U.S.A.) **4**, 1758 (1965).
827. Rittell, W.: Helv. Chim. Acta **45**, 2465 (1962).
828. Rittenberg, D., Shemin, D.: Ann. N. Y. Acad. Sci. **47**, 119 (1946).
829. Roberts, E., Rouser, G.: Anal. Chem. **30**, 1291 (1958).
830. Rocchi, R., Marchiori, F., Scoffone, E.: Gazz. chim. ital. **93**, 823 (1963).
831. Rockland, L. B., Blatt, J. L., Dunn, M. S.: Anal. Chem. **23**, 1142 (1951).
832. — Underwood, J. C.: Anal. Chem. **26**, 1557 (1954).
833. Roeske, R. W.: Chem. & Ind. **1959**, 1121.
834. Ross, D. L., Skinner, C. G., Shive, W.: J. Med. Pharm. Chem. **3**, 519 (1961).
835. Ross, A. F., Stanley, W. M.: J. Gen. Physiol. **22**, 165 (1938).
836. Rothe, M., Kunitz, F.-W.: Ann. Chem., Justus Liebigs **609**, 88 (1957).
837. Rovery, M., Fabre, C., Desnuelle, P.: Biochim. Biophys. Acta **12**, 547 (1953).
838. Rowlands, D. A., Young, G. T.: Biochem. J. **65**, 516 (1957).
839. Rudinger, J., Honzl, M., Zaoral: Collection Czechoslov. Chem. Commun. **21**, 202 (1956).
840. Ruhlmann, K., Grescke, W.: Angew. Chem. **73**, 113 (1961).
841. Russell, D. W.: Biochem. J. **87**, 1 (1963).
842. Ruttenberg, M. A., King, T. P., Craig, L. C.: Biochemistry (U.S.A.) **3**, 758 (1964).
843. Rydon, H. N., Smith, P. W. G.: Nature **169**, 922 (1952).
844. Ryle, A. P., Sanger, F.: Biochem. J. **60**, 535 (1955).
845. — — Smith, L. F., Kitai, R.: Biochem. J. **60**, 541 (1955).
846. Sachs, H., Brand, E.: J. Am. Chem. Soc. **75**, 4610 (1953).
847. Sakaguchi, S.: J. Biochem. (Tokyo) **5**, 25, 133 (1925).
848. Sakakibara, S., Inukai, N.: Bull. Chem. Soc. Japan **38**, 1979 (1965).
849. — Itoh, M.: Bull. Chem. Soc. Japan **40**, 656 (1967).
850. — Nobuhara, Y.: Japanese Patent 2710 (1967); cf. Chem. Abs. **68**, 13402 (1968).
851. — Shimonishi, Y.: Bull. Chem. Soc. Japan **38**, 1412 (1965).
852. Sakami, W., Toennis, G.: J. Biol. Chem. **144**, 203 (1942).
853. Sakiyama, F., Okawa, K., Yamakawa, T., Akabori, S.: Bull. Chem. Soc. Japan **31**, 926 (1958).
854. Salander, R. C., Piano, M., Patton, A. R.: Anal. Chem. **25**, 1252 (1953).
855. Sanger, F.: Biochem. J. **39**, 507 (1945).
856. — Advances in Protein Chem. **7**, 1 (1952).
857. — Biochem. J. **44**, 126 (1949).
858. — Thompson, E. O. P.: Biochim. Biophys. Acta **9**, 225 (1952).
859. — — Biochem. J. **53**, 353 (1953).
860. — — Biochem. J. **53**, 366 (1955).
861. — — Biochim. Biophys. Acta **71**, 468 (1963).
862. — Tuppy, H.: Biochem. J. **49**, 463 (1951).
863. — Lecture at Northwestern University, Chicago 1957.
864. Sarges, R., Witkop, B.: J. Am. Chem. Soc. **86**, 1862 (1963).
865. — — Biochemistry (U.S.A.) **4**, 2491 (1965).
866. Saroff, H. A., Karmen, A.: Anal. Biochem. **1**, 344 (1960).
867. — — Healy, S. A.: J. Chromatog. **9**, 122 (1962).
868. Satake, K., Okuyama, T., Ohashi, M., Shinoda, T.: J. Biochem. (Tokyo) **47**, 654 (1960).
869. — Sasakawa, S., Maruyama, T.: J. Biochem. (Tokyo) **53**, 516 (1963).
870. — Take, T., Matsuo, A., Tazaki, K., Hiraga, Y.: J. Biochem. (Tokyo) **60**, 2 (1966).
871. — Tanaka, M., Shino, H.: J. Biochem. (Tokyo) **50**, 6 (1961).

872. Savige, W. E.: Australian J. Chem. **14**, 664 (1961).
873. Scanes, F. S., Tozer, B. T.: Biochem. J. **63**, 282, 565 (1956).
874. Schachman, H. K.: In Ultracentrifugation in biochemistry. New York: Academic Press 1959.
875. — Edelstein, S. J.: Biochemistry (U.S.A.) **5**, 2681 (1966).
876. Schaffer, N. K., Lang, R. P., Simet, L., Drisko, R. W.: J. Biol. Chem. **230**, 185 (1958).
877. — Simet, L., Harshman, S., Engle, R. R., Drisko, R. W.: J. Biol. Chem. **225**, 197 (1957).
878. Schellenberger, A., Jeschkeit, H., Henkel, R., Lehmann, H.: Z. Chem. **7**, 191 (1967).
879. Schelling, V.: J. Biol. Chem. **96**, 17 (1932).
880. Schlack, P., Kumpf, W.: Z. physiol. Chem., Hoppe Seyler's **151**, 125 (1926).
881. Schlögl, K., Wessely, F., Woidich, H.: Monatsh. Chem. **87**, 425 (1956).
882. Schluck, P., Kumpf, W.: Z. physiol. Chem., Hoppe Seyler's **154**, 125 (1926).
883. Schlüssler, H., Zahn, H.: Chem. Ber. **95**, 1076 (1962).
884. Schnabel, E.: Ann. Chem., Justus Liebigs **659**, 168 (1962).
885. — Ann. Chem., Justus Liebigs **702**, 188 (1967).
886. — Zahn, H.: Ann. Chem., Justus Liebigs **614**, 141 (1958).
887. Schneider, F.: Z. physiol. Chem., Hoppe Seyler's **321**, 38 (1960).
888. Schoellman, N. G., Shaw, E.: Biochemistry (U.S.A.) **2**, 252 (1963).
889. Schönheimer, R.: Z. physiol. Chem., Hoppe Seyler's **154**, 203 (1926).
890. Schott, H. F., Larkin, J. B., Rockland, L. B., Dunn, M. S.: J. Org. Chem. **12**, 490 (1947).
891. Schramm, M.: Biochemistry (U.S.A.) **3**, 1231 (1964).
892. — Moore, S., Bigwood, E. J.: Biochem. J. **57**, 33 (1954).
893. Schröder, E.: Ann. Chem., Justus Liebigs **670**, 127 (1963).
894. — Ann. Chem., Justus Liebigs **681**, 241 (1965).
895. — Ann. Chem., Justus Liebigs **688**, 250 (1965).
896. — Klieger, E.: Ann. Chem., Justus Liebigs **673**, 196 (1964).
897. — — Ann. Chem., Justus Liebigs **673**, 208 (1964).
898. — Lübke, K.: In: The peptides, Vol. I and Vol. II. New York: Academic Press 1965.
899. — — **655**, 211 (1962).
900. — In: Peptides. Proc. European Peptide Symp., 8th, p. 245. Noordwijk, Neth., 1966.
901. Schroeder, W. A.: J. Am. Chem. Soc. **74**, 5118 (1952).
902. — In: Methods in enzymology (Colowick, S. P., Kaplan, N. O., Eds.), Vol. XI, p. 445. New York and London: Academic Press Inc. 1967.
903. — Cua, J. T., Matsuda, G., Fenninger, W. D.: Biochim. Biophys. Acta **63**, 532 (1963).
904. — Holmquist, W. R., Shelton, J. R.: Anal. Chem. **38**, 1281 (1966).
905. — Kay, L. M., le Gette, J., Honnen, L., Green, F. C.: J. Am. Chem. Soc. **76**, 3556 (1954).
906. — Jones, R. T., Cormick, J., McCalla, K.: Anal. Chem. **34**, 1570 (1962).
907. — Robberson, B.: Anal. Chem. **37**, 1583 (1965).
908. — Shelton, J. R., Robberson, B.: Biochim. Biophys. Acta **147**, 590 (1967).
909. — — Shelton, J. B., Cormick, J., Jones, R. T.: Biochemistry (U.S.A.) **2**, 992 (1963).
910. Schultz, J., Allison, H., Grice, M.: Biochemistry (U.S.A.) **1** 694 (1962).
911. Schwarz, H., Arakawa, K.: J. Am. Chem. Soc. **81**, 5691 (1959).
912. Schwyzer, R.: In: Peptides. Proc. European Peptide Symp., 8th, p. 1. Noordwijk, Neth., 1966.
913. — Helv. Chim. Acta **36**, 414 (1953).
914. — Helv. Chim. Acta **37**, 647 (1954).
915. — Helv. Chim. Acta **37**, 674 (1954).
916a. — Aung, Tun-Kyi: Helv. Chim. Acta **45**, 859 (1962).
916b. — Costopanagiotes, A., Sieber, P.: Chimia (Switz) **16**, 295 (1962).
917. — Dietrich, H.: Helv. Chim. Acta **44**, 2003 (1961).

918. — Feurer, M., Iselin, B., Kagi, H.: Helv. Chim. Acta **38**, 80 (1955).
919. — Iselin, B., Feurer, M.: Helv. Chim. Acta **38**, 69 (1955).
920. — — Kappeler, H., Riniker, B., Rittel, W., Zuber, H.: Helv. Chim. Acta **41**, 1273 (1958).
921. — — — — — — Helv. Chim. Acta **46**, 1975 (1963).
922. — — Rittel, W., Sieber, P.: Helv. Chim. Acta **39**, 872 (1956).
923. — Kappeler, H.: Helv. Chim. Acta **44**, 1991 (1961).
924. — Rittel, W.: Helv. Chim. Acta **44**, 159 (1961).
925. — — Costopanagiotis, A.: Helv. Chim. Acta **45**, 2473 (1962).
926. — — Kappeler, H., Iselin, B.: Angew. Chem. **72**, 915 (1960).
927. — Sieber, P.: Helv. Chim. Acta **40**, 624 (1957).
928. — — Helv. Chim. Acta **42**, 972 (1959).
929. — — Helv. Chim. Acta **43**, 1910 (1960).
930. — — Helv. Chim. Acta **49**, 134 (1966).
931. — — Kappeler, H.: Helv. Chim. Acta **42**, 2622 (1959).
932. — — Zatsko, K.: Helv. Chim. Acta **41**, 491 (1958).
933. — Surbeck-Wegmann, E., Dietrich, H.: Chimia (Switz) **14**, 366 (1960).
934. Scoffone, E., Fontana, A., Rocchi, R.: Biochem. Biophys. Res. Commun. **25**, 170 (1966).
935. — — — Biochemistry (U.S.A.) **7**, 971 (1968).
936. — Vianello, E., Lorenzini, A.: Gazz. chim. ital. **87**, 354 (1957).
937. Seiler, N., Wiechman, J.: Experientia **20**, 559 (1964).
938. Seki, T.: J. Biochem. (Tokyo) **47**, 253 (1960).
939. Sela, M., White, Jr., F. H., Anfinsen, C. B.: Science **125**, 691 (1957).
940. Shaw, J. L. D., McFarlane, W. D.: Can. J. Research **16B**, 361 (1938).
941. Sheehan, J. C., Chapman, D. W., Roth, R. W.: J. Am. Chem. Soc. **74**, 3822 (1952).
942. — Cruickshank, P. A., Boshart, G. L.: J. Org. Chem. **26**, 2525 (1961).
943. — Frank, V. S.: J. Am. Chem. Soc. **71**, 1856 (1949).
944. — Goodman, M., Hess, G. P.: J. Am. Chem. Soc. **78**, 1367 (1956).
945. — Hess, G. P.: J. Am. Chem. Soc. **77**, 1067 (1955).
946. — Hlavka, J. J.: J. Org. Chem. **21**, 439 (1956).
947. — Preston, J., Cruickshank, P. A.: J. Am. Chem. Soc. **87**, 2492 (1965).
948.— Yang, D. D. H.: J. Am. Chem. Soc. **80**, 1154 (1958).
949. Shields, J. E., McGregor, W. H., Carpenter, F. H.: J. Org. Chem. **26**, 1491 (1961).
950. Shinoda, T., Satake, K.: J. Biochem. (Tokyo) **50**, 293 (1961).
951. Sifferd, R. H., du Vigneaud, V.: J. Biol.Chem. **108**, 753 (1953).
952. Singer, T. P., Barron, E. S. G.: J. Biol. Chem. **157**, 241 (1945).
953. Sjöquist, J.: Acta Chem. Scand. **7**, 447 (1953).
954. — Biochim. Biophys. Acta **16**, 283 (1955).
955. — Arkiv Kemi **11**, 151 (1957).
956. — Biochim. Biophys. Acta **41**, 20 (1960).
957. — Arkiv Kemi **11**, 129 (1957).
958. Slobin, L. I., Carpenter, F. H.: Biochemistry (U.S.A.) **5**, 499 (1966).
959. Sluyterman, L. A.: Biochim. Biophys. Acta **25**, 402 (1957).
960. — Veenendaal, H. J.: Chem. Abstr. **44**, 4363 (1960).
961. Smillie, L. B., Hartley, B. S.: Biochem. J. **101**, 232 (1966).
962. — — Biochem. J. **105**, 1125 (1967).
963. Smith, E., Sheppard, Jr., H.: Nature **208**, 878 (1965).
964. Smith, E. L.: In: The enzymes (Summer, J. B., Myrbäck, K., Eds.), Vol. I, Part 2., p. 793. New York: Academic Press 1951.
965. — Hill, R. L.: In: The enzymes (Boyer, P. D., Lardy, H., Myrbäck, K., Eds.), Vol. IV., p. 37. New York: Academic Press 1960.
966. — Kimmel, J. R., Brown, D. M.: J. Biol. Chem. **207**, 533 (1954).
967. — Polglase, W. J.: J. Biol. Chem. **180**, 1209 (1949).
968. Smith, E. L., Spackman, D. H.: J. Biol. Chem. **212**, 271 (1955).
969. Smith, I.: Nature **169**, 60 (1952).
970. Smith, M. J. H., Taylor, K. W.: Biochem. J. **55**, xxx (1953).

971. Smithies, O.: Advances in Protein Chem. **14**, 65 (1959).
972. Smyth, D. G., Battaglia, F. C., Meschia, G.: J. Gen. Physiol. **44**, 889 (1961).
973. — Nagamatsu, A., Fruton, J. S.: J. Am. Chem. Soc. **82**, 4600 (1960).
974. — Stein, W. H., Moore, S.: J. Biol. Chem. **237**, 1845 (1962).
975. — — — J. Biol. Chem. **238**, 227 (1963).
976. Smyth, D. S., Utsumi, S.: Nature **216**, 332 (1967).
977. Snell, E. E.: Advances in Protein Chem. **2**, 85 (1945).
978. Snellman, K. O.: Netherlands Patent 6, 508, 512 (1965).
979. Sogami, M., Foster, J. F.: J. Biol. Chem. **237**, 2514 (1962).
980. Sokolovsky, M., Riordan, J. F., Vallee, B. L.: Biochemistry (U.S.A.) **5**, 3582 (1966).
981. — Sadeh, T., Patchornik, A.: J. Am. Chem. Soc. **86**, 1212 (1964).
982. — Vallee, B. L.: Biochemistry (U.S.A.) **5**, 3574 (1966).
983. — Wilchek, M., Patchornik, A.: J. Am. Chem. Soc. **86**, 1202 (1964).
984. Sondheimer, E., Holley, R. W.: J. Am. Chem. Soc. **76**, 2476 (1954).
985. Soru, E., Zaharia, O.: Arch. Biochem. Biophys. **119**, 498 (1967).
986. Spackman, D. H., Smith, E. L., Brown, D. M.: J. Biol. Chem. **212**, 255 (1955).
987. — Stein, W. H., Moore, S.: Anal. Chem. **30**, 1190 (1958).
988. Spande, T. F., Wilchek, M., Witkop, B.: J. Am. Chem. Soc. **90**, 3256 (1968).
989. Speck, J. C., Riehm, J. P., Evard, R.: Federation Proc. **20**, 220 (1961).
990. Spies, J. R., Chambers, D. C.: Anal. Chem. **21**, 1249 (1949).
991. Staab, H. A.: J. Am. Chem. Soc. **609**, 75 (1957).
992. Stacey, R. W., Waxman, B. D.: In: Computers in biochemical research. New York: Academic Press 1965.
993. Stalling, D. L.: Personal communication.
994. — Gille, G., Gehrke, C. W.: Anal. Biochem. **1967**, 18.
995. — Gehrke, C. W.: Biochem. Biophys. Res. Commun. **22**, 329 (1966).
996. Stark, G. R.: Biochemistry (U.S.A.) **7**, 1796 (1968).
997. — Federation Proc. **24**, 225 (1965).
998. — Stein, W. H., Moore, S.: J. Biol. Chem. **235**, 3177 (1960).
999. — Smyth, D. G.: J. Biol. Chem. **238**, 214 (1963).
1000. — In: Methods in enzymology (Hirs, C. H. W., Ed.), Vol. XI, p. 125. New York: Academic Press 1967.
1001. Steers, Jr., E., Craven, G. R., Anfinsen, C. B., Bethune, J. L.: J. Biol. Chem. **240**, 2478 (1965).
1002. Stegemann, H.: Z. physiol. Chem., Hoppe Seylers' **213**, 255 (1958).
1003. Stegnik, L. D., Vestling, C. S.: J. Biol. Chem. **241**, 4923 (1966).
1004. Stein, W. H., Moore, S., Bergmann, M.: J. Biol. Chem. **154**, 191 (1944).
1005. Stelakatos, G. C., Theodoropoulos, D. M., Zervas, L.: J. Am. Chem. Soc. **81**, 2884 (1959).
1006. Stern, R., de Luca, M., Mehler, A. H., McElroy, W. D.: Biochemistry (U.S.A.) **5**, 126 (1966).
1007. Stevens, F., Glazer, A. N., Smith, E. L.: J. Biol. Chem. **242**, 2764 (1967).
1008. Stewart, F. H. C.: Australian J. Chem. **19**, 1067 (1966).
1009. — Australian J. Chem. **19**, 1511 (1966).
1010. — Australian J. Chem. **19**, 2373 (1966).
1011. — Australian J. Chem. **20**, 365 (1967).
1012. — Australian J. Chem. **20**, 2243 (1967).
1013. — Australian J. Chem. **21**, 477 (1968).
1014. — Australian J. Chem. **20**, 1991 (1967).
1015. — Australian J. Chem. **21**, 1101 (1968).
1016. Stewart, J. M., Young, J. D., Benjamini, E., Shimizu, M., Leung, C. Y.: Biochemistry (U.S.A.) **5**, 3396 (1966).
1017. — Mizoguchi, T., Woolley, D. W.: J. Med. Chem. (in press).
1018. Stricks, W., Kolthoff, I. M.: J. Am. Chem. Soc. **73**, 4569 (1951).
1019. — Kolthoff, I. M., Kapoor, R. C.: J. Am. Chem. Soc. **77**, 2057 (1955).
1020. — — Tanaka, N.: Anal. Chem. **26**, 299 (1954).

1021. Studer, R. O., Lergier, W., Vogler, K.: Helv. Cheim. Acta **46**, 612 (1963).
1022. Suran, A. A.: Arch. Biochem. Biophys. **113**, 1 (1966).
1023. Svedberg, T., Pedersen, K. O.: In: The ultracentrifuge. Oxford: University Press 1940.
1024. Swallow, D. L., Abraham, E. P.: Biochem. J. **70**, 364 (1958).
1025. Swan, J. M.: Australian J. Sci. Research **A 5**, 711 (1952).
1026. — Nature **180**, 643 (1957).
1027. Swensen, A. D., Boyer, P. D.: J. Am. Chem. Soc. **79**, 2174 (1957).
1028. Synge, R. L. M.: Chem. Revs. **32**, 135 (1943).
1029. — Biochem. J. **39**, 351 (1945).
1030. Szent-Györgyi, A.: Arch. Biochem. Biophys. **42**, 305 (1953).
1031. Szewzuk, A., Mulczyk, M.: European J. Biochemistry **8**, 63 (1969).
1032. Tabone, J., Robert, D., Thomassey, S., Mamounas, N.: Bull. soc. chim. biol. **32**, 529 (1950).
1033. Takahashi, K.: J. Biol. Chem. **240**, PC4117 (1965).
1034. Tamura, Z., Nakajima, T.: Tanpakushitsu, Kakusan, Koso (Protein, Nucleic acid, Enzyme) **12**, 729 (in Japanese).
1035. Tanaka, M., Nakashima, T., Benson, A., Mower, H., Yasunobu, K. T.: Biochemistry (U.S.A.) **5**, 1666 (1966).
1036. Tanford, C.: In: Physical chemistry of macromolecules. New York: Wiley Publishing Co. 1961.
1037. — Kawahara, K., Lapanje, S.: J. Am. Chem. Soc. **89**, 729 (1967).
1038. Tang, J., Hartley, B. S.: Biochem. J. **102**, 593 (1967).
1039. Taniuchi, H., Anfinsen, C.: J. Biol. Chem. **241**, 4366 (1966).
1040. — — Sodja, A.: J. Biol. Chem. **242**, 4736 (1967).
1041. — — — J. Biol. Chem. **242**, 4752 (1967).
1042. Taschner, E., Biernat, J. F., Rzeszotarska, B., Wasielewski, Cz.: Ann. Chem., Justus Liebigs **646**, 123 (1961).
1043. — Chimiak, A., Bator, B., Sokolowska, T.: Ann. Chem., Justus Liebigs **646**, 134 (1961).
1044. — Liberek, B., Wasielewski, Cz., Biernat, J.: Angew. Chem. **71**, 743 (1959).
1045. — Wasielewski, Cz., Biernat, J. F.: Ann. Chem., Justus Liebigs **646**, 119 (1961).
1046. — — Sokolowska, T., Biernat, J. F.: Ann. Chem., Justus Liebigs **646**, 127 (1961).
1047. Tesser, G. I., Schwyzer, R.: Helv. Chem. Acta **49**, 1013 (1966).
1048. Theodoropoulos, D.: J. Org. Chem. **23**, 140 (1958).
1049. — Craig, L. C.: J. Org. Chem. **21**, 1379 (1956).
1050. Thompson, A. R.: Nature **169**, 495 (1952).
1051. — Biochem. J. **61**, 253 (1955).
1052. Thompson, E. O. P.: Biochim. Biophys. Acta **15**, 440 (1954).
1053. — O'Donnell, I. J.: Biochim. Biophys. Acta **53**, 447 (1961).
1054. Thomson, A. R.: In: 5th Colloquium in amino acid analysis, monograph 2, p. 116. Domont (France): Technicon International Div. 1967.
1055. — Gibbons, D. G., Eveleigh, J. W., Miles, B. J.: Biochem. J. **108**, 8p (1968).
1056. — Miles, B. J.: Nature **203**, 483 (1964).
1057. Tibbs, J.: Nature **168**, 910 (1951).
1058. Timasheff, S. N., Gorbunoff, M. J.: Ann. Rev. Biochem. **36**, 13 (1967).
1059. Tiselius, A.: Biochem. J. **31**, 1464 (1937).
1060. Titani, K., Narita, K., Okunuki, K.: J. Biochem. (Tokyo) **51**, 350 (1962).
1061. Toennies, G., Homiller, R. P.: J. Am. Chem. Soc. **64**, 3054 (1942).
1062. — Kolb, J. J.: Anal. Chem. **23**, 823 (1951).
1063. — — Nature **177**, 281 (1956).
1064. Toi, K., Bynum, E., Norris, E., Itano, H. A.: J. Biol. Chem. **242**, 1036 (1967).
1065. Tower, D. B., Peters, E. L., Wherrett, J. R.: J. Biol. Chem. **237**, 1861 (1962).
1066. Trautman, R.: In: Instrumental methods of experimental biology (Newman, D. W., Ed.), p. 211. New York: Mac Millan Co. 1964.
1067. Travis, J., McElroy, W. D.: Biochemistry (U.S.A.) **5**, 2170 (1966).
1068. Tristram, G. R.: Advances in Protein Chem. **5**, 83 (1949).

1069. Tsung, C. M., Fraenkel-Conrat, H.: Biochemistry (U.S.A.) **4**, 793 (1965).
1070. — — Biochemistry (U.S.A.) **5**, 2061 (1966).
1071. Turner, R. A., Schmerzler, G.: Biochim. Biophys. Acta **13**, 553 (1954).
1072. Udenfreind, S., Velick, S. F.: J. Biol. Chem. **190**, 733 (1951).
1073. Uliana, J. A., Doolittle, R. F.: Arch. Biochem. Biophys. **131**, 561 (1969).
1074. Vallee, B., Neurath, H.: J. Biol. Chem. **217**, 253 (1955).
1075. van Eys, J., Judd, J., Ford, J., Womak, W. B.: Biochemistry (U.S.A.) **3**, 1755 (1964).
1076. van Holde, K. E.: Fractions **1**, (1967).
1077. — Baldwin, R. L.: J. Phys. Chem. **62**, 734 (1958).
1078. van Orden, H. O., Smith, E. L.: J. Biol. Chem. **208**, 751 (1954).
1079. van Slyke, D. D.: J. Biol. Chem. **73**, 99 (1927).
1080. — J. Biol. Chem. **83**, 425 (1929).
1081. Vaughan, J. R.: J. Am. Chem. Soc. **73**, 3547 (1951).
1982. Vaughan, Jr., J. R., Osato, R. L.: J. Am. Chem. Soc. **74**, 676 (1952).
1083. — J. Am. Chem. Soc. **74**, 6137 (1952).
1084. — Eichler, J. A.: J. Am. Chem. Soc. **75**, 5556 (1953).
1085. Vaughan, J. R., Osato, R. L.: J. Am. Chem. Soc. **73**, 5553 (1951).
1086. Veber, D. F., Milkowski, J. D.: Denkewalter, R. G., Hirschmann, R.: Tetrahedron Letrs. (**26**), 3057 (1968).
1087. — Pfister, K., Hirschmann, R.: J. Med. Chem. **10**, 968 (1967).
1088. Velick, S. F., Udenfriend, S.: J. Biol. Chem. **191**, 233 (1951).
1089. Veronese, F. M.: Private communication.
1090. — Boccu, E., Fontana, A.: Gazz. chim. ital. (1968) (in press).
1091. Voisetnet, M. E.: Bull. soc. chim. **33**, 1198 (1905).
1092. Wachsmuth, E. D., Fritze, I., Pfleiderer, G.: Biochemistry (U.S.A.) **5**, 169 (1966).
1093. — — — Biochemistry (U.S.A.) **5**, 169, 175 (1966).
1094. Wagner, J., Winkler, G.: Z. anal. Chem. **183**, 1 (1961).
1095. Waldschmidt, E., Kuhn, K.: Chem. Ber. **84**, 381 (1951).
1096. Waley, S. G., Watson, J.: J. Chem. Soc. 2394. (1951).
1097. — — Biochem. J. **57**, 529 (1954).
1098. Wang, K. T., Wang, I. S. Y., Lin, A. L., Wang, C. S.: J. Chromatog. **26**, 323 (1967).
1099. Wang, S., Carpenter, F. H.: J. Biol. Chem. **240**, 1619 (1965).
1100. Ward, D. N., Coffey, J. A., Ray, D. B., Lamkin, W. M.: Anal. Biochem. **14**, 243 (1966).
1101. Wasielewski, C.: Roczniki Chem. **41**, 1915 (1967); cf. Chem. Abs. **68**, 87571n (1958).
1102. Weber, C. J.: J. Biol. Chem. **86**, 217 (1930).
1103. Weber, K., Watson, J. D.: Cold Spring Harbor Symposia Quant. Biol. **36**, 257 (1966).
1104. Weber, K. L., Konigsberg, W.: J. Biol. Chem. **242**, 3563 (1967).
1105. Weber, R. E., Engelhardt, D. L., Zinder, N. D.: Proc. Natl. Acad. Sci (US) **55**, 155 (1966).
1106. Weil, L., Seibles, T. S.: Arch. Biochem. Biophys. **95**, 470 (1961).
1107. — Telka, M.: Arch. Biochem. Biophys. **71**, 473 (1957).
1108. Weiss, M., Szobolew, N.: Biochem. Z. **58**, 119 (1913—14).
1109. Weitzel, G., Schneider, F., Fretzdorff, A.-M.: Z. physiol. Chem., Hoppe Seylers' **307**, 23 (1957).
1110. Wells, J. R. E.: Biochem. J. **97**, 228 (1965).
1111. — Anal. Biochem. **19**, 448 (1967).
1112. Weygand, F., Csendes, E.: Angew. Chem. **64**, 136 (1952).
1113. — Bull. soc. chim. biol. **43**, 1269 (1961).
1114. — Reiher, M.: Chem. Ber. **88**, 26 (1955).
1115. — Geiger, R.: Chem. Ber. **89**, 647 (1956).
1116. — Hoffman, D., Wunsch, E.: Z. Naturforsch. **21b**, 426 (1966).
1117. — Hunger, K.: Z. Naturforsch. **13b**, 50 (1958).
1118. — — Chem. Ber. **95**, 1 (1962).
1119. — Konig, W., Nintz, E., Hoffmann, D., Huber, P., Kahn, N., Prinz, W.: Z. Naturforsch. **21b**, 325 (1966).

1120. — Rinno, H.: Chem. Ber. **92**, 517 (1959).
1121. — Steglich, W.: Z. Naturforsch. **14b**, 472 (1959).
1122. — — Chem. Ber. **92**, 313 (1959).
1123. — — Pietta, P.: Tetrahedron Letrs. 3751, (**1966**).
1124. Whitaker, D. R., Strasdine, G. A.: Can. J. Biochem. Physiol. **41**, 1105 (1963).
1125. Whitaker, J. R.: Anal. Chem. **35**, 1950 (1963).
1126. White, Jr., F. H.: J. Biol. Chem. **235**, 383 (1960).
1127. White, J.: J. Biol. Chem. **106**, 141 (1934).
1128. Wieland, T., Birr, Ch.: Angew. Chem. (Int.) **5**, 310 (1966).
1129. — Heinke, B., Vogler, K., Morimoto, H.: Ann. Chem., Justus Liebigs **655**, 189 (1962).
1130. — Kern, W., Schring, R.: Ann. Chem., Justus Liebigs **569**, 117 (1950).
1131. — Schring, R.: Ann. Chem., Justus Liebigs **569**, 122 (1950).
1132. — Bernhard, H.: Ann. Chem., Justus Liebigs **572**, 190 (1951).
1133. — Schäfer, W., Bokelmann, E.: Ann. Chem., Justus Liebigs **573**, 99 (1951).
1134. — Schneider, G.: Ann. Chem., Justus Liebigs **580**, 159 (1953).
1135. — Vogeler, K.: Angew. Chem. **74**, 904 (1962).
1136. Wilcox, H. G., Heimberg, M.: Biochem. Biophys. Acta **152**, 424 (1968).
1137. Williams, K. T., Bevenue, A.: Science **113**, 582 (1951).
1138. Williams, J. W., van Holde, K. E., Baldwin, R. L., Fujita, H.: Chem. Revs. **58**, 715 (1958).
1139. Williams, Jr., R. C., Rajagopalan, T. G.: J. Biol. Chem. **241**, 4951 (1966).
1140. Winkler, S.: Z. physiol. Chem., Hoppe Seyler's **228**, 50 (1934).
1141. Winstead, J., Wold, F.: Biochemistry (U.S.A.) **3**, 791 (1964).
1142. Wintersberger, E., Cox, D. J., Neurath, H.: Biochemistry (U.S.A.) **1**, 1069 (1962).
1143. Witkop, B.: Advances in Protein Chem. **16**, 221 (1961).
1144. Wolman, Y., Kivity, S., Frankel, M.: Chem. Commun. **1967**, 629.
1145. — Ladkany, D., Frankel, M.: J. Chem. Soc. (c) **1967**, 689.
1146. Wood, J. L., du Vigneaud, V.: J. Biol. Chem. **130**, 109 (1939).
1147. Woodward, G. E., Fry, E. G.: J. Biol. Chem. **97**, 465 (1932).
1148a. Woodward, R. B., Olofson, R. A.: J. Am. Chem. Soc. **83**, 1107 (1961).
1148b. — — Mayer, H.: J. Am. Chem. Soc. **83**, 1010 (1961).
1149. — — — Tetrahedron Letrs. Suppl. **8**, 321 (1966).
1150. Woolley, D. W.: J. Biol. Chem. **172**, 71 (1948).
1151. Wunsch, E., Drees, F.: Chem. Ber. **99**, 110 (1966).
1152. — Fontana, A.: Chem. Ber. **101**, 323 (1968).
1153. — — Drees, F.: Z. Naturforsch. **22b**, 607 (1967).
1154. — Fries, G., Zurck, A.: Chem. Ber. **91**, 542 (1958).
1155. — Zwick, A.: Chem. Ber. **99**, 105 (1966).
1156. — — Wendlberger, G.: Chem. Ber. **100**, 173 (1967).
1157. — — Fontana, A.: Chem. Ber. **101**, 326 (1968).
1158. — — Jaeger, E.: Chem. Ber. **101**, 336 (1968).
1159. Yajima, H., Okada, Y., Kinomura, Y., Minami, H.: J. Am. Chem. Soc. **90**, 527 (1968).
1160. Yamashiro, D.: Nature **201**, 76 (1964).
1161. Yang, C. A.: Biochim. Biophys. Acta **133**, 346 (1967).
1162. Yonda, A., Filmen, D. L., Pate, H., Alonzo, N., Hirs, C. H. W.: Anal. Biochem. **10**, 53 (1965).
1163. Young, G. T.: Angew. Chem. **71**, 742 (1959).
1164. — J. Chem. Soc. **1963**, 1105.
1165. Young, J. D., Benjamini, E., Stewart, J. M., Leung, C. Y.: Biochemistry (U.S.A.) **6**, 1455 (1967).
1166. Youngs, G. C.: Anal. Chem. **31**, 1019 (1959).
1167. Yphantis, D. A.: Biochemistry (U.S.A.) **3**, 297 (1964).
1168. Zahn, H.: Naturwissenschaften **54**, 396 (1967).
1169. — Danho, W., Gutte, B.: Z. Naturforsch. **21b**, 763 (1966).
1170. — Diehl, J. F.: Angew. Chem. **69**, 135 (1957).

1171. — LAFRANCE, N. H.: Ann. Chem., Justus Liebigs **630**, 37 (1960).
1172. — — Z. physiol. Chem., Hoppe Seyler's **319**, 143 (1960).
1173. ZAORAL, M.: Collection Czechoslov. Chem. Commun. **27**, 1273 (1962).
1174. ZERVAS, L., BOROVAS, D., GAZIS, E.: J. Am. Chem. Soc. **85**, 3660 (1963).
1175. — HAMALIDIS, C.: J. Am. Chem. Soc. **87**, 99 (1965).
1176. — PHOTAKI, I.: J. Am. Chem. Soc. **84**, 3887 (1962).
1177. — — COSMATOS, A., BOROVAS, D.: J. Am. Chem. Soc. **87**, 4922 (1965).
1178. — — GHELIS, N.: J. Am. Chem. Soc. **85**, 1337 (1963).
1179. — THEODOROPOULOS, D. M.: J. Am. Chem. Soc. **78**, 1359 (1956).
1180. — WINITZ, M.: GREENSTEIN, J. P.: J. Org. Chem. **22**, 1515 (1957).
1181. ZLATKIS, A., ORO, J.: Anal. Chem. **30**, 1156 (1958).
1182. — — KIMBALL, A.: Anal. Chem. **32**, 162 (1960).
1183. ZOMZELY, C., MARCO, G., EMERY, E.: Anal. Chem. **34**, 1414 (1962).
1184. ZUBER, H.: Nature **201**, 613 (1964).
1185. — Abstr. 7th Int'l. Congr. Biochem., Tokyo, p. 541, 1967.

Subject Index

Molecular Biology, Biochemistry and Biophysics

Molekularbiologie, Biochemie und Biophysik

Volumes published:

Vol. 1 J. H. van't Hoff: Imagination in Science. Translated into English, with Notes and a General Introduction by G. F. Springer. With 1 portrait. VI, 18 pages. 1967. Soft cover DM 6,60/US $ 1.65

Vol. 2 K. Freudenberg and A. C. Neish: Constitution and Biosynthesis of Lignin. With 10 figures. IX, 129 pages. 1968. Cloth DM 28,—/US $ 7.—

Vol. 3 T. Robinson: The Biochemistry of Alkaloids. With 37 figures. X, 149 pages. 1968. Cloth DM 39,—/US $ 9.75

Vol. 4 A. S. Spirin and L. P. Gavrilova: The Ribosome. With 26 figures. X, 161 pages. 1969. Cloth DM 54,—/US $ 14.90

Vol. 5 B. Jirgensons: Optical Rotatory Dispersion of Proteins and Other Macromolecules. With 65 figures. XI, 166 pages. 1969. Cloth DM 46,—/US $ 12.70

Vol. 6 F. Egami and K. Nakamura: Microbial Ribonucleases. With 5 figures. IX, 90 pages. 1969. Cloth DM 28,—/US $ 7.—

Vol. 8 Protein Sequence Determination. A Sourcebook of Methods and Techniques. Edited by Saul B. Needleman. With 77 figures. XXI, 345 pages. 1970. Cloth DM 84,—/ US $ 23.10

Vol. 9 R. Grubb: The Genetic Markers of Human Immunoglobulins. With 8 figures. XII, 152 pages. 1970. Cloth DM 42,—/US $ 11.60

Forthcoming volume:

Vol. 7 F. Hawkes: Nucleic Acids and Cytology. With approx. 35 figures. Approx. 288 pages.

Volumes being planned:

Aida, Kô: Amino Acid Fermentation

Ando, T., M. Yamasaki and K. Suzuki: Protamines: Isolation, Characterization, Structure and Function

Anker, H. S.: Studies on the Mechanism and Control of Fatty Acid Biosynthesis

Antonini, E.: Kinetic Structures and Mechanism of Hemoproteins

Beermann, W.: Puffs and their Function in Developmental Biology

Breuer, H., and K. Dahm: Mode of Action of Estrogens

Brill, A. S.: The Role of Heavy Metals in Biological Oxidation Processes

Burns, R. C., and R. W. F. Hardy: Nitrogen Fixation in Bacteria and Higher Plants

Callan, H. G., and O. Hess: Lampbrush Chromosomes

Chapman, D.: Molecular Biological Aspects of Lipids in Membranes

Chichester, C. O., H. Yokoyama and T. O. M. Nakayama: Biosynthesis of Carotenoids

Dumonde, D. C.: Cell and Tissue Antigens

Dus, K. M.: Analysis of Proteins

Volumes being planned:

ELEY, D. D.: Biological Materials as Semi-Conductors
FENDER, D. H.: Problems of Visual Perception
GIBSON, I.: Paramecium Aurelia and Cytoplasmic Factors
GIBSON, Q. H.: Rapid Reactions in Biochemistry
GORINI, L. C., and R. A. ZIMMERMANN: Translation of Genetic Information into Cell Structure and Cell Components
GUILLEMIN, R.: Hypothalamic Releasing Factors
HAUGAARD, N., and E. HAUGAARD: Mechanism of Action of Insulin
HEINZ, E., and W. WILBRANDT: Biological Transports
HOFFMANN-OSTENHOF, O.: Biochemistry of the Inositols
HOFMANN, TH.: Chymotrypsin and other Serine Proteinases
JARDETZKY, O., and G. C. K. ROBERTS: Nuclear Magnetic Resonance in Molecular Biology
JOLLÈS, P., and A. PARAF: Biological and Chemical Basis of Adjuvants
KERSTEN, W., H. KERSTEN and D. VAZQUEZ: Antibiotics Inhibiting Nucleic Acid and Protein Synthesis
KLEINSCHMIDT, A. K., and D. LANG: A Manual for Electron Microscopy of Nucleic Acid Molecules
KLENK, E.: Cerebrosides, Gangliosides and other Glycolipids
LARNER, J.: Chemistry of Hereditary Disease
LUDWICK, T. M., J. E. GANDER and R. JENNESS: Biosynthesis of Milk
LUKENS, R. J.: Chemistry of Fungicidal Action
MATHEWS, M. B.: Molecular Evolution of Connective Tissue
MATTHAEI, J. H.: Protein Biosynthesis
MAURER, P. H.: Chemical and Structural Basis of Antigenicity
McMANUS, T. J.: Red Cell Ion Transport and Related Metabolic Processes
MEIENHOFER, J.: Synthesis of Biologically Active Peptides
NELSON, R. A.: The Complement System. Physicochemical Characterization and Biologic Reactivities
OKUNUKI, K.: Cytochromes
OSBORN, M. J., and L. ROTHFIELD: Bacterial Cell Wall Synthesis and Function
PAUL, K. G.: Peroxidases and Catalases
PORATH, J.: Modern Methods of Separation of Proteins and other Macromolecules
QUASTEL, J. H., and C. S. SUNG: Chemical Control of Cell Metabolism and Function
REEVES, P.: Bacteriocins
RODWELL, V. W., and R. SOMERVILLE: Metabolism of Amino Acids
RUNNSTRÖM, J.: Molecular Biology of the Fertilization Process
SCHANNE, O., and ELENA RUIZ DE CERETTI: Impedance Measurements in Biological Cells
SELIGER, H. H.: Bioluminescence
SÖLL, D., and U. L. RAJBHANDARY: Transfer-RNA
SHULMAN, S.: Tissue Specificity
STADTMAN, E. R.: Enzymatic Regulation of Metabolism
TS'O, O. P., and P. A. STRAAT: Nucleic Acids Polymerases
VINNIKOV, J. A.: The Molecular and Cellular Basis of the Reception
WERLE, E.: Plasma Kinines
YAGI, K.: Biochemistry of Flavins
ZWEIFACH, B. W.: Shock